POTATO

Improvement, Production and Utilization Technologies

An Approach to Doubling Farmers' Income

THE EDITORS

Dr. Jagesh Kumar Tiwari, Senior Scientist (Horticulture-Vegetable Science), ICAR-Indian Institute of Vegetable Research, Varanasi has over 14 years of research experience on potato at ICAR-CPRI, Shimla since 2008 to 2022. He is post-doctorate from University of Adelaide, Australia and International Potato Centre, Lima, Peru. He has bagged awards ICAR-Lal Bahadur Shastri Outstanding Young Scientist Award, ICAR Hari Om Ashram Trust Award, NAAS Associateship Award, and IAHS-DP Ghosh Memorial Young Scientist Award etc. and published over 100 research papers/reviews.

Dr. (Ms.) Tanuja Buckseth, Scientist (Senior Scale), Horticulture-Vegetable Science, ICAR-CPRI, Shimla has over eight years of research experience in seed potato production. She has contributed significantly in seed potato research for producing healthy, disease free potato seed stocks, strengthened the Hi-tech Potato seed production system (Micropropagation and aeroponic) by refining the various parameters. She has published over 40 research articles in reputed international & national journals and received awards/recognitions from several professional societies.

Dr. (Ms.) Dalamu, Senior Scientist (Horticulture-Vegetable Science), ICAR-CPRI, Shimla has over 10 years of research experience in potato breeding and varietal development through conventional and marker assisted selection for various biotic stresses and nutritional enhancement, potato DUS characterisation and varietal protection. She has published more than 40 research articles, book chapters and popular articles. She is life member of Indian Potato Association, Indian Society of Vegetable Science and Horticultural Society of India.

Dr. (Ms.) Aarti Bairwa, Scientist (Sr. Scale), Nematology, ICAR-CPRI, Shimla has over 9 years of research experience in potato nematodes and their different management aspects including development of potato cyst nematode resistant varieties. She has published more than 45 research papers, reviews, book chapters and technical/popular articles. She has received awards from different professional societies and is life member of Indian Potato Association.

Dr. N.K. Pandey, Ex-Director (Acting), ICAR-CPRI, Shimla has over 35 years of experience on potato in Agriculture economics and Transfer of Technology. He is entrusted with the demonstrating the potato technologies and human resource development to farmers/stakeholders and staff of the Institute. He has also assessed the impact of potato technologies developed by the Institute. Dr Pandey has over 150 research publications. He has been awarded ICAR Team Research award and CPRI Merit Certificate.

Dr. Manoj Kumar, Ex-Joint Director & Principal Scientist (Soil Science), ICAR-Central Potato Research Institute, Regional Station, Modipuram, Meerut (UP) has nearly 20 years of research experience on potato coordination & management and soil-nutrient management technologies. He has published over 200 research articles and received awards/ recognitions from several professional societies/ institutes. He is Fellow of Indian Potato Association, Shimla.

POTATO

Improvement, Production and Utilization Technologies

An Approach to Doubling Farmers' Income

– Editors –

Jagesh Kumar Tiwari

Tanuja Buckseth

Dalamu

Aarti Bairwa

N.K. Pandey

Manoj Kumar

2023

Daya Publishing House®

A Division of

Astral International Pvt. Ltd.

New Delhi – 110002

ISBN 9789359191669

Publisher's Note:

Every possible effort has been made to ensure that the information contained in this book is accurate at the time of going to press, and the publisher and authors cannot accept responsibility for any errors or omissions, however caused. No responsibility for loss or damage occasioned to any person acting, or refraining from action, as a result of the material in this publication can be accepted by the publisher or the authors. The publisher is not associated with any product or vendor mentioned in the book. The contents of this work are intended to further general scientific research, understanding and discussion only. Readers should consult with a specialist where appropriate.

Every effort has been made to trace the owners of copyright material used in this book, if any. The authors and the publisher will be grateful for any omission brought to their notice for acknowledgement in the future editions of the book.

Published by : **Daya Publishing House®**
A Division of
Astral International Pvt. Ltd.
– ISO 9001:2015 Certified Company –
4736/23, Ansari Road, Darya Ganj
New Delhi-110 002
Ph. 011-43549197, 8130496929
E-mail: info@astralint.com
Website: www.astralint.com

Printed at : **Replika Press Pvt Ltd**

Preface

Potato (*Solanum tuberosum* L.) is the third most important food crop in the world after rice and wheat in terms of human consumption. Potato is an economically important staple food because of its high yield potential, high nutritive value, easily digestible and wholesome food. The United Nations declared the year 2008 as the International Year of the Potato. Potato holds a great potential to secure both food and nutritional security at global level. In 2020, the global annual potato production is 359.07 million tonnes from 16.49 million hectares area with 21.76 tonnes per hectare productivity. India (51.30 mt) is the second largest producer of potato after China from an area of 2.15 m ha with productivity of 23.77 t/ha. Whereas, Uttar Pradesh is the top potato producer followed by West Bengal and Bihar.

The book "Potato: Improvement, Production and Utilization Technologies" is based on technologies developed at ICAR-Central Potato Research Institute (CPRI), Shimla, Himachal Pradesh, India. The book is based on the lectures and practicals conducted during the Winter School on "Innovations in Potato Improvement, Production and Utilization of Technologies for Doubling Farmers' Income" w.e.f. 18 January to 7 February, 2022 at ICAR-CPRI, Shimla (Course Director: Dr. Jagesh Kumar Tiwari). The book covers various chapters on potato improvement, production, protection, post-harvest management and social issues such as introduction, role of AICRP (potato), true potato seed technology, molecular breeding for biotic and abiotic stresses resistance/ tolerance, processing potato and quality parameters, nutrient management in potato, water management in seed and ware potato,potato germplasm, DUS characterization and plant variety & farmer's right, somatic hybridization, diploid F_1 hybrid TPS potato breeding, potato nematodes management, transgenics and genome editing in potato improvement, status of adoption of potato technologies and varieties in India, impact of climate change on potato: current scenario and adaptation strategies, soil and plant tests for fertilizer recommendation, post-harvest management of potatoes, application of proteomics and metabolomics, nutritional value and biochemical analysis, viral diseases and their management, fungal diseases and their management, soil and tuber borne diseases of potato and their management, seed potato production, production of disease-free planting material through hi-tech system including aeroponics, farm mechanization and drone technology, agro-techniques for quality seed potato production, functional genomics for improving nitrogen use efficiency, good agricultural practices, breeding potato varieties for different agro-climatic zones, economics, marketing and export, insect pest management, virus-vector interaction, protocols for analysis physio-biochemical parameters, ploidy estimation, micropropagation, soil and plant sample test, DNA fingerprinting and gene expression analysis and protoplast fusion in potato.

We believe this book is useful for potato farmers, stakeholders, researchers, academicians and students working towards doubling of farmers' income and also to enhance knowledge on potato research and development. We are grateful to the Competent Authority, Education Division, ICAR, Director, ICAR-CPRI and thanks to the authors for their contribution to this book.

Editors

Jagesh Kumar Tiwari

Tanuja Buckseth

Dalamu

Aarti Bairwa

N.K. Pandey

Manoj Kumar

Contents

Chapter 1

Potato: An Overview of National and International Perspective

N.K. Pandey, Dalamu and Jagesh Kumar Tiwari

ICAR-Central Potato Research Institute, Shimla, Himachal Pradesh

1. Potato: The Crop and the Food

Potato (*Solanum tuberosum* L.) is one of the most important food crop after rice and wheat, contributing to food and nutritional security in the world. This tuber crop of the family *Solanaceae* has about 200 wild species. It originated in the high Andean hills of South America, from where it was first introduced into Europe towards the end of 16th century through Spanish conquerors. There the potato developed as a temperate crop and was later distributed throughout the world largely as a consequence of the colonial expansion of European countries. It was introduced to India by early 17th century probably through British missionaries or Portuguese traders.

Potato is an annual, herbaceous, dicotyledonous and vegetatively propagated plant. It can also be propagated through botanical seed known as True Potato Seed (TPS). The potato tuber is a modified stem developed underground on a specialized structure called stolon. It contains all the characteristics of a normal stem like dormant bud (eye) and scaly leaf (eyebrow). Potato tuber is a bulky commodity which responds strongly to its prevailing environment thus needs proper storage. Potato is a highly nutritious, easily digestible, wholesome food containing carbohydrates, proteins, minerals, vitamins and high quality dietary fibre. A potato tuber contains 80 per cent water and 20 per cent dry matter consisting of 14 per cent starch, 2 per cent sugar, 2 per cent protein, 1 per cent minerals, 0.6 per cent fibre, 0.1 per cent fat, and vitamins B and C in adequate amount. Thus, potato provides more nutrition than cereals and vegetables. Keeping in view the shrinking cultivable land and burgeoning population in India, potato is a better alternative to deal with the situation.

2. Potato in India

In Europe the potato crop is grown in summer having long photoperiod of up to 14 hours and the crop duration of 140-180 days. The potato in Indian plains is, however, grown in completely contrasting situations. Nearly 85 per cent of the crop is grown during winters having short photoperiod (with about 10-11 hours sunshine) and the crop duration is also limited to 90-100 days because of short and mild winter. The mornings usually have fog, which further reduces the sunshine hours posing severe constraints on photosynthetic activity. Besides, the post-harvest period consists of long hot summer, which creates storage problems. All these problems called for suitable varieties and technologies for growing potatoes under the sub-tropical conditions of India. This necessitated initiating indigenous potato research and development programmes, and accordingly the Central Potato Research Institute (CPRI) came up in 1949 at Patna. The headquarters was later on shifted to Shimla in order to facilitate hybridization and maintenance of seed health. In 1971 the All India Coordinated Research Project (AICRP) on potato was initiated under the aegis of the Indian Council of Agricultural Research (ICAR) at the CPRI with an objective to coordinate potato research and development in the country across diverse agro-ecological regions. The success story of over seven decades of potato research in India is phenomenal. Compared to the area, production and productivity in 1949-50, the increase over this period is 9, 33 and 3.6 fold, respectively (Table 1.1).

Table 1.1: Area, Production and Yield of Potato in India

Year	*Area (million ha)*	*Production (million tones)*	*Yield (t/ha)*
1949-50	0.239	1.54	6.59
1959-60	0.362	2.73	7.55
1969-70	0.496	3.91	7.89
1979-80	0.685	8.33	12.15
1989-90	0.940	14.77	15.71
1999-00	1.340	25.00	18.44
2009-10	1.835	36.57	19.93
2010-11	1.860	42.34	22.72
2011-12	1.910	41.48	21.75
2012-13	1.992	45.34	22.78
2013-14	2.024	46.39	22.92
2014-15	2.076	48.0	23.20

Year	*Area (million ha)*	*Production (million tones)*	*Yield (t/ha)*
2015-16	2.117	43.41	20.50
2016-17	2.179	48.60	22.30
2017-18	2.142	51.31	23.95
2018-19	2.173	50.19	23.10
2019-20	2.158	51.30	23.77

Source: FAOSTAT 2022.

Table 1.2: State-wise Area, Production and Yield of Potato in India during 2017-18

States	*Area (1000 Hectares)*	*Production (1000 T)*	*Yield (t/ha)*
Uttar Pradesh	614.78	15555.53	25.30
West Bengal	427.50	12782.50	29.90
Bihar	304.78	7740.80	25.40
Gujarat	133.29	3806.95	28.56
Madhya Pradesh	136.29	3144.64	23.07
Punjab	98.52	2571.04	26.10
Haryana	34.72	897.58	25.85
Assam	102.87	720.97	7.01
Chattisgarh	44.87	694.61	15.48
Jharkhand	48.21	690.23	14.32
Others	195.91	2705.18	13.81
Total	2141.74	51310.00	23.96

Source: Horticulture statistics at a glance (2018).

It was only because of indigenously developed technologies that potato in India has shown spectacular growth in area, production and productivity during the last seven decades. The major achievements of potato research in India are as under:

3. Varietal Improvement

So far 66 potato varieties have been bred for different agro-climatic regions of the country with 32 varieties alone for north Indian plains. Varieties have also been developed for north Indian hills and other special problem areas *viz.*, Sikkim, north Bengal hills and south Indian hills. Of the 66 varieties developed, 19 possess multiple resistance to different biotic and abiotic stresses. Besides, twelve varieties are suitable for processing purposes. These are Kufri Chipsona-1, Kufri Chipsona-2, Kufri Chipsona-3, Kufri Chipsona-4, Kufri

Himsona, Kufri Frysona, Kufri Jyoti, Kufri Chandramukhi, Kufri Lauvkar, Kufri Surya. Kufri Sangam and Kufri FryoM. All these varieties fall in three maturity groups, *i.e.* early (70-80 days), medium (90-100 days) and late (110-120 days). The potato varieties developed by CPRI are grown not only in India but also in several neighbouring countries. The variety Kufri Chandramukhi is grown in Afghanistan, Kufri Jyoti in Nepal and Bhutan, and Kufri Sindhuri in Bangladesh and Nepal. Besides, five Indian hybrids are also commercially grown in Sri Lanka, Madagascar, Mexico and Philippines.

4. Seed Plot Technique

This technique was developed in 1970s to enable healthy seed potato production in the sub-tropical Indian plains under low aphid period. This technique aided by bio-technological approaches for virus elimination, micro-propagation and effective viral diagnostics has sustained the National Potato Seed Production Programme by producing about 2870 tonnes of breeder's seed annually (CPRI Annual report 2020). This breeder's seed is further multiplied to about 0.50 million tonnes of certified seed by the State Departments of Agriculture/Horticulture. Thus, the country saves about 484 million US dollars because most Asian countries like Pakistan, Bangladesh and even China continue to import seed potatoes from Europe. The decentralization of potato breeding from hills to plains in India through the seed plot technique enabled the development of varieties suited to different agro-climatic regions of the country. The area under seed potato production also increased by 12 times and enabled the availability of seed potato throughout the country in proper physiological state.

5. Tissue Culture

Efforts are being made to improve seed health standards and reduce the time required for production of breeder's seed by employing *in vitro* techniques of meristem culture and micro- propagation. Presently, about 33 per cent of Breeder's seed production (CPRI Annual report 2020) programme is fed annually by microtubers produced through tissue culture. It is proposed to produce 100 per cent of breeder's seed through tissue culture propagated material in the years to come.

6. Agro-techniques

The development of package of practices for potato production in different agro-climatic zones has helped in improving potato productivity in these zones. The potato crop is input intensive and requires optimum cultural practices for achieving higher productivity. Optimum cultural practices depend on delineated phenological phases of crop growth and development *viz.*, pre-emergence, emergence to tuber initiation, tuber initiation to tuber bulking and tuber bulking to termination of bulking. The cultural practices are adjusted in the Indian plains in a way so that tuber initiation and development coincide with the period when night temperature is less than 20 °C and day temperature is below 30 °C. The phenological phase of tuber initiation to tuber bulking is mainly conditioned by nutrition and moisture. For this purpose, fertilizer and irrigation requirement in different agro-climatic zones have been worked out through multi-locational trials under AICRP (Potato). Termination of tuber bulking coincides with onset of foliage senescence. By manipulating the nutrition and moisture, the foliage senescence is delayed for ensuring continuation of linear tuber bulking phase resulting in higher yield. Several profitable potato-based inter-cropping and crop rotations have also been identified for different regions of the country. Potato can be profitably intercropped with wheat, mustard and sugarcane. These cropping systems have helped in the maintenance of soil fertility and have improved the fertilizer economy, crop yield and gross returns. Besides, potato cultivation has also been mechanized in selected regions through the fabrication and development of cost-effective tools and implements.

7. Plant Protection

Effective management practices have been devised for the major potato diseases and insect-pests in India. Late blight is the most notorious disease of potato which occurs almost every year in the hills and plains. Besides chemical control measures, several late blight resistant varieties have been developed. Potato varieties have also been bred which possess resistance to wart and cyst nematodes. Cultural and biological control measures have also been developed to control the diseases and insect-pests. The development of late blight forecasting systems for hills and plains has enabled the early warning mechanism for the appearance of late blight disease.

8. Storage

In European countries, the potato crop is grown in summer and the main storage season is the cold winter. However, in India, 85 per cent of potato is produced in winter and stored during long hot summer. This requires storage of potatoes in cold stores at 2-4 °C, which involves substantial cost. It also leads to accumulation of reducing sugar in the potato tubers resulting in sweetening of potatoes.However, there are a number of traditional low-cost and non-refrigerated storage structures (essentially based on evaporative or passive evaporative cooling) in use in India with varying degrees of success. These traditional structures have been studied, validated and popularized for particular regions. In non-refrigerated storages, use of sprout suppressants has also been popularized to prevent excessive weight loss and shrinkage due to sprouting. The CIPC (isopropyl-N-chlorophenyl carbamate) is the most effective sprout inhibitor when applied @ 25 mg a.i. per kg tubers.

9. Processing and Value Addition

In addition to raw consumption, potatoes can be processed into several products like chips, french fries, cubes, granules and canned products. The primary determinants for potato processing include high dry matter and low reducing sugar content. A dry matter content of more than 20 per cent is desirable for chips, french fries and dehydrated products. Similarly, reducing sugar content in tubers up to 100 mg/100 g fresh weight is considered acceptable for processing. Twelve varieties *viz.*, Kufri Chipsona-1, Kufri Chipsona-2, Kufri Chipsona-3, Kufri Chipsona-4, Kufri Jyoti, Kufri Chandramukhi, Kufri Lauvkar, Kufri Surya, Kufri Himsona, Kufri Frysona, Kufri Sangam and Kufri FryoM have been developed for processing purposes. In India, potato processing in organised sector started about a decade ago, and the recent proliferation of this sector mainly results from the development of indigenous potato processing varieties, *viz.*, Kufri Chipsona-1, Kufri Chipsona-2, Kufri Chipsona-3 and Kufri Chipsona-4, Kufri FryoM by CPRI. These varieties are now being used by the industries for processing into chips and French-fries.

10. Computer Applications

Simulation modelling is now widely used in various disciplines to work out tactical decisions. CPRI has developed INFOCROP-POTATO model to simulate the potato growth and development, to determine the best growing period, to optimise management practices under different agro-ecological regions, and to forecast the accurate yield estimates. An expert system (Potato Pest Manager) has also been developed for decision support with respect to identification and management of diseases and insect-pests.

11. Transfer of Technology

Research achievements alone are not adequate to gauge the success of an agricultural system. The research information needs to be assessed and refined under various bio-physical and socio-economic situations through adaptive research before it is labelled as a technology. In this regard, the multi-locational trials under AICRP (Potato) and the TOT projects undertaken by CPRI such as Operational Research Project (ORP), Lab-to-Land Programme (LLP), Tribal Area Development (TAD) programme and Institution-Village Linkage Programme (IVLP) proved landmark in getting feedback from the field and development of appropriate technologies. Transfer of technology to the end users is a complex task which consists of a number of components and dimensions. One of the important components is proper linkage between technology generating system and the client system. In this regard, innovative approaches like need assessment, participatory planning and implementation, and direct scientist-farmer interface facilitated faster dissemination of technologies and consequent adoption by the farmers/clients. The CPRI has build up linkages with farmers through demonstrations, trainings, Kisan Melas, potato school on All India

Radio, supply of literatures and other extension activities. Besides, studies have been conducted to measure the socio-economic impact and constraints in transfer of potato technology.

12. Potato Export

Although India contributes 12.50 per cent to the total world potato production from 11.30 per cent of global area under potato cultivation, its 0.7 per cent share in world's potato export is quite insignificant. Indian potatoes are truly free from the prohibited disease like wart, black scruf, and pests like tuber moth and nematodes, which are the barometer for phytosanitary standards. India has also the natural advantage of exporting fresh table potatoes during January to June when supply from European countries dwindles. It can also supply fresh potatoes round the year because India has diverse agro-climates and potato is grown throughout the year in one or the other part of the country. Potato has a good future in India under the changed scenario of global economy. Globalisation has resulted in many developing countries becoming much more integrated into the international potato trade. With the phasing out of quantitative restrictions on agricultural commodities, the imports and exports of potato would be based on the differences in price and production cost between the importing and exporting countries involved. Due to low production cost in the country as a result of availability of cheap labour, India will have competitive advantage in the international potato trade.

13. Potato in the New Millennium

With the improvement in the living standard of people in India, the dietary habits will shift from cereals to vegetables. Under such a situation it is estimated that India will have to produce 122 million tonnes of potato by 2050 (Singh *et al.*, 2014). This target could be achieved only by improving the productivity level. The productivity of potato in India is quite low (23.77 t/ha) as compared to that of Kuwait (48.73 t/ha), New Zealand (50.74 t/ha) and USA (50.80 t/ha), top three counries wrt potato productivity (FAOSTAT 2022). This is due to shorter crop duration in India. There is a wide ranging variations in the agro-ecological setting of different parts of the country, which results in wide variations in the productivity levels of different states (Table 1.2). Therefore, all our efforts may be put in to develop location-specific and problem-specific varieties and technologies.

Most of the people in India have either no knowledge or wrong notions about the nutritive value of potato. With low fat (0.1 per cent) and calorie contents, it does not cause obesity. Due to misconception regarding potato consumption, the per capita consumption of potato in India is only about 26 kg/year. On the other hand, the per capita consumption in European nations are more than 100 kg/year. Hence, there is ample scope for improving the consumption of potatoes in India. For this purpose, a publicity campaign like eggs and milk needs to be launched through mass media such as television, radio and newspapers highlighting its nutritional value. Moreover, the possibility of using surplus potatoes as animal feed also needs to be explored.

The surplus potatoes in a season are stored in cold stores at 2-4 °C in the country. This makes stored potatoes just unfit for processing and loses preference for table purposes due to accumulation of sugar content. To avoid sweetening potato are required to be stored at 2-4 °C. Only seed potatoes should be cold stored at 2-4 °C. This would release atleast 60 per cent of cold storage space that can be converted to store potatoes for processing and table purposes at 10-12 °C with CIPC treatment leading to considerable savings on energy and storage costs.

Processing is a fast growing sector in the potato world economy. Due to increased urbanization, rise in per capita income and expanding tourism, the demand for processed potato products in India and international market has risen at a fast pace. However, in India, processing of potatoes constitutes less than 2 per cent of the total annual production as compared to 60 per cent in USA, 47 per cent in the Netherlands and 22 per cent in China. Hence, there is great scope to expand the potato processing industries in India and also to diversify the processing to produce flour, cubes, granules, flakes and starch. Under the changed global scenario, the potato production and utilisation pattern is changing very fast. These changes harbour many opportunities which could be tapped through effective extension system. The use of modern information and communication technologies (ICT) to create awareness is highly pertinent in the contemporary times. This would enable us to reach directly to the end users by eliminating the intermediate channels which create distortion of information. Efforts are also needed to devise market-based extension strategies in order to promote entrepreneurship among potato growers with regard to potato production and marketing.

Chapter 2

Role of AICRP (Potato) and True Potato Seed (TPS) Technology in Potato

Vinod Kumar, Dalamu and S. Sood

ICAR-Central Potato Research Institute, Shimla, Himachal Pradesh

1. AICRP (Potato)

Potato is one of the major non-cereal food crops and currently ranks 3rd in terms of human consumption in the world after rice and wheat. The productivity of potato in India was quite low (9.97 t/ha) during the period of inception of AICRP (1970-71). Since there was ample scope of improving the productivity of potato in sub-tropical climates, the All India Coordinated Research Project (AICRP) on Potato was initiated during 4th Five Year Plan in the year 1970-71. The main objectives of the project were to carry out region specific trials on evaluation of new hybrids, agronomic practices, potato based cropping systems, control measures for various diseases and pests, potato storage practices and identification of methods for reducing post-harvest losses. Presently, it has a nationwide network of 25 centers located in distinct agro-climatic regions of the country. Seven of these centers are based at CPRI regional stations, 17 in different State Agricultural Universities (SAUs) and 1 voluntary center at Ranichauri in Uttarakhand. The area, production and productivity has touched a hike of 903, 3210 and 355 per cent respectively during 2018-19 compared to the year of establishment of this project that is a direct reflection of the success of this project and AICRP (P) on potato industry in the country.

2. Objective of the Scheme

The overall objective of AICRP (Potato) is to "develop high yielding varieties with desirable attributes for diversified end usage aid improved production technologies for potato in different agro-climatic regions" which could be accomplished through the following specific objectives:

- ☆ Identification of improved high yielding potato varieties according to consumer preferences and user groups and different target environments.
- ☆ Identification of sustainable agronomic practices and profitable potato based cropping systems for different regions.
- ☆ Monitoring of diseases/pests and other problems of potato in varied agro-climates and develop appropriate measures for their control/ management.
- ☆ Identification of areas suitable for production of disease-free potato seed and develop seed production technology.
- ☆ Development of technologies for minimizing post-harvest losses.
- ☆ Development of linkages with Agricultural Universities, State Departments of Agriculture/Horticulture, KVKs and other developmental/extension agencies.

Table 2.1: Details of AICRP (Potato) Centers Since its Inception

Location/Institute/University		*Year of Establishment*	*Current Status*
Headquarters			
Shimla	PC Unit, CPRI, Shimla	1971 (IV Plan)	Continued
Regular centers			
Bhubaneshwar	OUAT, Bhubaneshwar	1971 (IV Plan)	Continued
Chhindwara	JNKVV, Jabalpur	1971 (IV Plan)	Continued
Deesa	SDAU, SK Nagar	1971 (IV Plan)	Continued
Dharwad	UAS, Dharwad	1993 (VIII Plan)	Continued
Dholi	RPCAU, Pusa	1994 (VIII Plan)	Continued
Faizabad	NDUAT, Faizabad	1987 (VII Plan)	Continued
Gwalior	CPRI Regional Station	2006 (X Plan)	Continued
Hassan	UHS, Bagalkot	1972 (IV Plan)	Continued
Jalandhar	CPRI Regional Station	1993 (VIII Plan)	Continued
Jorhat	AAU, Jorhat	1971 (IV Plan)	Continued
Hisar	CCSHAU, Hisar	1975 (V Plan)	Continued
Kalyani	BCKV, Kalyani	1975 (V Plan)	Continued
Kanpur	CSAUAT, Kanpur	2009 (XI Plan)	Continued
Kota	AU, Kota	1971 (IV Plan)	Continued
Kufri	CPRI Regional Station	1976 (V Plan)	Continued
Modipuram	CPRI Regional Station	1976 (V Plan)	Continued
Ootacamund	CPRI Regional Station	1987 (VII Plan)	Continued
Pantnagar	GBPUAT, Pantnagar	1975 (V Plan)	Continued

Location/Institute/University		*Year of Establishment*	*Current Status*
Patna	CPRI Regional Station	1971 (IV Plan)	Continued
Pune	MPKVV, Rahauri	2009 (XI Plan)	Continued
Passighat	CAU, Imphal	2009 (XI Plan)	Continued
Raipur	IGKV, Raipur	1999 (IX Plan)	Continued
Shillong	CPRI Regional Station	1971 (IV Plan)	Continued
Srinagar	SKUAST (K), Srinagar	1972 (IV Plan)	Continued
Voluntary centers			
Ranichauri	UH and F, Ranichauri	1984 (VI Plan)	Continued

3. On-going Experiments

The current relative priorities under the project are:

3.1 Crop Improvement

Varieties for early bulking; nutrient and water use efficiency; late blight resistance; processing; suitable for cultivation in non-traditional seasons and areas; and suitable for growing under organic/low input regimes.

3.2 Crop Production

Integrated nutrient and water management; potato based intensive cropping systems; agro-techniques for increasing total factor productivity (TFP); agro-techniques for non-traditional areas and seasons; organic farming; and development of low input production technologies.

Crop Protection: Surveys/monitoring of potato pests; integrated pest management; late blight forecasting and management; and effective management of whiteflies.

4. Achievements of AICRP Potato since Inception

- ☆ 1231 hybrids evaluated and 45 were recommended for commercial release.
- ☆ Three TPS population namely HPS-I/13, TPS-C3 and 92-PT-27 were also recommended.
- ☆ 125 region-specific potato production recommendations were brought out.
- ☆ 77 plant protection measures were recommended.
- ☆ 4 recommendations were also brought out on potato seed production.
- ☆ Three software for decision support were developed
  - ❖ "INDO-BLIGHTCAST" for forecasting late blight appearance,
  - ❖ "Potato Growing Season Descriptor" for crop scheduling decisions

- "Potato Empirical Nitrogen Management Advisory System" for precision nitrogen recommendations.

5. True Potato Seed (TPS) Technology

Sexual or botanical seed of potato commonly called "True Potato Seed" (TPS) is a radical alternative to seed tubers for raising a commercial potato crop. Low multiplication rate, high storage and transportation costs, carry-over of pathogens, and physiological degeneration are some of the constraints associated with the use of seed tubers. Inevitably, performance declines as viruses accumulate due to which an elaborate and complex system of disease-free seed production is necessary. Costs of healthy seed tubers may account for 50 per cent of the total production costs. Compared to seed tubers, TPS offers several advantages, such as low cost of planting material, reduced transmission of pathogens/pests, and convenience as well as inexpensiveness of storage and transport *etc.* With rare but important exceptions (*e.g.* potato spindle tuber viroid), TPS is indeed free of virtually all diseases including systemically transmitted viruses. By using TPS, a considerable amount of valuable potato stocks that is used as seed can be saved for consumption as food.

The possibility of raising commercial crop from TPS was first explored in India in late 1940s. The attempts, however, were not successful because efforts were mainly made to use self-seeds of cultivar Phulwa that flowers naturally under short days when potato crop is grown in plains. The produce of this population was highly heterogeneous for most of the economic characters and also had low yields due to inbreeding depression (Ramanujam 1954). Renewed interest and formal research efforts to exploit the TPS as low cost alternative for potato production in developing countries were initiated by International Potato Center (CIP) in 1977 (Mendoza 1984).

6. Production of Hybrid TPS

The advantage of TPS technology has extended the use of botanical seed from mere breeding to raising of commercial crop. For this, production of TPS in large quantities becomes imperative. Generally, potato flowers under long photoperiod conditions. Such conditions are available in hills of our country where the crop is grown during summer. The TPS production in hills is however hampered owing to the regular occurrence of late blight and premature foliage killing of susceptible parents. Therefore, efforts were done to induce flowering under short day conditions. The main steps involved in hybrid TPS production:

6.1 Planting of Hybridization Block

Cut seed pieces or whole tubers of desirable parents are planted at 50 x 25 cm. Male and female lines should be planted in separate blocks under artificially extended photoperiod with the help of high density Sodium Vapour Lamps of

100/250 W per 100 sq.m. The extra light is given from germination till berry setting for 5 to 6 hours immediately after sunset. Male lines should be planted one week earlier and two to three batches to ensure continuous supply of pollen.

6.2 Hybridization

Trim the female flower bunches to retain 6-8 large size buds per bunch. Collect right stage flowers from the male parents in the morning and remove anthers for storing overnight. Extract the pollen next morning by shaking anthers in a nylon tea sieve to separate the pollen and store in a refrigerator at 6-10 °C if not required immediately. However, the use of fresh pollen every day is more effective. Pollinate each receptive stigma twice by dipped in pollen at 8 hourly intervals.

6.3 Harvesting of Berries and Seed Extraction

Harvest berries 6-7 weeks after pollination and allow them to ripen at room temperature till they become soft. Macerate the berries by hand or using reverse screw juice extractor into pulp. Treat the pulp with 10 per cent HCl and stir for 20 minutes to separate the seeds from debris. Wash the seeds with water 3 to 4 times to remove the acid. Treat the seeds with 0.05 per cent sodium hypochlorite for 10 minutes for surface disinfecting. Dry the clean seeds in shade followed by sun drying to reduce moisture contents to 5-6 per cent. Pack the seeds in double polythene bags and store over calcium chloride as desiccant under temperature below 20 °C in a refrigerator.

7. Use of TPS

TPS can be sown directly in the field or in nursery beds, and seedling grown from TPS can be left in place or transplanted. Seedling tubers (*i.e.* tubers harvested from TPS seedlings) can either be sold as ware potatoes or used as seed tubers. Since plants grown from TPS generally produce more and smaller tubers per stem than plants from seed tubers, seedling tuber production is especially suitable to form the basis of a seed tuber system.

7.1 Direct Seeding

Potato true seeds are small and seedlings rather delicate, and as such make severe demands on horticultural skills. Thus, raising the crop from direct seeding (Figure 2.1) prone to risks involving failure due to poor germination and mortality of seedlings. This method is thus rarely recommended for raising a commercial potato crop. However, under experimental conditions direct sowing of 200 g TPS per ha and an emergence of 75 per cent in well-controlled field conditions was reported to produce a yield of 28-30 tons of seedling tubers with an average weight of 25-30 g (Renia 1995).

Figure 2.1: Direct seeding tps crop.

7.2 Seedling Transplants

Raising seedlings in nursery beds and transplanting them to field is the possible alternative of TPS use for potato production (Figure 2.2). Nursery beds allow proper water and soil management, and ensure rapid germination and seedling growth. Sowing in nursery beds also permits off-season production as nursery beds can be protected against extreme temperatures and watered

Figure 2.2: TPS Seedlings Transplant Crop.

optimally. This shortens the growing period of the crop in the field. Transplanting requires good field conditions, assured irrigation and carries considerable labor costs. Direct seeding or transplanting for ware tuber production seems to have potential only in areas where consumers accept small tubers.

7.3 Seedling Tubers

Seedlings raised in the nursery beds can be allowed to grow to maturity in the nursery beds itself to produce seedling tubers (Figure 2.3). Seedling tuber production in nursery beds or otherwise well-controlled environments is the most feasible way of using TPS at present. Maximum yields of seedling tubers in nursery beds are obtained at plant densities of 80-100 seedlings per m^2 resulting in seedling tubers of 1 to over 40 g with an average size of 10-15 g. Only 750 kg of seed tubers of an average tuber size of 15 g suffice to plant one hectare of potatoes, compared with 2-3 tons for 40-60 g seed tubers. This means that 6.0-7.5 g of TPS (0.60-0.75 mg per seed, 80 per cent emergence) and 80 m^2 of nursery bed can produce seedling tubers for 1 ha potato crop.

Figure 2.3: TPS Seedlings in Nursery Bed and Seedling Tubers.

The use of seedling tubers or later-generation tubers from TPS- varieties is culturally similar to the use of tubers from conventional cultivars (seed rate, initial crop development, number of tubers per stem *etc.*). Also the yield potential of seedling tubers and later generations of selected TPS varieties compares well with that of clonal cultivars. The variation in performance of the tubers derived from selected TPS varieties is not significantly larger than the tubers from a clonal cultivar.

8. TPS Quality

The quality of TPS is determined by genetic constitution of the parents, stage of seed development, nutritional status of mother plant and biochemical components of the seed. Seed vigour generally defined as the superior performance of a genotype after planting compared to the same genotype or other genotypes under defined experimental conditions is a good indicator of quality of seed. Upadhya *et al.* (1981) described an additional quality component of TPS concerning the type of embryo shape of the seed. Seeds with a cercinnate and inverted "U" shaped embryos are heavier than seeds with other types of embryos (rod shaped, aborted *etc.*) and give a higher percentage of germination and more vigorous seedlings with higher yield. However, we found that more than 90 per cent seeds of both open and hybrid families have cercinnate and inverted "U" shaped embryos and these two classes of embryos were at par to each other for various seedling characters (Gopal 2004). By eliminating the poor vigour seedlings before transplanting the seeds with undesired embryos get eliminated and help in improving the productivity of the seedlings.

9. Advantages of TPS

- ☆ TPS requirement for planting one hectare land cost 20 times less than the seed tubers (TPS costs approx. Rs. 3000-3500/whereas, seed tubers costs RS. 60,000-65,000/for one-hectare land)
- ☆ Several viruses, bacteria, fungi and nematodes which contaminate seed tubers get filtered during the process of pollination and fertilization. TPS is thus disease free and saves investment on seed health testing
- ☆ TPS crop comprises of segregating population which gives multiline effect providing better safe guards against disease epidemics
- ☆ TPS can be stored at room temperature under dry conditions without the loss of viability for many years
- ☆ Unlike seed tubers which are bulky, there is no transportation problem with TPS
- ☆ By using TPS a considerable amount of valuable potato stocks which are used as seed can be consumed as table potato. Currently the country is using 3.1 million tonnes seed tubers which by 2020 will be 6 million tonnes
- ☆ Tuber seed production for present area under potato requires 40,000 ha land whereas for the production of TPS for the same area only 2000 ha land is required
- ☆ Disease free potato tuber seed can be produced only in northern plains and north-western hills, whereas, TPS can be produced in most parts of the country
- ☆ In tuber seed propagation varieties are likely to undergo pathological and physiological degeneration but in hybrid TPS the hybrid vigour is ensured.

10. Constraints in Adopting TPS Technology

Raising of crop through TPS is labour intensive, needs constant attention and is more prone to damage by climatic vagaries like high temperature, heavy rains and draught *etc.* than the crop raised from seed tubers. TPS crop also takes about 20-25 days more for maturity than the conventional potato crop raised from seed tubers. Limited choice of hybrid TPS populations is also coming in the way of adoption of this technology. TPS being a segregating progeny, its produce is not so uniform as of the conventional potato crop and hence has lower market value.

11. Conclusion

If we summarise the experience gained with this technology, we can safely conclude that direct seeding (drilling) as method to raise a TPS crop has failed, transplants were better, but seedling tubers, still better. Technically, a crop raised from seedling tubers cannot be regarded as a TPS crop. The only practical way to raise a truly TPS crop is through seedling transplants. Further, for production of seedling tubers too, raising a successful seedling crop either in nursery beds or as transplants in field is mandatory. In its present state a successful crop of TPS seedlings can be raised at the most in restricted niches where small farmers grow potatoes with irrigation and intensive management in small plots. Those who have experience of raising vegetables' seedlings and enough family labor could better manage potato seedlings crop.

Breeding for hardiness and early bulking using conventional as well as non-conventional approaches is thus absolutely essential if this technology is to be popularized on a larger scale. On the other hand, a little variability within the produce of a TPS crop might not matter particularly for consumers in developing countries. Rather this might even be biologically advantageous for achieving good disease resistance. Identification of parents producing open-pollinated seeds under varying thermo-photoperiodic conditions and having reasonably uniform tuber progenies with high tuber yield is another aspect that needs to be further researched upon. This will help farmers to produce their own seed at a low cost and getting good returns for their efforts and time spent on production of seedling tubers and then raising normal potato crop from the seedling tubers. All this needs sustained efforts to overcome various intrinsic problems associated with TPS technology. The past long experience with this technology, however, makes one pessimistic about its future.

Chapter 3

Molecular Breeding for Biotic Stresses Resistance in Potato

Vinay Bhardwaj

ICAR-Central Potato Research Institute, Shimla, Himachal Pradesh

1. Introduction

The productivity of potato crop is constantly threatened by assorted biotic and abiotic stresses, which affects the quality and quantity of tubers, the main food storage organ. Among various biotic stresses, fungi (*Phytophthora infestans, Alternaria solani, Rhizoctonia solani, Verticillium dahliae*), viruses (PVY, PLRV, PVX, PVM, PVA, PVS), bacteria (*Erwinia species, Streptomyces scabies, Pseudomonas solanacearum*) and nematodes (*Globodera pallida* and *Globodera rostochiensis*) poses serious losses and remains as major impediment in potato production worldwide. Management of biotic stresses include both protective approaches such as breeding for resistance and therapeutic approach like chemical control. Employing durable resistant varieties is an important contrivance in the management strategy of biotic stresses. This reduces recurring chemical fungicide sprays, thus minimising the impact on environment, besides rendering economic benefits. The sub-sections in the chapter summarises conceptual information on significant biotic stresses, detailing their area of distribution, races/pathotype, mechanism and sources of resistance, screening methodology, economic importance and various breeding strategies adopted worldwide to scuffle the diseases caused by respective pathogens, with special state of past, present and future panorama of Indian breeding programme.

2. Late Blight

Late blight is caused by *infestans* which is highly variable and therefore, breeding varieties against it has been a seesaw story. Pathological variants (races) could be detected universally after the resistance genes from *S. demissum* were transferred to the commercial cultivars. In 1909, R.N. Salaman demonstrated the heritable nature of resistance to late blight in wild species *S. edinense*. Resistance to blight can occur both in foliage and tubers. However, breeders have largely neglected the latter. Broadly resistance can be grouped into two types: (i) race-specific resistance (also called vertical resistance or major gene resistance, qualitative or discontinuous resistance) and (ii) race non-specific (also called horizontal resistance or minor gene resistance, field resistance, polygenic resistance, quantitative resistance or partial resistance).

The race-specific resistance based on gene-for-gene relationship was initially identified in hexaploid (2n=6x=72), wild species *S. demissum*. It is expressed in the form of hypersensitive response of the tissue to all races of *P. infestans* that did not possess the corresponding virulence to the resistance genes (R-genes). Specific resistance is conditioned by a series of major dominant genes each of which is brought into action by distinct pathotypes; currently eleven such genes (ex-demissum) are recognized. *S. stoloniferum* has also been found to possess similar, if not identical, resistance genes. However, the R-genes wherever deployed were defeated in due course of time. In India, the process of transfer of R-genes from *S. demissum* background was started in mid-fifties and the first set of late blight resistant varieties was released for commercial cultivation in 1968. Of these, cv. Kufri Jyoti (possessing R-genes 3.4.7) became most popular which is still grown in several parts of the country. Since then a lot of varieties carrying R-genes have been bred and deployed across the country. In cv. Kufri Jyoti, matching virulences (3.4.7) were detected immediately after 5-6 years of its cultivation both in North-eastern and North-western hills. Both frequency of matching virulences, their combinations and disease increased rapidly making it completely susceptible by 1988. In India, this problem has been avoided by making adjustments in screening methodology. The seedlings in are challenge-inoculated with the most complex race (8-9 gene complex). The seedlings showing either complete susceptibility or immunity are discarded. The selected seedlings possess both R-gene resistances coupled with a high degree of field resistance. Race non-specific resistance is a quantitative and multifaceted trait, probably governed by many genes, it is, therefore, difficult to analyse in Mendelian ratios. Field resistance to late blight operates mainly through four factors, *viz.*, infection efficiency, incubation period, colonization rate and sporulation efficiency. Many host factors, environmental aspects, edaphic, nutritional and climatic have an effect on these four components of resistance. Besides, components of field resistance to tuber blight include the depth in the soil at which the tubers are produced, the ease with which the spores are washed down from the canopy into the soil, the rapidity of periderm formation and the resistance to wounding. Although, at the phenotypic level, both types of resistances can be easily identified, yet at the genotypic level these are almost similar. Genetic analysis of resistance to late blight using DNA markers showed that major genes for resistance (R-gene) are closely linked to the factors controlling quantitative resistance suggesting that there is no real difference between qualitative and quantitative resistance to late blight as far as the nature of the genes involved. The differences observed at the phenotypic level may be the result of various allelic and non-allelic interactions. Thus a complex picture emerges, which renders both selection for and evaluation of blight resistance a slow process.

Several wild *Solanum* species possesses high degree of resistance to late blight. Species like *S. bulbocastanum*, *S. demissum*, and *S. stoloniferum* had clones, which possessed low infection frequency. Besides clones of *S. bulbocastanum* and *S. demissum* developed only small lesions, whereas clones of *S. stoloniferum* possess high degree of resistance to tissue colonization. Umaerus demonstrated that *S. demissum* is a treasure of resistance. Besides R- genes, it also possesses field resistance, which operates primarily through low infection frequency at seedling stage. In German-Dutch potato collection nothing encouraging was found in *S. tuberosum* ssp. *andigena* and the primitive cultivars except an accession of *S. phureja*. However, resistance was detected in wild diploid Mexican species like *S. pinnatisectum*, *S. bulbocastanum*, *S. polyadenium* and *S. verrucosum*. It was also detected in Bolivian and Aregentinian species, including *S. chacoense*, *S. berthaulti*, *S. microdontum* and *S. vernei*. Although, initially *S. tuberosum* ssp. *andigena* was thought to be devoid of any resistance to late blight but, there has been renewed interest in it as andigena potatoes respond to selection for resistance and produce a number of highly resistant clones. Resistance has also been reported in the Russian diploid sources, including *S. polytrichon*, *S. simplicifolium* and *S. microdontum*, which showed resistance to tuber blight as well.

3. Viruses

Viral diseases are an important constraint for potato crop because of their systemic distribution in the host and are mainly responsible for the degeneration of seed stocks. At least twelve viral diseases are known to infect potato crop in India and elsewhere. Among them, viruses PVX, PVY, PVS, PVA, PVM and leaf roll (PLRV) are important. Introducing resistant cultivars is one of the most efficient ways of reducing the losses caused by viruses. Resistance genes to different potato viruses have been identified in many wild potato species. Some of these genes have been incorporated in many of the recently released potato cultivars.

The nature of resistance against viruses is of several types: i) tolerance, ii) resistance to infection, iii) hypersensitivity usually giving field immunity, and iv) extreme resistance or immunity. Tolerance to viruses in potatoes is usually considered a dangerous type of resistance. Resistance to infection can be defined as the type in which only a small percentage of infection appears in the field and is governed by polygenes. Hypersensitivity and immunity are on the other hand due to mostly single dominant genes. Out of the four types of resistance, immunity gives almost complete elimination of virus and is preferable over other types. However, in recent years, more emphasis is being given to vector resistance where the resistance sought is against the vectors (aphids and other vectors), the carrier of viruses and not against viruses themselves. There are three main groups of strains of PVY *viz.*, PVY ^{o} (common strains), PVY^{N} (tobacco veinal necrosis strains) and PVY^{C} (stipple streak strains). The strain-specific resistance is controlled by the resistance gene *NY* while the extreme resistance is controlled by the gene *RY*. *Ny* genes are found in large number of cultivars including Pentland Crown, Pentland Ivory, King Edward and Cana and in hybrids derived from wild species like *S. chacoense*, *S. demissum* and *S. microdontum*. The sources of *Ry* gene are *S. stoloniferum*, *S. hougasii* and *S. tuberosum* ssp. *andigena*. Accordingly the genes are named as *Rysto*, *Ryhou* and *Ryadg*. *Ry* is inherited as a single dominant gene hence easy to breed.

Mild, moderate and severe strains of PVA occur. They differ in severity of symptoms produced in potato cultivars. The gene *Na*, present in many cultivars, protects the plant from infection under natural pressure from PVA by means of a hypersensitive response. The gene *Na* is linked to gene *Nx*tbr, which controls the resistance to PVX. PVX strains can be separated according to their serological reaction into two main pathotypes: 1 and HB. Cockerham (1970) identified several genes conferring hypersensitivity *viz.*, *Nxtbr*, *Nbtbr*, *Nxchc* and *Rx^{n}acl*. Similarly, genes conferring extreme resistance are known to occur in several species (*Rxadg*, *Rxacl* and *Rxscr*). Strains of PLRV differ in the severity of the symptoms they produce on potato and on test plants. Resistance to PLRV has been found to be oligogenic and has been detected in *S. brevidens* and *S. tuberosum*. Transfer of resistance genes from *S. brevidens* has been achieved through protoplast fusion but from *S. tuberosum* it has been possible only through bridge crosses. Among the hexaploid somatic hybrids derived from *S. tuberosum* and *S. brevidens* by protoplast fusion, some hybrids with high PLRV resistance were obtained.

The development of cultivars with multiple virus resistance remains a challenge for the breeders. This may be because the breeder must select for many important characters, therefore, introducing even a few genes for resistance to viruses becomes a difficult task. At CPRI, parental lines having virus resistance in duplex/triplex/tetraplex form have been developed. The progeny of triplex/tetraplex parents are being crossed with nulliplex parents to produce almost cent per cent immune clones. After the early evaluation for viral resistance, an evaluation for horticultural traits could be done at the later clonal generations.

4. Bacterial Wilt

Bacterial wilt or brown rot, first reported in India in 1892, is the most destructive of all bacterial diseases. Incidence of bacterial wilt is wide spread in all mid hill regions of the country and pockets of Assam, Meghalaya and Maharasthra. The disease damages the crop in two different ways - premature wilting of standing crop and rotting of tubers in fields, transit and stores. It is primarily seed borne, but survives equally well in soil. Host resistance is hard to find because of lack of co-evolution of the host and the bacterium, high variability in the bacterium and instability of the host resistance.

Resistance to bacterial wilt is a partially dominant character and is more of a polygenic type. Inheritance of resistance and its expression is complex and both additive and non-additive gene actions are involved, but the latter component is more important. A gene-for-gene relationship is not applicable to bacterial wilt. Certain genes other than those 'for resistance alone' have turned out to have the novel (pleiotropic) effects in conferring the resistance once the potato plant has come into contact with pathogen under a certain set of environmental conditions. These genes were eventually called 'genes for resistance' once a certain level of resistance was detected. The major or minor status of these genes depends on the particular genotype of the pathogen, and the particular environmental conditions that influence their expression. Attempts to transfer resistance from wild *Solanum* sp. into common potato result in excessive recombination, resulting in breakdown upon intercrossing. Non-strain specificity and race- cultivar specificity are the common features required for resistance to bacterial wilt. Thus, the host genotype x pathogen genotype interaction in potato-*R. solanacearum* system seems to be artifactual. Both the host and pathogen are sensitive to environmental changes. Therefore, host genotype x pathogen genotype interaction may also be a result of host genotype x environment and/or pathogen x environment interaction.

The resistance in the clones of species like *S. phureja* mainly and a few other *Solanum* species have been exploited extensively in the South American countries, but the Indian isolates of the bacterium has proved to be strongly virulent making these sources ineffective. Resistance in *S. phureja*, the only species where it has been studied in detail, is strain- and temperature- specific and breaks down under the warm climates. Nematode injury also leads to its break down. A collection of nearly 500 clones of *Solanum* species which carry low to moderate degree of resistance *i.e. S. phureja, S. microdontum, S. canasense, S. stenotomum, S. pinnetisectum, S. sparsipilum, S. kurtzianum, S. jamesii, S. polytrichon, S. vernei, S. acaule* and *S. stoloniferum* and inter-specific hybrids between a number of above species and also resistant varieties developed so far in other parts of the world, *viz.*, Prisca, Cruza, Caxamarca, Molenera and Ampola were screened against different isolates of the pathogen. All these cultivars/cultures proved susceptible to Indian isolates except for the one clone of diploid *S. microdontum* showing a moderate level of resistance. The efforts made to transfer the useful resistance from this source into the *tuberosum* background *via* dihaploids resulted in the development of two promising meiotic tetraploids. However, in field tests these were also found to be susceptible.

5. Wart

Wart disease of potato caused by a phycomycetous fungus, *Synchytrium endobioticum* (Schilb) was first reported from India in 1953 in North Bengal Hills. To avoid its further spread to other parts of the country, this area was brought under domestic quarantine in 1959. This has helped in containing this dreaded disease in Darjeeling district only. The disease causes cauliflower like growths on tubers, stolons and stem bases. The heavy infection of disease causes rottage of entire produce and results in total loss of crop. Cultivation of wart immune varieties on a long-term basis is the only viable alternative. The resistance genes are available in a number of varieties of *S. tuberosum*. Besides, a number of wild species like *S. boliviense, S acaule, S microdontum, S. demissum, S. sparsipilum, S. polytrichon, S. simplicifolium, S. chacoence* f.sp. *boergerii, S. vernei* and *S. spegazzinii* are known to have resistance to the disease. Monogenic dominant mode of inheritance for at least the control of necrotic response has been proved. However, modifying genes are also present which condition the nature and extent of response. Systematic breeding programme for wart resistance started in 1964. The crosses between wart immune/resistant *tuberosum* parents result in recovery of quite high percentage of resistant clones than between resistant x susceptible parents. Using Adina x Ultimus, several late blight resistant and wart immune hybrids were developed. One of them was released for commercial cultivation under the name Kufri Sherpa in 1983, which did not become popular because of its poor keeping quality, unattractive dull white skin, round tubers with medium deep eyes. Indian cultivars, *viz.*, Kufri Jyoti, Kufri Chamatkar, Kufri Muthu, Kufri Sheetman, Kufri Bahar, Kufri Khasigaro and Kufri Kumar are immune to the race of *S. endobioticum* prevalent in the Darjeeling hills. These cultivars except Kufri Jyoti also didn't establish in the area because of local preference for varieties with red skin tubers. To develop a red tuber variety, Pimpernel was used as one of the parents. One hybrid from cross SLB/Z 405a x Pimpernel was selected and has since been released as cultivar Kufri Kanchan. This variety is immune to wart and possesses high degree of field resistance to late blight.

6. Nematodes

About 90 species of nematodes belonging to 38 genera have been reported to be associated with potatoes. Among these, the root knot nematodes and potato cyst nematodes have been recognized as the major pests. At least nine species of Root-knot nematode (*Meloidogyne* spp.) are known to infect potatoes. Among these, *M. incognita* is the most important throughout the world followed by *M. javanica*. The dominant root-knot nematode species affecting potato both in hills and plains is *Meloidogyne incognita* while *M. javanica* infestation is restricted to mid hills and plains.Heavily infested plants are stunted with yellowish leaves while no visible symptoms of nematode injury are seen under low infestation levels. The galls on potato roots are small and often go unnoticed. Wart-like structures are formed due to tuber infection, which reduces the commercial value and keeping quality of tubers. The second stage juveniles (hatched out from the egg masses laid by females) infect the young roots resulting in the formation of giant cells and the formation of syncytium and development of the nematode in the roots causes formation of galls or root knots. The nematode infection on tubers is characterized by the formation of typical wart like pimples on the outer skin.

Potato cyst nematodes (PCN), *Globodera* spp. also known as Golden nematode or potato root eelworms, is considered as one of the major pests throughout the world. Quarantine or regulatory actions are imposed against them in most countries. In India, the PCN was first detected in 1961 by Jones at Ootacamund in Tamil Nadu. The efforts to locate the source of resistance to cyst nematodes though began in 1968, a systematic breeding programme at CPRI was started in 1971. Over 2000 genotypes comprising group tuberosum, group andigena and tuber bearing *Solanum* species were screened and resistance was located in 20 accessions of 14 wild tuber bearing species *S. ehrenberjii, S. vernei, S. chacoense, S. phureja, S. demissum, S. gourlayi, S. microdontum, S. sucrense, S. tarijense, S. acaule, S. fendleri, S. multidissectum, S. oplocense, S. sparsipilum* and some accessions of *S. tuberosum* ssp. *adigena*. Simultaneously, efforts were also made to procure resistant breeding lines to both the species from the Netherlands and USDA. A parental line VTn^2 62.33.3 (*S. tuberosum* x *S. vernei* hybrid) received from the Netherlands, having resistance to both the species, was extensively used in crosses with late blight resistant cultivar Kufri Jyoti resulting in release of hybrid Kufri Swarna for cultivation in cyst nematode infested areas of Nilgiri hills.

Chapter 4

Breeding for Processing Potato and Quality Parameters

V.K. Gupta

ICAR-CPRI, Regional Station, Modipuram, Meerut, Uttar Pradesh

1. Introduction

Potato (*Solanum tuberosum* L.) is 3rd most important food crop in the world after rice and wheat in terms of human consumption and popularly known as the 'king of vegetables'. India is the second biggest potato producer and almost a third of all potato is harvested in China and India. Availability of indigenous varieties, good quality seeds and right package of agronomic practices triggered a revolution in potato production causing very fast growth in area, production and productivity. Potato production increased more than 31 times since 1950, due to both increased area and yield. The per capita potato consumption in India has risen from 12 kg/year in the early nineties to about 24 kg/year during 2018-19. Increased production, however, brought in a peculiar problem of glut situation when supply exceeded demand resulting in price collapse. In India till 1989, all the varieties were bred for consumption as fresh potato. Therefore, despite large quantities of potato available in the market, the availability of potatoes for processing was poor. Long day adapted European processing varieties were not successful in India, as they were not suited to short winter days. Some of the varieties like Kufri Chandramukhi, Kufri Jyoti and Kufri Lauvkar were used by the chipping industry from the produce of crop grown in warmer areas where night temperatures are relatively high (around).

Efforts to introduce European and north American processing and table varieties has not been successful in the country mainly because these varieties were bred for long day conditions whereas, the crop in India is grown under short winter days. Keeping in view the demand of quality raw material for processing, ICAR-Central Potato Research Institute launched a breeding programme in 1990 for developing indigenous processing varieties.

2. Indian Breeding Programme for Processing Potatoes

Under this programme, crossing were made at ICAR-Central Potato Research Station, Kufri (HP) in hills (31 ° N 77 ° E; 2402 m above msl) and raising and evaluating of segregating progenies and selection of suitable genotypes was done at ICAR-Central Potato Research Institute, Campus, Modipuram (U.P.) in the plains (29 ° N 76 ° E; 222 m above msl). Assessment of segregating populations for as many characters as possible in the initial stages was done to eliminate undesirable genotypes facilitating better assessment of the selected genotypes in subsequent generations (easily identifiable characters were considered for this purpose). Extensive evaluation of genotypes was done in subsequent generations for tuber dry matter, reducing sugars, chip colour and yield. Finally the selected hybrids are tested under industrial processing conditions.

Generally, any plant breeding program is based on two fundamental steps, (1) production of genetic variability through either conventional or innovative methods, and (2) efficient selection within the variability created. There are two methods to create variability, sexual hybridization and genetic engineering. Similarly, breeding varieties for value addition comprises of following major steps:

- ☆ Identification and evaluation of parental lines with desirable attributes
- ☆ Combining the desired attributes with high yielding base variety through hybridization or genetic manipulations
- ☆ Evaluation of segregating population/clones and selection for stability of yield and desired processing traits

Inheritance of processing especially related to quality traits is poorly defined and genetically least researched. The available information on various quality traits of interest is presented below:

3. Morphological Attributes

3.1 Tuber Shape

Tuber shape is a varietal character but it is also influenced by environmental conditions and cultural practices. Inheritance of the tuber shape is yet not finally concluded due to contradictory results of the researchers. Some researchers considered long tubers as being dominant to round tubers, but others suggested that round shape was dominant to either oval or long oval.

3.2 Growth Cracks

The causes of growth cracks are not well understood. It is attributed to fluctuating water stress or changes in tuber growth rates. In addition to the abiotic stresses, there is also a heritable component, which does not result in clearly delineated groups, suggesting that a number of minor genes may be involved. Therefore, for selection of this character one has to rely on observations in various environments over a number of years, as is presently practiced.

3.3 Hollow Heart

Hollow heart is a physiological defect resulting from an internal cavity of varying dimensions. It is found more frequently in the larger tubers. Development of hollow heart is often associated with periods of rapid tuber growth, which may have been preceded by a period of moisture or nutritional stress. The expression of this trait is affected by various environmental and genetical factors.

3.4 Greening

Potato tubers exposed to light in the field, during or after the harvest develop a green pigmentation, initially at the surface and subsequently throughout the whole tuber. This condition is caused by the formation of chlorophyll. Greening of potatoes is often associated with an increased level of glycoalkaloids, which impart a bitter taste to the tubers and are poisonous. Tuber greening is inherited in a quantitative manner with most of the variation being additive. However, broad sense heritability was sufficiently large to permit effective selection within the potato breeding programmes.

4. Biochemical Attributes

4.1 Dry Matter

Starch constitutes about 70 per cent of the dry matter (DM) content of potato tubers. There is considerable variation among cultivars for DM content and this trait is also influenced by the environment during the growing season as well as storage. Dry matter has been associated with many other quality parameters such as texture, suitability for processing and susceptibility to mechanical damage. It is polygenically controlled and responds to selection, but the environmental effects are so large that much effort is required to achieve significantly higher dry matter content that are currently available in the known varieties. Too high DM content also badly affects certain culinary properties, such as disintegration on boiling (sloughing), so that selection for high DM content *per se* is seldom a breeding objective. Moreover, high DM has a negative correlation with total tuber yield.

4.2 Reducing Sugars

Starch is the major carbohydrate source in potato tubers, yet small but varying amounts of sugars, namely sucrose, glucose and fructose are also present. Reducing sugar content of 0.1 per cent per 100 g fresh weight is ideal for processing into chips and higher than 0.33 per cent is unacceptable. Sugar content of tubers varies considerably among seasons, locations and also varieties. The chip colour has high heritability thus allowing fairly reliable prediction of parental values based on progeny means.

4.3 Glycoalkaloids

Potato tubers contain small quantities of naturally occurring steroidal glycoalkaloids, a class of potentially toxic compounds, found throughout the family Solanaceae. Cultivars vary with respect to their inherent glycoalkaloid content; at lower levels it is suggested that they may enhance potato flavour, but at higher concentration (above 15 mg 100 g $^{-1}$ fresh wt.) they impart bitterness and levels above 20 mg 100 g^{-1} fresh wt. are considered unsuitable for human consumption resulting in symptoms typically associated with food poisoning. There are conflicting reports available in the literature about the inheritance of this trait. Some reports suggest that it is inherited in a dominant manner with a couple of genes controlling this trait while others indicate its polygenic inheritance. Nevertheless, it is important that initial content of parental material may be assessed before starting a breeding programme particularly with wild species.

5. Indian Processing Varieties

Concerted breeding efforts for development of processing varieties by CPRI has led to develop and release of eight processing varieties namely Kufri Chipsona-1 and Kufri Chipsona-2 (1998), Kufri Chipsona-3 (2006), Kufri Himsona (2007), Kufri Frysona (2009), Kufri Chipsona-4 (2010), Kufri FryoM (2018) and advanced stage hybrid MP/6-39 (recommended for release). These varieties are gradually occupying large areas of various regions for meeting the requirement of quality potatoes for processing. There is however, need to continue to breed better varieties for different purposes, especially early-medium maturing varieties for warmer areas of country where growing period is shrinking and for vast Indo-Gangetic plains. The salient features of the processing varieties are given below:

Kufri Chipsona-1 (MP/90-83)

The variety is a selection from the progeny of the cross CP2416 x MS/78-79. The female parent CP2416 is the Mexican genotype MEX 750826, while the male parent is from Indian potato breeding programme which involved Kufri Jyoti and EM/H-1601 in its parentage. The variety produces white-cream ovoid tuber with shallow eyes and white-cream flesh. The variety produces ~30-35 t/ha yields under north Indian plains. Its dry matter content is 20-23 per cent and reducing sugar content is 10-75 mg/100 g fresh weight. The variety yields processing grade tubers to the tune of 60-75 per cent. The variety has good late blight resistance and is well adapted to Indo-Gangetic plains. The tubers possess excellent keeping quality. The variety is suitable for making chips, French fries and flakes.

Kufri Chipsona-2 (MP/91-G)

The variety is a selection from the progeny of the cross CP2346 (F-6) x QB/B 92-4. The female parent F-6 is an accession received from International Potato Centre, Lima, Peru, while the male parent is an indigenous advanced stage hybrid developed from Kufri Red and Navajo (from USA). The variety produces white-cream round tuber with medium eyes and cream flesh. The variety produces nearly 30-32 t/ha yields under north Indian plains. The tubers of this variety have reducing sugars in the tune of 30-100 mg/100 g fresh weight and dry matter content is 22-24 per cent.The variety has high degree of late blight resistance. The processing grade yield is 65-75 per cent and is highly suitable for chip making.

Kufri Chipsona-3 (MP/97-583)

The variety is a selection from the progeny of the cross MP/91-86 x Kufri Chipsona-2. The female parent MP/91-86 is a promising selection from the processing breeding programme. The variety produces white-cream ovoid tuber with shallow eyes and white flesh. The variety produces ~30-35 t/ha yields under north Indian plains. The tubers of this variety have reducing sugar content of 10-100 mg/100 g fresh weight and dry matter content 20-23 per cent. The variety has moderate resistance to late blight and is suitable for chips and flakes preparation. The variety yields reasonably good process grade tubers to the tune of 70-80 per cent.

Kufri Himsona (MP/97-644)

The variety is a selection from the progeny of the cross MP/92-35 x Kufri Chipsona-2. The female parent MP/92-35 is a selection from processing breeding programme. The variety produces white-cream ovoid tuber with shallow eyes and white flesh. The variety produces ~15-20 t/ha yields under hilly regions of the country. The variety produces nearly 30-35 t tuber yield when grown in north-western/central plains The tubers have dry matter content of 21-24 per cent and reducing sugars 10-80 mg/100 g fresh weight. The variety has a high degree of resistance to late blight. The variety is suitable for chips and flakes making

Kufri Chipsona-4 (MP/01-916)

The variety is selection from the progeny of cross Atlantic x MP/92-35. It produces high yield with higher proportion of chip grade tubers. It has early maturity with field resistance to late blight, thus helping farmers in saving on costly fungicides. The variety produces white-cream round-ovoid tuber with shallow eyes and white flesh. The variety produces ~18-22 t/ha yields during kharif crop in Karnataka and 30-35 t/ha tuber yield in rabi crop in plains. It is suitable for preparation of chips owing to its round-ovoid shape, high dry matter (>20 per cent) and low reducing sugars (40-80 mg/100 g fresh weight). It will fill the void of a suitable chipping variety from Karnataka, West Bengal and Madhya Pradesh where processors are in need of variety combining high tuber yield and high level of late blight resistance. Long dormancy and good keeping quality will help storage of this variety for longer period thus ensuring round the year availability of raw material to chipping industry.

Kufri Frysona (MP/98-71)

The variety is selection from the progeny of MP/90-30 x MP/90-94. It produces attractive white-cream long oblong tubers with shallow eyes and white flesh. The variety produces nearly 30-35 t/ha tuber yield in rabi crop in plains of the country. It possesses very good field resistance against late blight disease and has reasonably good frost tolerance. It is a good keeper under country store conditions and possess longer tuber dormancy period of more than 8 weeks. It possesses high tuber dry matter (22 per cent), low reducing sugars (30-80 mg/100 g fresh weight) and very good quality French fries can be prepared. The industrial testing has shown the superiority of this hybrid for French fry making in terms of taste, texture and colour.

Kufri FryoM (MP/04-578)

The variety is selection from the progeny of Kufri Chipsona-1 x MP/92-35. It produces attractive white long oblong tubers with shallow eyes and white flesh. The variety produces nearly 30-35 t/ha tuber yield in plains of the country. Tubers do not show deformities like cracking or hollow heart. It has field resistance against late blight disease. It is a good keeper under country store conditions and possess longer tuber dormancy period of more than 10 weeks. It possesses 20 per cent tuber dry matter, low reducing sugars (50-90 mg/100 g fresh weight) and very good quality French fries can be prepared.

Kufri Sangam (MP/6-39)

It is main-season dual purpose variety having medium maturity with high tuber yield (35-35 t/ha), field resistance to late blight, very good keeping/culinary quality and suitable for growing in central plains (processing and table purpose) and northern plains (table purpose). It produces attractive white-cream ovoid uniform tubers with shallow eyes and white-cream flesh. It possesses field resistance to late blight. The variety has been for recommended for release in 37th group meeting of AICRP (Potato) held during 3-4th September, 2019 at JNKVV, Jabalpur, Madhya Pradesh.

6. Future Thrusts

To keep the pace with the future needs of growing processing industry in India, suitable cultivars needs to be developed with the desirable traits which are expected are:

- Processing Varieties for growing under early to medium crop conditions to fit well in popular cropping systems,
- Varieties can withstand high temperature during tuberization and harvesting
- Stable processing quality after long term storage
- Management of acrylamide content
- Management of dry-rot, root knot nematode, fungal and bacterial diseases
- Development of varieties for resistance to low temperature sweetening for storing potatoes at 4–6 °C to control the development of diseases, weight loss and sprouting in store, with reduced reliance on sprout inhibiting chemicals is need of the hour.

Chapter 5

Nutrient Management in Potato Cultivation

Manoj Kumar

ICAR-CPRI, Regional Station, Modipuram, Meerut, Uttar Pradesh

1. Introduction

Precise and timely application of inputs are becoming domineering need in present day agriculture to make farming more efficient and environment friendly. A persistent balanced supply of nutrient elements is essential for normal plant growth. Nutrient management today is handling a complex dynamics of integrating soil, crop, weather, and management practices to achieve the best balance for maximizing profit while contributing to the conservation of our resources. It aims to achieve the goals of optimizing nutrient use efficiency, yields, crop quality, and economic returns, while reducing off-site transport of nutrients that may impact the environment. Several approaches have been advanced suiting to different production systems and with advancement of science several smart management options has progressed. Important issues to be addressed for smart management of nutrients are achievable optimum yields and/or crop quality, the available management options (including sources, timing and method of application, and timing of nutrients using a budget based on all sources and sinks active at the site) and the management of soil, water, and crop to minimize the off-site transport of nutrients from nutrient leaching out of the root zone, surface runoff, and volatilization (or other gas exchanges). In addition in plant system, nutrient interactions also needs to be governed for better production level and quality. Several approaches like soil test crop response, targeted yield equations, site specific nutrient management, real time N management using chlorophyll index, diagnosis and recommendation integrated system (DRIS), variable rate application technology, use of GIS and remote sensing, controlled release formulations for targeted delivery of inputs through nano or polymeric materials is of tremendous interest in achieving smart soil nutrient management. Recent times it has become almost imperative to develop recommendations using decision support systems based on various models and fine tune in season nutrient application depending upon real time information on crop health.

2. Nutrient Requirements of Potato

Nutrient requirement of potato crop which has to be fed from external sources in the form of fertilizer and manure depends principally on two factors, what is the yield level of crop (tuber) and nutrient supplying capacity of soil. In addition, it is further fine tuned by uptake efficiency of cultivars and management factors. In potato based cropping systems potato being most productive crop on per unit time basis becomes pivotal from productivity and profitability of the whole system. Potato has shallow root system compared to other crops limiting its foraging capacity in the soil. On the other hand, uptake of nutrients by potato per unit area and time is high due to fast rate of early growth and tuber bulking In view of these facts, direct application of fertilizers to potato is essential. Potato responds most favourably to direct application of fertilizer nutrients and organic manures compared to other crops in the sequence. In a cropping system, fertilizer requirement of potato has to be met on priority and any available farmyard manure (FYM) should be devoted to potato only.

A mature crop of potato yielding about 30 t/ha of tubers, on an average removes 120-150 kg N, 10-15 kg P and 120-140 kg K. The Indian soils are generally deficient in organic matter thus unable to release N at a rate required to maintain adequate N supply to the growing plant. Uptake of N ceases in the later part of crop growth, therefore, most of the N has to be supplied to potato in first 70 days, as the uptake is normally more than the average rate of 2 kg N/ha/day. During the early growth period, most N goes to foliage part particularly in leaves and rate of N accumulation is faster than dry matter production as a result N content in leaves goes as high as 6 per cent or even higher. Subsequently, as the growth takes place and development stage advances, with increased dry matter production, stolen formation, tuber initiation and tuber bulking, dilution of N takes place in foliage part and N is translocated towards tuber. At harvest, foliage N remains around 2 per cent with some variation depending upon cultivars and N supply level. Tuber N content is normally lower than foliage even at harvest on dry weight basis. Similarly, P content in plant tissues decreases with increasing developmental stage of crop from 0.3 to 0.4 per cent to around 0.2 per cent at harvest. Whereas, K content in plant leaves may go as high as 10 per cent in leaflets during active growth period (30 to 40 days after planting) which further decreases to around 2 per cent at harvest.

In addition to nutrient uptake, one important factor determining nutrient external supply need of crop is soil status of these nutrients including organic matter. In the Indo-gangatic northern plains of India soils are mostly poor in organic matter content. It's content is below 0.5 per cent in most soils of Punjab and western part of UP and it rarely goes beyond 0.7 per cent in soils of Bihar depending upon soil texture. On contrary, in hill soils of north western and north eastern hills organic matter content is normally more than 1 per cent. However, due to slow rate of mineralisation it is not able to meet the N requirement of crop. Therefore, application of nitrogen in form of fertilizers and manures becomes indispensable. Unlike OM, most potato growing soils in plains has shown build up of P. Continuous application of P to all the crops in sequence and only partial removal by crop uptake than applied P has attributed for build up of P in soils. Large numbers of soil samples analysed from different part of plains intensively growing potato has revealed P content above 20 ppm and there is likely to be no or little response to P application in these soils. However, in acid soils of NW and NE hills available P is normally below the critical level due to severe fixation of P by sesquioxide at lower pH and it's application responses by increasing tuber yield.

The story of K is not the same. It's availability is basically governed by soil mineralogy. In plains most soils are having micaceous phyllosilicate as dominant mineral which is a rich source of K. Therefore, soils are generally high in total K but only fraction of it is in available form because of dynamic equilibrium between fixed, non-exchangeable and exchangeable K. With high crop intensity, high K removal and due to negative balance in most of potato based cropping systems there has been considerable mining of this nutrient and the soils are becoming deficient in K with time. It is likely that at most places potato responses very well to its application. On the other hand, in acid soils which are dominant in kaolinite or sesquioxide like mineral are inherently poor in potassium and therefore it application is must for successful production of potato.

Unlike major nutrients, micronutrients are required in only infinitesimal amounts. For example, one hectare of potato crop contains about 200 g Zn, 800 g Mn, 1300 g Fe and 50 g Cu. Even though these elements are needed in traces, many soils do not supply them in needed amounts required for healthy growth and optimal yield of potato.

3. General Recommendations

The optimum rate of NPK fertilization in different zones based on tuber yield response in trials conducted in different potato growing agro-climatic zones in India is given (Table 5.1). Fertilizer management of potato crop is usually based on averages on regional basis known as package of practices. These recommendations take into account the general growing conditions of the regions including soil health, climate/weather and cultivars grown in that part. Processing varieties of potato Kufri Chipsona 1 and Kufri Chipsona 3 and the likes grown specifically to produce large processing grade tubers require higher rate of fertilization by 33 to 50 per cent than recommended. However, this approach give most crude recommendation as taking single value for thousands of hectare crop area of any agro-eco region is too much simplification of facts to be exact or precise. Such an approach results in suboptimal or excessive application of inputs resulting in inefficient utilization of costly fertilizers on one hand and risks environmental degradation on the other.

Table 5.1: Optimum Fertilizer Requirement of Potato Crop in different Potato Frowing Zones of India

Potato Growing Zones	*Nitrogen*		*Phosphorus (P_2O_5)*		*Potassium (K_2O)*	
	Rate (kg/ha)	*Response (q/ha)*	*Rate (kg/ha)*	*Response (q/ha)*	*Rate (kg/ha)*	*Response (q/ha)*
North-western plains	180	110	80	26	120	57
West-central plains	190	127	80	29	120	34
North-eastern plains	200	166	80	25	160	67
Plateau region	115	85	50	21	110	36
North-western hills	130	148	120	112	100	64
North-eastern hills	100	85	160	66	130	64
SWB* hills	100	85	150	52	110	52
Southern hills	115	120	150	35	110	38
Mean	**141**	**116**	**109**	**46**	**120**	**51**

* Sikkim and West Bengal hills.

3.1 Soil Test Basis

Soil testing can in general be used for assessing the suitability of soil for potato or other crops cultivation. However, more specifically soil testing is important for establishing the nutrient supplying capacity of soil and developing the fertilizer manure application schedule for any crop. While simply testing the soil pH, EC, OC, texture can give idea of soil being suitable for any given crops. For assessing nutrient supplying capacity of soil and quantification of amount, source, method and time of external nutrient application, there is need to do very precise chemical analysis. The analysis must suit to particular soil type and agro-climatic situations. The test values may not indicate quantitatively the amount of nutrient present in the soil but it must correlated well with the uptake and yield levels. For potato crop important parameter needed to be tested for judging nutrient supplying capacity involve soil organic matter, available nitrogen, phosphorus, potassium, secondary nutrients (Ca, Mg, S) and micro-nutrients.

3.2 Organic Matter

Organic matter is important as it improves soil fertility (nutrient supply) and has a great influence on the soil's physical environment and population of micro flora and fauna in the soil. In addition, soils rich in organic matter are known to be sufficient in micronutrients *viz.*, Zn, Cu Fe and Mn. The test for estimation of organic matter content in the soil by Walkley and Black method where in organic matter is oxidized with chromic acid-sulphuric acid is frequently used in soil testing laboratories.

3.3 Nitrogen, Potassium and Phosphorus

Nitrogen, Potassium and Phosphorus are the three major nutrients required in terms of their removal and uptake by the potato tubers and haulms and thus are to be applied in substantial quantities. In potato crop nitrogen is the first limiting nutrient in all type of soils. Likewise, potassium is the second major limiting nutrients in alluvial, red soils and brown hill soils whereas, phosphorus is the second major limiting nutrient in acidic soils. The soil test choice, therefore, depends upon the type of soil. It is more valid in case of P whose availability depends upon the form of P present in the soil.

3.3.1 Nitrogen

Organic carbon has been a very good indicator of available N, as major part of N in soil is present in organic form which has to be mineralized in NH_4^+ and NO_3^- from before being absorbed by potato plants. NO_3^-N being present as a tiny fraction of total N gives an idea at what rate organic N is being mineralized for uptake by plant. Chemical methods based on oxidation of organic matter are available for assessing the soil N availability. Some of the tests currently used for judging N availability to potatoes are total N, $KMnO_4$-oxidisable N, $K_2Cr_2O_7$-oxidisable N and organic matter. The tests involving mild chemical oxidation of organic N which correlates well with N uptake is widely being used for judging the N availability. Although, total N gives a true picture of soil N but in soil testing it is seldom used as total N estimation by Kjeldahl method is a time consuming process. $KMnO_4$-oxidisable- N (Subbiah and Asija 1956) is based on oxidation of soil organic matter in alkaline $KMnO_4$ medium (NaOH) and is commonly used in soil testing laboratories in the country. Likewise, $K_2Cr_2O_7$-oxidisable N (Sharma and Sud, 1981) is based on oxidation of organic matter under acidic conditions using H_2SO_4 and holds good for all type of soils. In heavy soils of Patna (with considerable amount of clay and organic matter) in eastern Indo-Gangetic plains, the potato response to N was significantly related to NO_3^- content in the soil.

3.3.2 Phosphorus

Plant requirement is mainly met by $H_2PO_4^-$ ions in acidic soils and HPO_4^{2-} and $H_2PO_4^-$ ions in neutral and alkaline soils. Therefore, soil pH plays a vital role in governing the availability of P to the plant. Phosphorus is retained in soil as insoluble P compounds of Fe, Al and Mn in acidic soils and Ca and Mg in alkaline soils, hence the choice of method for estimating P depends upon the quantitative amount of each fraction present in the soil and also pH of the soil.

Bray P (0.03 N NH_4F in 0.25 N HCl) method is able to differentiate absorbed and acid soluble forms of P and thus useful for soils having high amount of Al and Fe compounds. On the other hand, Olsen's method (0.5 M $NaHCO_3$, pH 8.3) extracts phosphate from Fe and Al only partly and low Ca activity prevents the precipitation of liberated phosphate as calcium phosphate. It has been found to be a good index of judging P availability in alluvial and black soils containing Ca and Mg forms of P. For acidic hill soils, organic carbon test is equally effective in predicting P and K availability to potato.

3.3.3 Potassium

The dynamic equilibrium between different form of K (fixed, non-exchangeable and exchangeable) in the soil at any time governs the K availability to the plant. The choice of method for assessing the K availability from the soil depends upon the soil type, soil capacity to release K and nature of crop species, its ability to utilize K from soil pool and fertilizer/manure. A wide range of chemical extractants have been used in the soil testing laboratories with each method having its own advantage. The validity of each test further depends upon its ability to correlate crop response to the amount of K extracted by that method. In potato, most frequent tests used is 1N ammonium acetate and it has been found to be good index for judging K availability to potatoes in acidic (North western and eastern hills), alluvial (Indo-Gangetic plains) and black (Plateau) soils. Exchangeable K determined by Morgan method is equally good for alluvial soils. The Q/I parameters *i.e.* intensity of K in relation to its capacity in the soil has also been found to be positively associated with K availability to the plant.

For the soil test of S, $CaCl_2$ or Ca $(H_2PO_4)_2$ containing 500 ppm P are most commonly used as extractant for measuring S availability to the potato crop. Availability of SO_4^-S from organic S bound with soil organic matter is closely associated with various soil tests. For other secondary elements *i.e.* calcium and magnesium, the most appropriate method is 1N ammonium acetate having pH 7.0 where in the two cations Ca^{2+} and Mg^{2+} are completely exchanged by NH_4^+ ions from ammonium acetate in absence of excess $CaCO_3$. The amount of Ca and Mg can be best estimated volumetrically by EDTA method. The former can also be conveniently determined by flame photometer.

3.3.4 Micronutrients

The Di-ethylene tri-amine penta-acetic acid (DTPA) test for micronutrients availability in the soil *viz.*, Zn, Fe, Mn and Cu has found wide application in potato. Among other methods for micronutrients availability in soils, use of 0.1N HCl and 0.05N HCl+0.025 N H_2SO_4 have also been used for estimating Zn and Cu held by soil organic matter. Trehan and Grewal (1985) found that NH_4HCO_3-DTPA test has advantage over other methods as it is able to estimate K, Ca and Mg in the same extract in addition to micronutrients.

4. Interpretation of Soil Test Value

Moderation in rate of fertilization is possible on the basis of soil test value (STV). Conventionally, based on soil test value, soil is classified in three (Low, Medium and High), five (very low, low, medium, high and very high) or more classes of nutrient supplying capacity and accordingly, external doses of nutrients are increased or decreased. Soil test based fertilizer recommendation results in fertilizer economy through balanced application and enhances tuber yield. Soil tests are more reliable, convenient and efficient as they are performed before planting the crop to decide on rate of fertilization precluding nutrient imbalances in the crop. The safe level is the nutrient concentration in the soil is maintained appreciably above the critical deficiency concentration for optimum potato production.

Another, approach in the interpretation of soil test values involve determining a fixed value below which crop is likely to response nutrient application. These estimates are called critical limits. For this purpose most widely used method was developed by Cate and Nelson (1956 and 1971) in which per cent yield or per cent Bray yield is plotted in scattered diagram against soil test values within a particular soil group but varying in soil nutrient availability indices. Further, points are tried to be divided into two groups responsive and non-responsive. This can be done graphically (Figure 5.1) or statistically. In statistical technique the soils can be classified into more than two classes *viz.*, low, medium and high ranges of nutrient level.

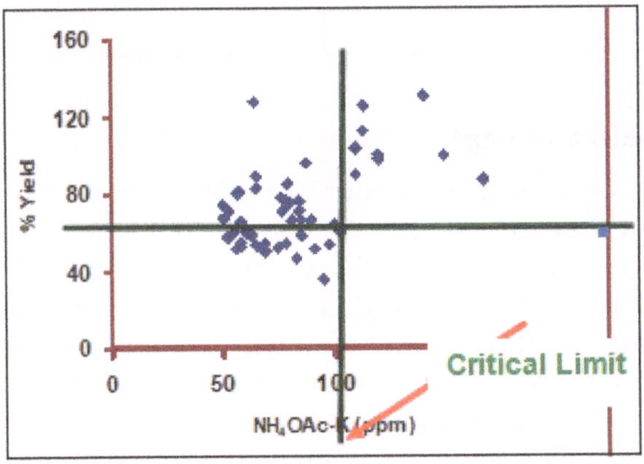

Figure 5.1: Graphical Representation of Soil Test Value for available K against per cent Yield.

Critical limit concept has practically been more used in micronutrients as their uptake is always in very small quantity. It is advisable to group soils into responsive (deficient) and non-responsive. The critical limits and methods to ameliorate the deficiency is presented in Tables 5.2 and 5.3.

Table 5.2: Critical Deficiency Limits of Micronutrients in Soil and Plant for Potato Crop

Micronutrient	*DTPA-Extractable Micronutrients in soil (ppm)*	*Critical Concentrations of Micronutrients in 4th Leaves (ppm)*			
		Days after Planting			
		30	*40*	*50*	*60*
Zinc (Total)	0.75	27.5	25.0	18.0	16.0
Copper (Total)	0.32	8.5	8.5	6.0	4.5
Iron (Ferrous form)	6.60	—	80.0	83.0	73.0

Table 5.3: Doses of Micronutrient Application for Correction of their Deficiency in Potato

Micronutrient	*Soil Application (kg/ha)*	*Spray Application (g/100 lit. water)*	*Tuber Soaking Treatment (g/100 lit. water)*
Zinc sulphate	25	200	50
Ferrous sulphate	50	300	75
Manganese sulphate	25	300	50
Copper sulphate	25	200	50
Ammonium molybdate	2	100	20
Sodium borate	2	100	20

5. Targeted Yield Concept

Target yield concept takes account of total nutrient requirement of the crop and nutrient supplying capacity of the soil. In this approach, it is assumed that there is a linear relationship between increasing yield level and nutrient uptake by the crop and for obtaining a particular yield, the plant must up take a definite amount of nutrients. The total requirement of nutrient for a given yield level is calculated on the basis nutrient concentration. The fertilizer requirement is estimated taking into consideration the contribution from soil available nutrients. The needed parameters to be experimentally obtained for a given soil type, crop and agro-climatic conditions for formulating fertilizer recommendation for targeted yield are nutrient requirement of crop for unit quantity of produce, the percent contribution from soil available nutrients to total uptake and the percent contribution from the applied fertilizer nutrients to total uptake.

The yield targeting equations are reasonably good in predicting nutrient requirement, though, it has some limitations. It must be used for similar soils with similar range of soil test values, occurring in a particular agro-eco-region, yield target chosen should not be unduly high or low and should be within the range of experimental yields obtained and recommended agronomic practices should be followed while raising crops.

The target yield approach matches fertilizer application with soil available nutrients and crop requirements as they vary across the field. Precision fertilization on soil test and target yield based fertilizer prescription equations increase the fertilizer use efficiency as compared to the recommended fertilization of potato.

6. Site-Specific Nutrient Management

Site-Specific Nutrient Management (SSNM) is an approach for feeding crop with nutrients as and when needed. In this approach there is possibility to dynamically adjust fertilizer use to optimally fill the deficit between the nutrient needs crop and the nutrient supply from indigenous sources, including soil, crop residues, and manures. In this approach mainly three nutrient elements (N P and K) need of crop is taken into account adjusted to season and location specific situations. The demand for N is strongly related to growth and development stage of crop. In order to achieve high yield, potato must not starve N supply at different stages up to tuber bulking. The SSNM approach envisages fertilizer N application in more number of doses to ensure the supply of sufficient N in synchronisation with the crop need at the same time not putting pressure on environment. Most convenient way can be to meet N crop need by rapidly assessing leaf N content, which is closely related to photosynthetic rate and biomass production and is a sensitive indicator of the N demand during the growing season. A chlorophyll meter like SPAD can provide a quick estimate of the leaf N status can effectively be used for real time N management. To make it more cost effective leaf colour chart has been developed for rice crop. This needs to be developed for high N requiring crop like potato also. For optimum use of P and K to overcome their deficiencies, to avoid the mining of soil and to allow best N management approaches based on decision support system with underlying the scientific principles of models, whereby, fertilizer P_2O_5 and K_2O rates are obtained from an estimate of attainable yield target (obtained from over fertilized reference plots, OFR) which must be realistically attainable and either the P- or K-limited yield determined by the nutrient omission plot technique (nutrient omission plot).

6.1 Nutrient Omission Plot Technique

In Nutrient omission plot technique, one plot is grown with abundant fertilizer supplements (NPK plot or NPK plus micronutrient plot) and the yield thus achieved is used to calculate the full demand for P and K. The attained yield can serve as a yield target. Other plots, without added P fertilizer or without added K provides an estimate of P-limited and K limited yield. The yields are then used to estimate optimal rate fertilizer P_2O_5 and K_2O rates, which overcome

P and K deficiencies and include sufficient P and K to prevent depletion of soil fertility arising from their long-term removal.

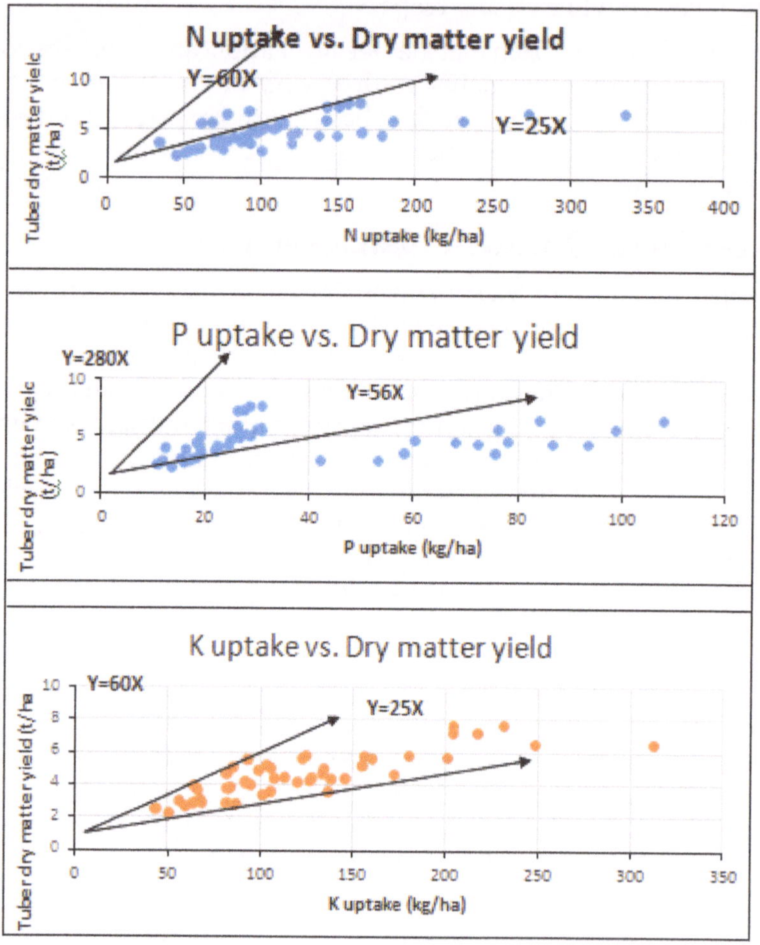

Figure 5.2: The Relationship between Tuber Dry Matter Yield and NPK Uptake. The upper and lower lines indicate tuber yields with maximum dilution and accumulation of nutrients.

6.2 Model Based Approaches

One recent approach on fertilizer recommendation for potato was based on QUEFT (Quantitative Evaluation of the Fertility of Tropical Soils) model. The model was calibrated for SSNM (site-specific nutrient management) of NPK in potato crop. For calibration of QUEFTS model, extensive review of literature was carried out and the data were collected and compiled on initial nutrient status, nutrient applied, uptake and yield under different experiments carried out at different locations during last 15 years. For working out the relationship between yield and nutrient supply, four steps were followed in the model; determination of indigenous nutrient supply, determination of the uptake of N, P and K as functions of potential supply of N, P and K (from soil + fertilizer), estimation of yield ranges as functions of actual uptakes of N, P and K when they are maximally accumulated and maximally diluted, and estimating the final yield by combining three yield ranges (one each for N, P and K) considering NPK interactions.

6.3 Real Time N Management

The environmental concern about NO_3^- pollution from heavily fertilized crops has led to development of chlorophyll meter based technique for efficient N fertilization. The SPAD chlorophyll meter is one of the most commonly used tool for nondestructive diagnosis of leaf chlorophyll content which in term is supposed to be the function of N content in leaf, provided other management variable are not much different. Minolta SPAD 502 measures light transmission in the red (650 nm) and near infrared (940 nm) parts of the spectrum to estimate leaf chlorophyll content. In crops like rice and wheat the SPAD has been extensively used for in season real time fine tuning of N application for its quantity and time of application. At many places it has been shown that crop based approaches to manage N is an improvement over soil based approaches, particularly for in season decision making on N application. Studies have found the critical limits of SPAD value was 48.03, 44.80 and 36.89 at 25, 40 and 60 DAP, respectively and the best period to detect N deficiency in potato for corrective N-fertilization was found to be 33-47 DAP. However, further studies under rainfed conditions in hills and with different cultivars in plains it was observed that SPAD values are dependent on rainfall, irrigation schedule, other nutrient status of soil and other seasonal variations (radiation *etc.*). Therefore, fixing a critical SPAD value for N application may not hold good always. To improve upon this, use of an over fertilized reference plot (OFR) is advocated where the crop does not show any sign of N stress. The top dressing with certain amount of N fertilizer to the test crop is done when SPAD value reaches 95 to 90 per cent of that of over fertilized reference plots.

7. GIS, Remote Sensing and Precision Farming

Remote sensing and GIS based technologies are being utilized increasingly and effectively for sustainable agricultural development and management. Recent developments in precision agricultural technologies are also very exciting. Remote sensing has been used in soil mapping, terrain analyses, crop stress, yield mapping and estimation of soil organic matter. Spatial mapping of available nutrient in potato growing pocket has very well been done in many parts of India using remote sensing and GIS tools. This involves classification of satellite imagery to delineate potato growing area and then strategically sampling soil from these areas using GPS so that entire pocket is covered in systematic manner. Further these soil samples are analyzed for desired soil properties and interpolation of these values is done using geo-statistical tools like kriging *etc.* to cover unsampled area and geo-spatial maps are developed for different nutrients. These maps can be classified giving reasonable range of soil properties for getting limited number of management zone and further recommendation map can be developed for the entire pockets.

Further remote sensing in combination with GIS at high resolution is being extensively used in precision farming because of its capacity to monitor spatial variability. Grid sampling at certain spacing, analysis of these samples for desired properties interpolation of these analyses results to unsampled locations by geo-statistical technique and further classification of these maps into limited number of management zones are important steps in precision farming. Another approach to make it more economical with some compromise on precision is to divide whole region into production level management zone, sub-region of a field having similar yield limiting factors and nutrient and other inputs are applied accordingly.

In Precision Farming, however, the input variable such as fertilizer are applied in right amount at right place and right time (variable rate application technology) as per the demand of the crop rather than prophylactic application. Thus it helps to improve nutrient use efficiency, economy and sustainable use of natural resources with minimizing wastage. Variable rate technology (VRAT) is application of input taking account of spatial variability of soil so that no part of the field is over fertilized or under fertilized. It is well known that soil properties are not uniform across the land escape and therefore, uniform application of fertilizer cannot give maximum efficiency. There are variety of VRAT available that can be grouped broadly into two types, namely GPS/Map based or sensor based. In map based methods application rate of input/fertilizer is based on digital prescription map which has been developed based on soil information, crop information, remotely sense images and numerous other informations sources. These systems have ability to determine location of machine in field and relate the position to the prescription map to decide the application rate. Thus, the concentration of input changes as applicator move through the field based on information stored in map.

While, in sensor based systems input is dispense at the rate sensor signals, based on soil properties and crop characteristics while on go. It uses basically a ground-based soil or crop canopy reflectance sensing, technically a type of remote sensing, is used to assess crop/soil condition and determine nutrient input recommendations. It does not required map or positioning systems. Based on instant information, a control system calculate the input/nutrient need of soil and transfer the information to controller/dispensing unit which delivered the input to the location measured by the sensor. Unlike aerial or satellite sensing, ground-based sensing need not be compromised by clouds and the sensors can be attached directly to an applicator so that the fertilization can be accomplished within seconds of crop sensing. Both these systems have limitations and advantages, therefore, some systems have also been developed to take advantage of both the methods.

8. Conclusion

Nutrient management in potato has evolved through phases and a paradigm shift has taken place. No more its' only objective is increasing yield, but it has to address the complex dynamics of integrating soil, crop, weather, and management practices to achieve the best balance for maximizing profit while contributing to the conservation of our resources. Therefore, it need to achieve the optimum nutrient use efficiency, yields, crop quality, and economic returns, while reducing off-site transport of nutrients that impact the environment. From region-based approach to model based expert system and decision support systems the nutrient management has progressed a long way. Further, we are progressing towards precision farming. However, challenge still remains how these recommendations get adopted at farmers' field.

Chapter 6

Water Management in Seed and Ware Potato Production

Name Singh

ICAR-CPRI, Regional Station, Modipuram, Meerut, Uttar Pradesh

1. Introduction

Potato (*Solanum tuberosum* L.) has a sparse and shallow rooting system and nearly 70 per cent of total water is used by the crop from upper 30 cm soil layer. It requires 400-600 mm water, which depends upon various factors *viz.*, climatic conditions, soil type/texture, growing season, duration and purpose of crop and irrigation methods *etc.* Water requirement of potato is quite high and frequent irrigations are necessary to maintain optimum soil moisture throughout its growth. Excess or limited availability of water, light textured soils having low water retention capacity and shallow rooting system of plants are some major factors which adversely affect the productivity. The yield and quality both depends upon a proper balance between soil air and soil moisture, which can be achieved only by adopting a sound water management program. Efficiency of water could be enhanced if water losses are minimized. Evaporation of water can be checked by use of mulches whereas percolation losses by resorting to drip and sprinkler irrigation. Water needs of crop constitute the amount of water required to mature the crop, encompassing consumptive water use (evapo-transpiration plus water required to build up plant tissues) and economically unavoidable water losses in the form of deep percolation and surface run off.

2. Techniques Used for Irrigation Scheduling

One pre-planting irrigation is usually given for land preparation and one pre-emergence light irrigation to ensure quick and uniform plant emergence in field. The various irrigation approaches have been used for working out water requirement of crop:

2.1 Time Interval Approach

In this approach different depths of water were applied at different intervals for irrigation scheduling. Studies based on delta interval approach revealed that about 45-65 cm water was needed by crop, and irrigations were required to be applied at 7-10 days interval.

2.2 Soil Moisture/Tension Approach

According to this concept, irrigations were scheduled whenever soil moisture tension increased to defined levels, and sufficient water was applied at each irrigation to bring back the soil moisture in the root zone to the field capacity. Tensiometers were used to monitor the change in soil moisture tension. The highest potato yields were obtained by frequent irrigations scheduled at low soil moisture of 0.2 to 0.3 atmosphere tension at 15-22 cm soil depth. With this regime at least 63 per cent of available soil water remained in the root zone before irrigation. This corresponded to the irrigation interval of 7-12 days.

2.3 Climatological Approach

The role of climate in governing water needs of crop was also recognized and criteria based on evapo-transpiration and evaporation was utilized for irrigation scheduling. Experiments designed on the basis of CPE values revealed that potato yields in sandy soils were highest when irrigations were applied at 15-20 mm CPE, on sandy loam soils at 25 mm CPE, on clay loam soils at 40 mm CPE and in loamy sand soils at 20 mm CPE. However, in another approach, irrigations applied at IW:CPE ratio of 2.0 produced yield comparable to that based on 0.25 bar soil moisture tension, but, former approach saved 12 cm water in loamy sand soils of Ludhiana.

2.4 Leaf Water Potential Approach

Plant is the integrating component of soil water status and atmospheric conditions. As such water status of the plant in terms of leaf water potential may serve as a sound basis for scheduling of irrigation. The research work carried out on this aspect in tarai soils of Pantnagar has shown that the irrigation at -2.5 bar (Ψ1) gave comparable results with the approaches based on days interval, soil moisture depletion and CPE values, but more work is required on this aspect in order to draw practical conclusions.

2.5 Stress Day Index Approach

Hukkery *et al.* advocated that SDI approach for timing of irrigation for potato by subjecting the crop at different stages of its growth to different pre-defined level of soil moisture stress. Stress day index was formulated as: SDI = CS x SD

Where CS is the crop susceptibility factor indicating the magnitude of yield reduction due to stress at a given stage, when compared without stress. SD is the stress day factor indicating the degree of imposed stress which may be in terms of soil or plant water deficit. SDI concept not only reveals the critical stages of crop growth but also gives quantitative information on yield reduction due to a given stress. When SDI and CS are known for different stages of growth, the SD for scheduling irrigation (SDI/CS =SD) could be worked out. However, a strong negative co-relation of the yield with the values of SDI was observed.

3. Critical Stages of Crop

Studies conducted by with-holding irrigation at one or the other growth period without given any consideration to actual soil moisture content, have revealed controversial results. According to some reports, the crop does not tolerate any stress at any growth stage, while in others, some stages are reported to be more critical than others. However, most of the reports, have indicated that stolonization followed by tuberization are most critical stages of potato.

4. Irrigation under Specific Situations

In North-western parts of India, where frost is a problem, the crop in such areas may have to be irrigated even at alternate days to prevent frost damage. The last irrigation to the main crop could be timed 10-12 days before the haulms die down or are killed deliberately. Early crop is harvested green and generally irrigated till harvest, although it is desirable to withhold the last irrigation a few days before harvest particularly if the produce has to be transported to the long distances. Lifting the crop when the soil is too moist impairs the transportability and keeping quality of the produce.

5. Need for Irrigation

Potato crop grown in hilly areas is mostly rain fed. Total rainfall received during crop season is adequate or more than crop water needs, but crop suffers from drought or excess water as the rains are not equally distributed. In certain areas like The Nilgiris hills, the rainfall is low during winter (Jan-May). Since this period is generally dry and irrigation is often required. Whereas, in plains and plateau areas, the potato grown in winter and rainfall is insufficient to meet the water requirement of crop. Hence, light and frequent irrigations are necessary to obtain the sustainable yield.

6. Water Management in Plains

About 80 per cent of potato cultivation in India is concentrated in north-western plains where it is grown under assured irrigation in *rabi* season (October-February). The irrigation method used in potato crop must ensure the adequate and uniform application of water with minimum losses as deep percolation and runoff. The water applied at each irrigation should be just sufficient to restore the soil moisture in the plant root zone at field capacity or to make up the soil moisture deficit. The different methods used for applying irrigation to potato are given below:

6.1 Furrow Irrigation

Among various methods *viz.*, surface, sub-surface, drip and sprinkler methods, the furrow irrigation method is most common and popular for irrigating the potato crop. The field is laid out in alternate ridges and furrows, and water should be applied in furrows. Under mechanized cultivation, the crop is irrigated in long furrows spaced widely at a distance of about 60 cm, where mechanization is not followed, the irrigations can be given in short furrows spaced at 45-50 cm apart. Whatever the form of furrow irrigation, water should never overflow the ridges as this seals the surface soil pores, thereby raises the bulk density and leads to anaerobic conditions. Such adverse effects result in seed tuber decay at planting especially in plains when soil temperatures are high. -Field channels of about 70-100 cm width serve bilaterally a set of 6-8 furrows in each bed. The ridges may be wetted only up to about two-third of height leaving the rest to be moistened gradually by capillary movement of water. Generally, 4-5 cm water should be applied at each irrigation. However, the irrigation efficiency under furrow method is low 40-50 per cent.

6.2 Micro-Irrigation Methods

The modern methods improve the efficiency of water and nutrients by increasing availability to the plants, they also improve hydro-thermal regimes and physical conditions of the soil by maintaining the proper balance between soil air and soil water in the root zone for better plant growth and development which results in to higher quality produce. With the increasing water scarcity and recurring droughts in many parts of country, the use of drip and sprinkler irrigation have become extremely important.

6.2.1 Drip or Trickle Irrigation

Drip method involves slow application of water in to root zone which provides scope to utilize water and nutrients efficiently. Moisture availability is always equal to field capacity in surface soil throughout growth period. Under drip irrigation about 40-50 per cent less water is required, it also enhances the production and keeps productivity at high level (20-30 per cent) in comparison to furrow irrigation. The raised bed two/three rows planting are beneficial. Increase tuber yield is contributed by increase in tuber size and numbers. Besides, the drip method provides several advantages, including no water runoff from field, precise application of water, ability to apply nutrients through fertigation, minimum leaching losses of water and nutrients with reduced canopy moisture and risk of foliar diseases. Earthing-up practice could also be dispensed. However, the initial cost consisting of pipes, filters, pump, fertilizer tank, laterals and drippers *etc.* is high, and technical know-how is required to install and run the system efficiently.

6.2.2 Sprinkler Irrigation

The sprinkler method gives uniform distribution of water without compacting soil surface, reducing water losses in terms of deep percolation and surface run off. It could also be used in special situations *viz.*, in undulating topography, extremely sandy soils with very high water intake rate or fine-textured soils with very low intake rate and costly and scare water supply. It is also beneficial in frosty nights as a preventive measure against freezing injury to the plants. Uniformity of water is assured only by overlapping pattern of sprays from different size nozzles. Nitrogen applied directly as foliar spray, helps in better growth and development. System is expected to economize water by 30-40 per cent with higher quality produce. Under sprinkler irrigation the closer spacing opted advantageously. Raised bed two/three rows planting are beneficial. However, the initial cost consisting of pipes, filters, pump, fertilizer tank, risers and sprinkler heads *etc.* is high, and technical know-how is required to install and run the system efficiently.

7. Scheduling of Micro-irrigation

Irrigation scheduling depends on moisture retention capacity of soil, growth stages, atmospheric factors and rooting patterns of potato cultivars. The proper irrigation scheduling is most important aspect for efficient utilization of available water resources. One pre-planting irrigation is usually applied for land preparation and one pre-emergence light irrigation can be given to ensure quick and uniform emergence. Thereafter, irrigation depth and time for drip and sprinkler methods can be worked out by using the following formula:

$$\textbf{Application rate (mm/hour)} = \frac{\text{Discharge of water (lph)}}{\text{Spacing between two drippers/sprinklers (m)} \times \text{Spacing between tworows of system (m)}}$$

Drip irrigation should be applied at alternate day. Experimental results advocated that integral drip laterals with inline/inbuilt/flat dippers (16 mm OD/30 cm/21 ph) placed on each ridge at 60 cm corresponds to operate for 35-45 minutes at 125 per cent CPE for each irrigation in single row planting. Whereas, for paired/triple row raised bed planting, the system should run for 1¼-1½ hours at each irrigation depending upon prevailing soil moisture in the field (Sood and Singh, 2008 and 2014). Ex situ Conservation

For sprinkler method, weekly two irrigations at 150 per cent of CPE are needed for sustainable yield. For example, the sprinkler system (*e.g.* 560 lph discharge/10 m x 10 m, distance between two sprinklers and two lateral lines)

corresponds to operate for 2.0-2½ hours at each irrigation depending upon the prevailing soil moisture in the field. However, the application rate and time for each irrigation may vary with the specifications of micro-irrigation.

7.1 Fertigation through Micro-irrigation

Micro-irrigation is most effective method to supply water to the plants with nutrients Potato crop raised with drip irrigation/fertigation, one third dose of recommended fertilizers (60:27:33 kg NPK/ha) through CAN or AS, SSP or DAP and MOP, respectively should be applied as basal at planting. The remaining two third dose of fertilizers *i.e.*120:53:67 kg NPK/ha should be given through fertigation in eight equal splits, twice in a week starting from plant emergence (21 DAP) to 50-55 days of planting (@ 15, 6.63 and 8.37 kg N, P_2O_5 and K_2O/ha through urea, di-ammonium phosphate and muriate of potash, resp.). The recommended fertilizer doses may vary for region to region and purpose of crop grown (seed, ware and processing *etc.*). For fertigation fill the fertilizer tank with 35 liters water and add 15 kg fertilizer and close the lid of tank tightly and then run the system for at least half an hour. After fertigation, the system needs to be run for 5 minutes.

Under sprinkler irrigation/fertigation, one third dose of nitrogen *i.e.* 60 kg N/ha with full dose of phosphorus (80 kg P_2O_5/ha) and potash (100 kg K_2O/ha) through CAN or AS, SSP or DAP and MOP, respectively should be applied at planting. The remaining two third nitrogen *i.e.*120 kg N/ha should be given through fertigation in eight splits, twice in a week starting from plant emergence (21 DAP) to 50-55 days of planting (@ 15 kg N/ha through urea). For fertigation, fill the tank with 35 liters of water and add 15 kg urea and close the lid of tank tightly and then run the system for at least 35-45 minutes (Sood and Singh, 2008 and 2014).

7.2 Advantages of Fertigation

- ✰ **Maximizing crop and soil productivity:** Higher yields and quality produce, and the marginal lands, like sandy soils, rocky soils, shallow soils and salt affected soils can be successfully put into cultivation.
- ✰ **Maximizing fertilizer efficiency:** Accurate and uniform application of fertilizers, the amounts and concentrations of nutrients are adapted to the plant needs, increases the availability and uptake of nutrients, and reduces nutrients losses by leaching and/or volatilization.
- ✰ **Minimizing production costs**: Large savings on time, transportation, labor and fuel costs.

8. Water Management in Hills

Potato is grown as a rain fed crop during summer from March/April to Sept/October in hilly areas. The total rainfall is generally in excess of evapo-transpiration, but, the crop usually suffers from drought before or onset of monsoon and from the excess moisture during later stages of growth. Studies have shown that there is surplus of 84 cm water per annum. This much quantity of water if conserved or stored and recycled during crop period by using modern irrigation techniques would greatly help in mitigating the drought problem for its effective use- i) increasing soil water storage capacity, ii) conservation of soil moisture, iii) in situ water harvesting, iv) water harvesting in farm reservoirs/ ponds, and v) proper drainage system *etc.*

9. Efficient and Economic Use of Water

Potato is an efficient user of water, but its needs are also inflated by unproductive water losses. A number of techniques have been developed to reduce water losses and to improve water use efficiency, as described below:

9.1 Use of Mulches

The potato fields have to be irrigated frequently as mentioned earlier, and are therefore, subject to high rates of evaporation especially before full vegetative growth covering the soil surface. Reduction in evaporation during early stages of growth either through tillage or mulching results in increased yield and thereby increases water use efficiency. Use of paddy straw mulch should be encouraged as it is reported to increase WUE by 15 per cent under Punjab conditions.

9.2 Use of Anti-transpirants

Studies on use of anti-transpirant *viz.*, abscisic acid have shown that it has only a transitory effect in reducing transpiration. As such this approach may not hold good in field, at least in potato.

9.3 Cultural Factors

It appears quite workable to improve water efficiency by stimulating growth through improved practices *viz.*, i) manuring and fertilizer use, ii) weed control, and iii) adoption of optimal cultural practices. Positive interaction of irrigation and nitrogen increased water use efficiency. Compared to nitrogen, water use efficiency was not affected by phosphate and potassic fertilizers. Weeds compete with crop not only for water but also for nutrient and space, therefore, they reduce the growth and yield, Hence, the weed control by cultural or chemical methods not only improves water use efficiency but also leads to water economy.

9.4 Inter-cropping

Even with the best management practices, some water is likely to go below the root zone. Part of this water could be recovered by wheat inter-sown in furrows, soon after potato are earthed up. This amounts to *in-situ* recirculation of water that might have percolated down.

9.5 Alternate Furrow Irrigation

Under conditions of water scarcity, 38 per cent saving in water was achieved by applying water to alternate furrows in turns at the expense of 11 per cent decrease in yield. This practice seems promising where water is a major limiting factor in potato production.

9.6 Stages of Water Stress

Knowledge of critical stages of water stress is essential for a judicious use of irrigation water. In potato, stolonization and early tuberization stages are most critical. An inadequate supply of water during these stages affects the yield more than the water stress at other stages. Therefore, plants should be irrigate at 0.3 bar tension or at 15-20 mm CPE during critical stages of growth. However, the plant could experience mild water stress (0.5 bar tension) during early and late crop stages and this approach can be used to save about 30 per cent water.

9.7 Improved Methods of Irrigation

As discussed earlier, drip and sprinkler methods of irrigation can economize irrigation water by 30-50 per cent. These micro-irrigation methods could be advocated to the progressive potato growers, because initial cost of installation is high.

Chapter 7

Potato Germplasm, DUS Characterization and Plant Variety and Farmer's Right

Dalamu, Babita Chaudhary, J.K. Tiwari, Vinod Kumar and A.K. Sharma

ICAR-Central Potato Research Institute, Kufri-Fagu Unit, Kufri, Himachal Pradesh

1. Abstract

Potato is the third most important food crop after rice and wheat. Potato was introduced in India towards beginning of seventeenth century by the Portuguese traders or by British missionaries. Today India is the second largest potato producer in the world. Indian potato gene bank is largest in Asian continent. Potato genetic resources at CPRI, Shimla constitutes more than 4,600 accessions belonging to cultivated species as well as wild or semi-wild species sourced from 30 countries. The major source of this collection has been the International Potato Center (CIP), Lima, Peru and the USA Potato Gene bank, Sturgeon Bay, Wisconsin. Potato germplasm are conserved either *in vitro* at CPRI, Shimla or *in vivo* as tubers at Kufri and Jalandhar and as True potato seeds (TPS). Potato germplasm are presently evaluated for 12 different parameters constituting adaptability wrt yield and maturity, biotic stress resistance, abiotic stress resistance and/or tolerance and nutritional and quality traits. Evaluation results and availability of various accessions at different conservation sites of cultivated as well as wild species are updated in computer databases and are available in form of inventories that serves for use in future basic and applied potato improvement research. In the era of agriculture globalization, protection of plant variety, the rights of breeder as well as recognizing farmers and their right is very important for sustainable growth and development of agriculture sector.

Considering this Government of India enacted Protection of Plant Varieties and Farmers Rights (PPV and FR) Act, 2001 by adopting sui generis system.

2. Introduction

Plant genetic resources play an ever growing role on world food security and economic development. Potato originated in the highlands of southern Peru over 10,000 years ago. Today, potatoes are grown in 149 countries from latitudes 65 °N to 50 °S and at altitudes from sea level to 4000 m. Potato crop belongs to the genus Solanum L., section Petota Dumort., which comprises 111 wild and cultivated species (Spooner *et al.*, 2014). In this, the cultivated species are four *viz.*, *S. tuberosum*, with two cultivar groups; Andigenum Group (diploids, triploids, and tetraploids) and the Chilotanum Group (tetraploids); *S. ajanhuiri* (diploids); *S. juzepczukii* (triploids) and *S. curtilobum* (pentaploids). The distributions of these species are from the southwestern United States to central Chile to Argentina. Wild potatoes are prevalent at sea level amongst the strand vegetation of Chilean beaches, dry deciduous forests in Mexico, scrub and cactus vegetation and forests in Mexico and USA, cool rain forests in the eastern Andeans up to high altitude Andean grasslands to an altitude of 4,500 m. The cultivated species have originated from the Andean highlands, most likely around Lake Titicaca, which lies between Peru and Bolivia up to 4400 m elevation. Germplasm management of any crop including potato is based on four major components *i.e.* collection, conservation, evaluation and documentation for their efficient utilization in improvement activities.

3. Potato Germplasm Collection

The collection of both wild and cultivated potato genetic resources has been carried out last several decades and is a continuous process. Initially, potato germplasm collection expeditions in its centers of origin and diversity were initiated by the Russians of Vavilov Institute in the 1920s (Hawkes 1990). Numerous collection expeditions were carried out particularly in Latin America by collectors from Europe, Asia, North and Latin America. The International Potato Center (CIP) has organized nearly 300 methodical exploration missions in the regions of greatest genetic diversity of wild and cultivated potatoes. As a result of these collections, present potato collection constitutes a total of 98, 285 potato accessions conserved *ex-situ* in 174 gene banks worldwide (FAO, 2010) that constitutes 24, 500-29, 500 unique potato accessions (Ellis *et al.*, 2020). INRA-RENNES, France has the major share of 11 per cent germplasm followed by VIR, Russian Federation 9 per cent, CIP, Lima, Peru 8 per cent, IPK, Germany and NR6, USA 5 per cent each, NIAS, Japan, CORPOICA, Colombia and CPRI, India 3 per cent each and remaining potato gene banks collections constituting 53 per cent. Presently, ICAR-CPRI, India has over 4600 accessions mainly procured from 30 countries (Table 7.1; CPRI Annual report 2020). However, there are certain gaps in potato genepool specifically for wild potato species that have yet to be tapped in *ex situ* germplasm collections. This includes over sampling of easy access areas, limited resources hampering multiple visits to the same site, species biasness wrt skewed distribution of number of accessions per species *etc.*

Table 7.1: Potato Germplasm Collection at CPRI, Shimla

Nature of Genotype	*No. of Accessions*			
	Tuber	*In vitro*	*TPS*	*Total*
Tuberosum (cultivars/parental lines) Indian				
Cultivars bred at CPRI	58	60	-	60
Advanced hybrids	86	50	-	96
Indigenous varieties	51	107	-	107
Indigenous samples	97	42	-	97
Exotic	1837	2700	-	2950
Andigena	701	77	-	762
Wild/semi cultivated species	119 (40 species)	130 (29 species)	294 (70 species)	540 (125 species)
Total				4612

4. Conservation

4.1 *Ex situ* Conservation

Ex situ Conservation is defined as the conservation of the components of biological diversity outside their natural habitat. The target genetic resources are collected transferred and stored away from area they were sourced, in botanical gardens, seeds genebanks, field genebanks, in vitro, DNA and pollen storage. Methods of *ex situ* conservation of potato germplasm are germplasm specific. Wild species and native cultivars are generally conserved as populations *i.e.* in the form of true seeds that are stored at 2-14 °C and relative humidity of 30 to 95 per cent. Botanically true seeds method of conservation of potato genetic resources have the advantages of low maintenance cost, longer conservation period and elimination of systemic viruses like potato spindle tuber viroid. On the other hand germplasm of cultivated species are conserved vegetatively. This method has advantages of readily availability for use. However vegetative conservation requires adequate space, funding, are more vulnerable to loss by various biotic and abiotic factors and natural disasters. In vitro or tissue culture conservation are also followed for potato germplasm esp for cultivated potatoes and for some elite wild species clones. In vitro method has advantages of maintenance in sterile and pathogen-free conditions with some disadvantages of induction of somaclonal variations during subculturing and high running costs. Cryopreservation is a new concept for long-term conservation of biological materials in liquid nitrogen at -196 °C. Cryopreservation is not a common practice yet, certain potato genebanks worldover have established cryo-banks of *in vitro*-grown shoot tips. The most successful methods for potato cryopreservation are dimethyl sulfoxide (DMSO) droplet vitrification and Plant Vitrification Solution (PVS2) droplet vitrification methods. However, the overall revival rate of cryopreserved shoot tips over a broad range of genotypes has been reported upto 51 per cent.

At CPRI, Shimla, India potato germplasm is being conserved in vegetative form in field gene banks at two locations: Hills of Kufri, Himachal Pradesh and plains of Jalandhar, Punjab. This method is followed for conservation of cultivated (tuberosum and andigena) accessions for ease of evaluation for adaptability as well as resistance/tolerance to various biotic and abiotic factors. Andigena collections along with wild species are maintained in true seed form. True seeds are produced by selfing and/or sibmating. In tissue culture conservation nodal cuttings are micropropagated on MS media containing 20 g/l sugar and 40 g/l sorbitol under a 16 hour photoperiod at 4-6 °C for minimal growth. Under such condition, potato plantlets can be conserved for up to 24 months without subculturing. Droplet vitrification of cryopreservation is being followed for conservation of the meristem tips or axillary buds.

4.2 On-farm/*In situ* Conservation

Many farmers, especially in areas of crop origin and diversity, have continued cultivating, managing and developing their plant material and thus generating crop diversity. Maintenance of crops under traditional farming systems is relevant as the varieties continue to evolve and adapt under different pressures, creating novel diversity and continuum of new variants that can persist under specific growing conditions. CIP carries out *in situ* conservation of potato genetic resources with farmers and local organizations in the center of origin of potatoes. The relationship of the ex situ conservation carried out by official institutions and the in situ conservation on farms need be strengthened for actual conservation of potato diversity in future.

5. Evaluation

Potato crop is affected by many pests and pathogens and is susceptible to numerous abiotic stresses that cause significant reductions in yields as well as in tuber quality. Several resistance genes have been discovered in both wild and cultivated potato species although relatively very few species have been used in the breeding of successful modern cultivars. Genetic diversity is the basic material for germplasm evaluation and further use in enhancement. Genetic diversity per se has no value in breeding, allelic variants alone or in combination confers desirable attributes. At CPRI, potato germplasm (cultivated) are evaluated for traits namely adaptability (temperate long days and sub-tropical short days), reaction to late blight, viruses, leaf hopper and mites, potato cyst nematodes, keeping quality, chipping and physico-chemical constituents, protein and micronutrient content, water use efficiency and foliage maturity. Evaluation of germplasm is to be furthered upto assessment of the parental values through combining ability studies. Wild species *S. bulbocastanum, S. etuberosum, S. tarijense, S. paucijugum, S. pinnatisectum, S. chacoense, S. sparsipilum, S. neorossii, S. schenckii, S. commersonii, S. demissum, S. albornozii, S. polyadenium, S. albicans, S. immite, S. andreanum, S. cardiophyllum, S. chomatophilum, S. microdontum, S. acaule, S. palustre etc.* are source of resistance to multiple biotic stresses. However, prezygotic barriers mostly in the form of stylar interactions, unilateral incompatibility, male sterility are barriers for efficient use of potato wild relatives.

6. Documentation

Comprehensive documentation is an integral part of germplasm conservation for its effective utilization. Since 1951, the information on potato germplasm has been published on regular basis. The information on passport data, morphological characteristics and reaction to various biotic and abiotic stresses of tuberosum as well as andigena germplasm has been published in the form of germplasm catalogues. Globally, Genesys, the global gateway to genetic resources (https://www.genesys-pgr.org/welcome), provides descriptor and trait information from multiple genebanks in a single database, which facilitates finding information on potato accessions worldwide.

7. Potato DUS Characterization

Protection of plant varieties, rights of farmer's and plant breeder's is required under the preview of agricultural globalization as well as an encouragement for further development of new improved plant varieties. Government of India had sought for establishment of an effective system for this and enacted "The Protection of Plant Varieties and Farmers' Rights (PPV and FR) Act, 2001" adopting sui generis system. This act is in compliance with International Union for the Protection of New Varieties of Plants (UPOV), 1978 and also has provisions to protect the interests of plant breeders, individual farmers and of farming community. The legislation provides to implement Trade Related Intellectual Property Rights that supports the specific socio-economic interests of all the stakeholders including private, public sectors and research institutions, and of farmers. For the implementation of the provisions of the Act the Department of Agriculture, Cooperation and Farmers Welfare, Ministry of Agriculture and Farmers Welfare established the Protection of Plant Varieties and Farmers' Rights Authority on 11th November, 2005 headquartered at NAS Complex, New Delhi. The authority is headed by The Chairperson who is the Chief Executive of the Authority. The authority comprise of fifteen members in addition to Chairperson, as notified by the Government of India. Eight of them are exofficio members on behalf of various Departments/Ministries, three from State Agricultural Universities and the State Governments, one representative each for farmers, tribal organization, seed industry and women organization associated with agricultural activities nominated by the Central Government.

The Registrar General is the ex-officio Member Secretary of the Authority.

8. Potato DUS Descriptors, Experimental Material and Layout

The guidelines for conduct of test for DUS on Potato (*Solanum tuberosum* L.) were standardized and published on 3rd November, 2009. Potato DUS test is based on 51 morphological descriptors in which canopy constitutes 2 traits, stem parameters are 8, 9 foliar characters, 17 floral, 9 tubers and 6 sprout traits. The observations of floral characters are recorded in hill location at Kufri, HP while remaining traits are recorded at plain regions in the Modipuram, UP and Jalandhar, Punjab. Up to now 17 potato varieties from public and 10 from private sector have been protected/registered under New (10), Extant Notified (15) and Variety of Common Knowledge (2) groups. The potato DUS experimental plot constitutes bed size of 4.8 m^2 with 4 rows of 2 m length and 60 cm row to row and 20 cm plant to plant distance per entry/variety. Total replications are three composing of 120 plants each. The observations are recorded in 30 plants or 30 plant parts *i.e.* 10 plants per replication excluding those in the border rows. The stages of observation are 30 days after withdrawal from cold storage for sprout traits, canopy, leaf and stem traits are recorded 50 days after planting, floral traits when 50 per cent of flowers open *i.e.* at 65 days after planting, plant maturity at 90 days after planting and tuber traits 115 days after planting. Uniformity of traits is recorded on the plot as a whole by visual assessment in a single observation on group of plants or parts of plants with maximum off-types not more than two. The DUS test experimental material must be free of any quarantine pest and pathogens. One hundred fifty fully matured, skin cured medium sized (3.5-5 cm diameter) healthy tubers are required for DUS testing trial.

The DUS test fee for potato is 48, 000 INR. Period of protection of potato varieties (annual crops in general) is 15 years. The duration of test for protection in *New category* requires two years independent similar growing seasons testing while for varieties to be protected in *Variety of Common Knowledge* one year DUS testing is required. No DUS test is done for varieties notified under Indian Seed Act, 1966 and is registered in *Extant category*.

8.1 Grouping Characters

Grouping of varieties facilitate the assessment of distinctiveness. Traits that do not vary, or vary only slightly, within a variety with its various states are fairly evenly distributed across all the varieties in the collection are suitable for grouping. In potato predominant colour of lightsprout, predominant colour of stem, floral corolla colour and predominant colour of tuber skin are the four grouping characters.

8.2 List of Potato DUS Descriptors and their States

1. Lightsprout - Predominant colour: White-green; Pink; Red-purple;

Purple; Blue (Figure 7.1)

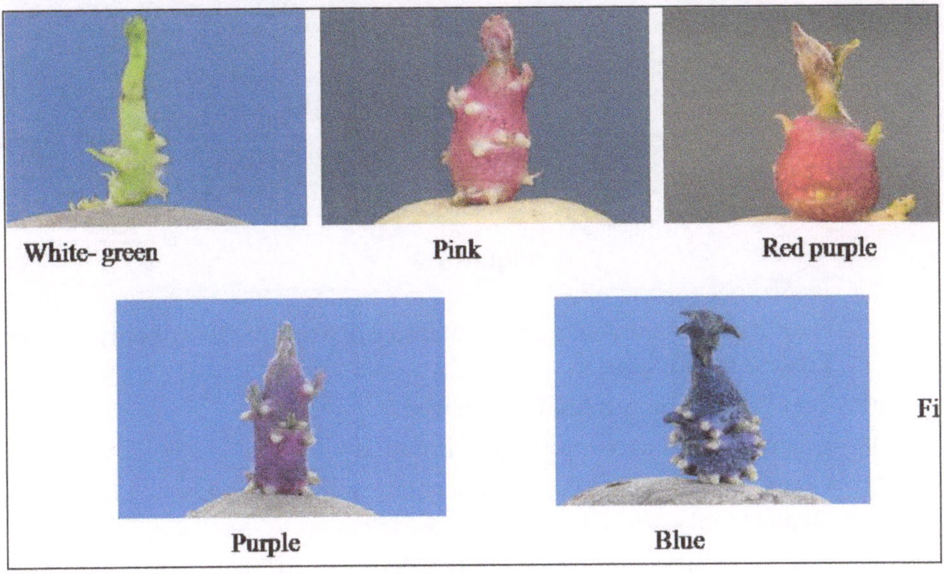

Figure 7.1

2. Lightsprout – Shape: Spherical; Conical; Cylindrical
3. Lightsprout - Intensity of anthocyanin colouration at base of sprout: Light; Medium; Dark
4. Lightsprout - Intensity of anthocyanin colouration at sprout tip: Light; Medium; Dark
5. Lightsprout - Pubescence base: Absent; Weak; Strong
6. Lightsprout - Length of apical sprout: Small; Medium; Long
7. Stem Solidity: Solid; Hollow
8. Stem Cross section: Round; Angular
9. Stem- Height of main stem: Small; Medium; Tall
10. Stem Predominant colour: Green; Red-brown; Purple; Dark purple
11. Stem -Secondary colouration: Absent; Green; Red-brown; Purple; Dark purple
12. Stem-Distribution of secondary colour: Absent; Only at base; Only at lower node; Throughout lightly scattered; Throughout highly scattered
13. Stem- Wing: Poorly developed; Highly developed
14. Stem- Wing type: Straight; Wavy
15. Leaf Structure: Open; Intermediate; Close
16. Leaf- Anthocyanin colouration of rachis: Absent (A); Present (B) (Figure 7.2)

Figure 7.2

17. Leaf- Anthocyanin colouration of mid rib: Absent; Present only at base; Present throughout
18. Leaf Length: Small; Medium; Large
19. Leaf Width: Narrow; Medium; Broad
20. Leaf Lateral shape: Narrow lanceolate; Lanceolate; Ovate lanceolate; Ovate; Oval
21. Leaf- Waviness of margin: Weak; Medium; Strong
22. Leaf -Glossiness of upper side: Weak; Medium; Strong
23. Leaf- Pubescence of blade: Absent; Present
24. Flower- Anthocyanin colouration of bud: Absent; Present
25. Flower- Anthocyanin colouration of floral stalk: Absent; Weak; Medium; Strong
26. Flower- Anthocyanin colouration of pedical articulation: Absent; Present
27. Flower- Pedical articulation position: Below the middle; At the middle; Above the middle
28. Flower- Corolla colour: White; Red-violet; Blue-violet
29. Flower- Corolla diameter: Small; Medium; Large
30. Flower- Inflorescence size: Small; Medium; Large
31. Flower- Anthocyanin colouration of outer side in white flower: Absent; Present
32. Flower- Intensity of anthocyanin colouration of corolla on inner side: Absent; Weak; Medium; Strong
33. Flower- Anther colour: Greenish-yellow; Yellow; Orange (Figure 7.3)
34. Flower- Anther cone type: Normal; Irregular (Figure 7.4)
35. Flower- Pistil type:Normal; Irregular
36. Flower- Stylar length in comparison to stamen column: Shorter; Equal; Longer
37. Flower- Stigma shape: Round; Lobed

Figure 7.3

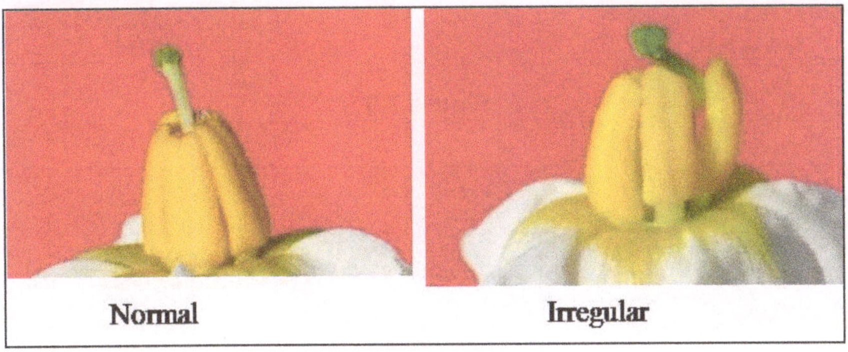

Figure 7.4

38. Flower- Stigma lobe: Unilobed; Bilobed; Trilobed
39. Flower- Premature bud dropping: Absent; Present
40. Flower- Intensity of flowering: Absent; Sparse; Medium; Profuse
41. Tuber- Predominant skin colour: Whitish-cream; Yellow; Orange; Brown; Pink; Red; Reddish-purple; Purple; Dark purple black
42. Tuber- Secondary skin colour: Absent; Whitish-cream; Russeted; Yellow; Pink; Red; Purple; Dark purple black
43. Tuber- Distribution of secondary skin colour: Absent; Confined to eyes; Present on eyebrows only; Spectacled; Splashed; Stippled (Figure 7.5)
44. Tuber- Skin type: Smooth; Rough
45. Tuber- Shape: Flattened; Round; Ovoid; Oblong; Pear shaped; Long -oblong; Reniform; Irregular
46. Tuber - Depth of eye:Protruding; Shallow; Medium deep; Deep
47. Tuber- Predominant colour of flesh: White; Cream; Yellow; Reddish purple; Dark purple
48. Tuber- Secondary colour of flesh: Absent (A); White (B); Cream (C); Yellow (D); Reddish purple (E); Dark purple (F) (Figure 7.6)

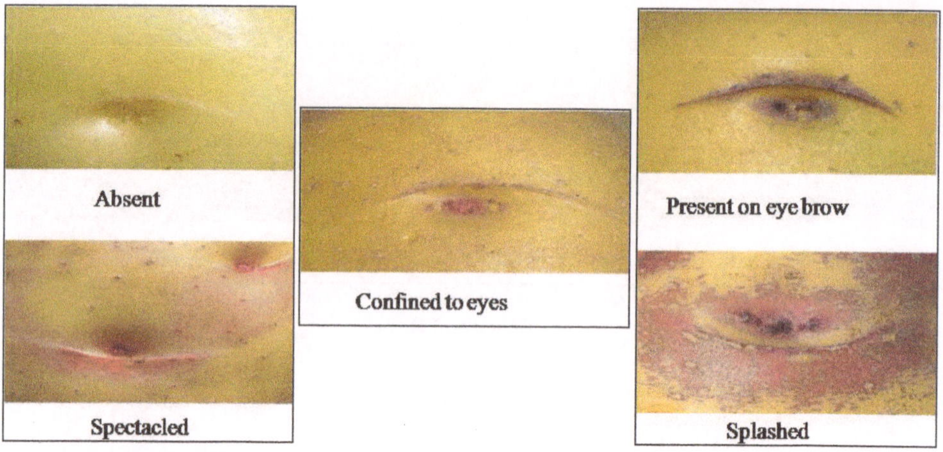

Figure 7.5

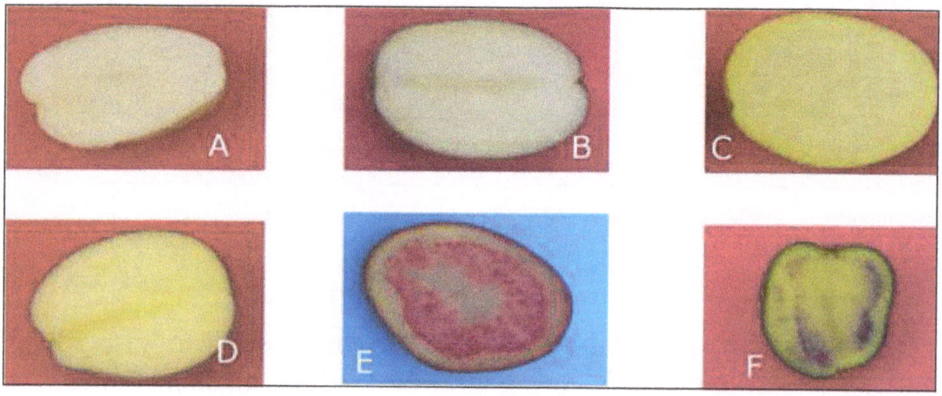

Figure 7.6

49. Tuber- Distribution of secondary colour of flesh: Outer cortex (A); Inner cortex(B); Outer medulla; Inner medulla; Vascular ring (C); Mottled (D) (Figure 7.7)
50. Canopy- Plant foliage structure: Compact; Semi-compact; Open (Figure 7.8)
51. Canopy- Time of maturity: Early; Medium; Late

9. Farmer's Right under the PPV and FR Act, 2001

The Act caters the rights of both the plant breeders and farmers. It asserts the necessity and importance of recognizing and protecting the rights of farmers with respect to the their contribution in conserving, developing and selecting varieties, improving and making available Plant Genetic Resources

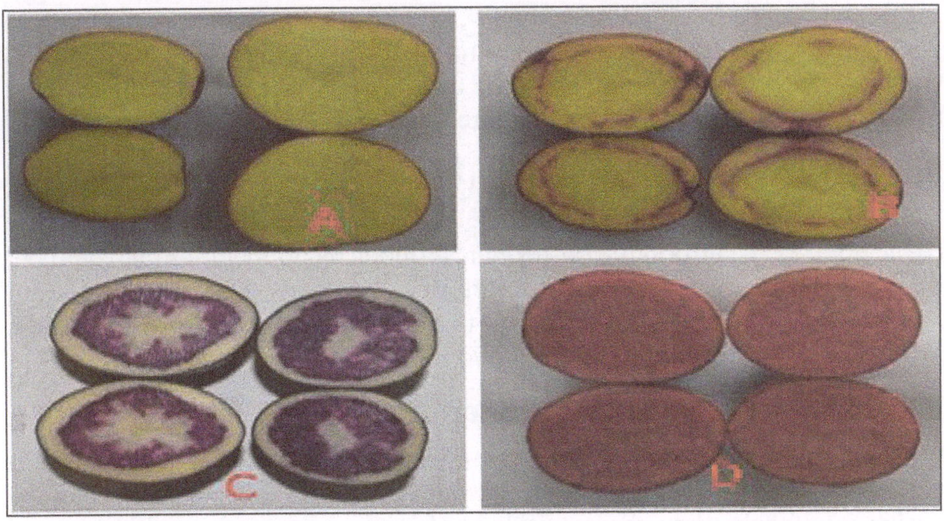

Figure 7.7

Figure 7.8

for the breeding new plant varieties. The Act protects the new farmers variety developed by them and value added to wild species or traditional varieties/ landraces through selection and identification for their economic traits. The farmer's rights considered their roles as users, conservers and breeders. Farmers are granted nine specific rights as follows:

9.1 Access to Seed (Section 39(1)

Farmers can save, use, sow, re-sow, exchange, share or sell their farm produce, including seed of protected varieties. However, they are not entitled to sell the branded seed of a variety protected under the Act.

9.2 Benefit Sharing (Section 26)

Farmers who provide Plant Genetic Resources to breeders for developing new varieties shall have right to receive a fair share of benefit from the commercial gains of the registered varieties. The Act combines the provisions for access and benefit-sharing along with Plant Breeder's Rights. Biological Diversity Act, 2002 permits use of genetic resources in breeding activities. PPV and FR Act necessities breeders to mention the geographical origin and nomenclature of the genetic resources used in the parentage of their new variety.

9.3 Compensation (Section 39(2)

Seed material of the registered varieties must be sold with the full description of its agronomic performance under recommended management practices. When such seed fails to perform at farmer's field under standard cultivation condition, he/she is eligible to claim compensation from the breeder through the intervention of the PPV and FR Authority.

9.4 Reasonable Seed Price (Section 47)

Farmers have the right to access seed of registered varieties at a reasonable price. When such condition is not met, the breeder's exclusive right over the variety is withheld under the provision of compulsory licensing, and is obligated to license the seed production, distribution and sales of the variety to a competent legal entity. Compulsory licensing of protected varieties ensures adequate seed supply to farmers at fair price.

9.5 Farmer's Recognition and Reward for Contributing to Conservation (Section 39(i)(iii) and Section 45(2)(C)

Farmers engaged in conservation and improvement and made significant contributions in providing genetic resources for breeding, get recognition and rewards from the national gene fund. The recognition and rewards are in form of Plant Genome Saviour Community awards given to farming communities and individual farmers for their contribution to *in-situ* and on farm conservation. Every year five Plant Genome Saviour Community Awards of Rupees ten lakhs each along with citation and memento to be conferred to the farming communities for their contribution in the conservation of Plant Genetic Resources. The Authority also gives ten Plant Genome Saviour Farmer Reward of Rupees one Lakh fifty thousand each with citation and memento and twenty Plant Genome Saviour Farmer Recognition of one lakh to the farmers engaged in the conservation of the Genetic Resources of the landraces and wild relatives of economics plants and their improvement through selection and conservation.

9.6 Registration of Farmers' Varieties (Section 39(1)(iii)

The PPV and FR Act permits registration of farmer's varieties that fulfill the criteria for distinctness, uniformity, stability and has proper denomination/ nomenclture. Farmer's varieties do not require the novelty trait. The registered farmer's varieties are entitled to all Plant Breeders Rights.

9.7 Protection from Innocent Infringement (Section 42)

The act considers of the old unrestrained rights that the farmers had over the seed of all varieties and the poor legal literacy among the farming community. In case of any infringement of the act if a farmer proves before court that he or she was not aware of the existence of any such act he or she will not be punished.

In addition to farmer's right, breeders have exclusive rights to produce, sell, market, distribute, import or export the protected variety and can appoint agent/licensee and may exercise for civil remedy in case of infringement of rights. Researcher can use any of the registered variety under the Act for conduct of experimental research as parental line for developing new variety. However repeated use needs prior permission of the registered breeder.

9.8 Prior Authorization for the Commercialization of Essentially Derived Varieties (Section 28 (6)

Farmers need to provide prior authorization for the commercialization of any essentially derived variety where farmers varieties are used by a third party as source material for the development of such essentially derived variety. This facilitates farmers to negotiate for authorization with the breeder with respect to royalties and benefit-sharing.

9.9 Exemption from Registration Fees and Payments for Farmers (Section 44)

Farmers have the privilege of complete exemption of any kind of fees or other payments. This includes fees of tests for distinctness, uniformity and stability, annual and renewal fee for registered farmer's variety and other services rendered by the PPV and FR Authority as well as for legal proceedings related to infringement or other cases in courts and tribunals.

Chapter 8

Somatic Hybridization in Potato Improvement

Jagesh K. Tiwari, Rasna Zinta, S.K. Luthra, Vinod Kumar, Tanuja Buckseth, Dalamu and R.K. Singh

ICAR-Central Potato Research Institute, Shimla, Himachal Pradesh

1. Introduction

Gene transfer is the basis for almost all crop improvement including potato. Conventionally, this is achieved through sexual hybridization; this rather limits the range of species from which gene flow can occur into a crop species. Wild species have contributed remarkably to the success of latter, they allowed the crops to retain their commercial status. As a result plant breeders have sought to utilize an increasing number of wild species as a source of valuable genes ranging from disease resistance to grain yield, and produce quality. But many sources of useful genes cannot be included in crop improvement programme primarily because of sexual incompatibilities. Genetic transformation, a focussed and direct gene transfer approach, require identification, isolation and cloning of the concerned genes. Further it is expensive and technically most exacting, although it may represent the ultimate strategy. However, some characters of interest may be govern by two or more and yet unknown genes; transfer of such characters through genetic transformation may pose many difficulties. Finally transfer of cytoplasmic organells, *viz.*, chloroplast and mitochondria may often be desired objectives; this, however is not possible through genetic transformation, while it can readily achieved by somatic hybridization.

Wild and cultivated species of potato have been effectively used in potato breeding but represent only a tiny fraction of the available potato biodiversity. Utilization of the wild tuber-bearing diploid species has been remained untapped potential source for transferring resistance trait into common potato (Bradshaw *et al.*, 2006). Wild tuber-bearing *Solanum* species are widely distributed from southwestern USA to central Argentina and Chile. This extensive geographical range has resulted in types adapted to a broad range of climatic and soil conditions. In the course of evolution, these plants have also developed resistance/tolerance to different pathogens and pests. Much of this effort has involved the examination of wild species for various resistance traits related to potato. This trait is particularly attractive to breeders to widen the potato genetic base, but the barrier between the cultivated potato and the many wild species has proved a difficult task, even when unconventional crossing methods are used (Orczyk *et al.*, 2003).

Many useful genes derives from wild sources cannot be transferred through conventional technique because of sexual incompatibilities are primarily due to difference in ploidy and endosperm balance number (EBN) (Spooner and Salas 2006). It is extremely difficult to cross 1 EBN wild species directly with common cultivated 4 EBN potato. Limited success has been obtained by utilizing bridging species but the incompatibility of 1 EBN wild species has generally prevented the use of this particularly valuable trait. However, modern research and new techniques have made it possible to expand considerably the genetic resources available for use in breeding programs. A few methods have now become available to overcome this problem. These methods include: manipulation of ploidy and endosperm balance number (EBN), bridge crosses, mentor pollination and embryo rescue, hormone treatment, and reciprocal crosses (Jansky 2006). Somatic hybridization, which removes prezygotic and some postzygotic barriers, can likewise surmount the barrier between cultivated and wild species. Somatic hybridization can provide a means of bypassing sexual incompatibility between *Solanum* species, leading to fertile plants that can be used directly in breeding programs.

Somatic hybridization generates functional combinations of large sets of genetic material, which makes it similar to sexual hybridization. This method can also be used to overcome limitations of genetic transformation. Many of the important traits are predominantly polygenic such as late blight resistance and thus unavailable as isolated and characterized sequences that are ready for genetic transformation. Therefore, efficient methods of transformation are yet to be available for multiple genes that are expressed in a coordinated manner (Orczyk *et al.*, 2003). On the other hand, somatic hybrids obtained directly after fusion contain all organelles from the cytoplasms of both parents. Somatic hybridization via protoplast isolation, electrofusion and regeneration is a useful tool to transfer polygenic traits such as late blight resistance in a single step. It enables a development of tetraploid somatic hybrid between diploid wild species and dihaploid of common potato. As a result, tetraploid somatic hybrids may be utilized in conventional breeding for late blight resistance and improvement of other traits. Thus, production of somatic hybrids between tetraploid 4EBN *S. tuberosum* and diploid 1EBN wild species has been envisaged for imparting durable resistance to late blight. In consequence, aim of this somatic fusion technology is creditable to enrich the cultivated potato gene pool by incorporating genes from a new exotic wild species, in order to enhance resistance to late blight disease.

Hence, somatic hybridization is the technique enables to transfer agronomically important traits by bypassing such sexual barriers, besides the conventional and recombinant-DNA technologies approaches. Despite these crossing-barriers, many researchers have used this technique and subsequently produced somatic hybrids with cultivated potato. Production of hybrid plants through the fusion of protoplasts of two different plant species/varieties is called somatic hybridization, and such hybrids are called somatic hybrids. Therefore, somatic hybridization can be resorted to only when the following two criteria are satisfied: i) isolation of protoplast in large quantity and ii) totipotency of the isolated protoplasts.

2. Procedures

The protoplast fusion by electric field requires sufficient amount of suitable plant material for protoplast isolation and their culture after electrofusion for obtaining plant regeneration. The procedure involves the following stages.

- ✰ *In vitro* culture of donor plants
- ✰ Protoplast isolation from leaf meshophyll tissues
- ✰ Verification of protoplast viability and protoplast fusion
- ✰ Protoplast fusion by electric field
- ✰ Regeneration and culture fusion products
- ✰ Characterization of putative somatic hybrids

2.1 Prepare Materials

- ✰ Three-week-old *in vitro*- grown microplants, raised from single nodal cutting

2.2 Microplant Incubation Condition

- ✰ 16 h photoperiod/40 μ mol m^{-2} s^{-1}/20 °C.

2.3 Pre-isolation (Protoplast) Incubation

- ✰ 48 h/dark/20 °C

2.4 Protoplast Isolation

- ✰ Mince young leaf tissues (1-2 g) in a ϕ 90 mm Petri dish containing digestion solution: 10 ml digestion solution for 1 g tissue.

2.5 Incubation (for protoplast isolation)

- ✰ 16 h/dark/25 °C/optional: gyratory shaking at 40-50 rpm; not exceeding 50 rpm.

2.6 Post-isolation Handling

- ✰ Add 0.3 M KCl (sterile) to the digestion medium/solution containing released protoplasts (protoplast suspension) in a 1:1 ratio (KCl: solution). For example, add 15 ml M KCl to 15 ml digestion medium/ solution.
- ✰ Filter the suspension through 40 μ nylon mesh, and collect in centrifuge tubes; 60 μ can be used but debris and/or undigested tissues will be much.

2.7 Protoplast Purification

- ✰ Centrifuge the filtrate at 50 x g (50 RCF) for 5 min.
- ✰ Discard supernatent and then resuspend the pellets in 10 (or 9) ml of 0.6 M sucrose (sterile).
- ✰ Overlay 1 ml of 0.3 M KCl onto this protoplast suspension.
- ✰ Centrifuge at 50 x g (50 RCF) for 5 min.
- ✰ Recover the protoplast (green upper portion/live protoplast) from sucrose: KCl interface
- ✰ Dilute the recovered protoplasts with 10 ml of 0.3 M KCl.
- ✰ Centrifuge at 50 x g (50 RCF) for 5 min to form pellet of the protoplast.
- ✰ Resuspend the pellets in 1-2 ml 0.5 M mannitol (sterile), centrifuge at 50 RCF) for 5 min., then retain pellets and discard supernatants
- ✰ Finally resuspend the pellets in 400-500 μl 0. 5M mannitol (sterile) to a final density of $1x10^6$ protoplast/ml for electrofusion.

2.8 Electrofusion Medium

- ✰ 0.5 M mannitol (sterile)/sterile-filtered (0.2 μ)/pH 7-7.3/adjust pH with 0.1 N NaOH

2.9 Symmetric Fusion

- ✰ 1:1 of each species

2.10 Electrofusion Settings

Chamber	BTX Microslide Model 453/3.2 mm gap
Alignment amplitude	16 V
Alignment time	25-30 s
Alignment field strength	50 V cm^{-1}
Electrofusion amplitude	260 V

DC pulse width	60 µs
Number of pulses	2
Electrofusion field strength	812 V cm^{-1}

2.11 Post-fusion Culture

- Dispense 50 µl Na-Alginate in each box of castor rack and mix well the 50 µl fusion products into it
- Add 2-3 ml Solution No 3 in each box and incubate at RT inside laminar for 30 min followed by add 2-3 ml Solution No 2 and incubate same for 1.30-2 h
- Remove Sol 2 and 3 by Pasteur pipette and add 5 ml VKMG (VKM Glucose) and tightly wrapped with parafilm.
- Incubate the castor racks under dark at 25 °C for the regeneration of calli.

3. Applications

Symmetric protoplast fusion approaches involving diploid *Solanum* species in combination with dihaploid *S. tuberosum* have been essentially used to develop tetraploid somatic hybrids potato having desirable introgression from wild relatives (Table 8.1). In the current years, application of this technology has been observed widely for the production of multiple resistant somatic hybrids. Interspecific potato somatic hybrid between commercial cultivars of potato *S. tuberosum* Agave and Delikat and wild diploid species *S. cardiophyllum* (1 EBN) has been produced for resistances to Colorado potato beetle, foliage blight and PVY (Thieme *et al.*, 2010). In addition, somatic hybrids between a diploid potato clone DG 81-68 susceptible to *P. infestans* and a resistant diploid tuber-bearing species *Solanum* x *michoacanum* were generated (Szczerbakowa *et al.*, 2010). Polzerová *et al.* (2011) have developed interspecific somatic hybrids between wild diploid species *S. pinnatisectum* (1 EBN) and *S. tuberosum* for the late blight resistance in potato. Following the successful production, somatic hybrids have been applied in the potato breeding for the development of advance progenies for transferring the resistance trait. For example, Thieme *et al.* (2008) have developed novel somatic hybrids and their fertile BC_1 progenies having resistances to late blight, Colorado potato beetle and PVY from a diploid wild species *S. tarnii* into common potato.

Interspecific potato somatic hybrids between 1 EBN wild *Solanum* species *S. pinnatisectum* (+) dihaploid *S. tuberosum*, and ii) S. *etuberosum* (+) dihaploid *S. tuberosum* have been produced following optimized protocol of the protoplast isolation, electrofusion and regeneration of plantlets (Sarkar *et al.*, 2011; Tiwari *et al.*, 2010, 2011). The somatic hybrids *S. pinnatisectum* (+) *S. tuberosum* have

Table 8.1: Symmetric Protoplast Fusion in Potato for Incorporation of Desirable Traits from Wild Diploid Species into *S. tuberosum*

Diploid Wild Species	*Trait(s) Transferred*	*References*
S. acaule	PVX resistance	Yamada *et al.*, 1997
	Bacterial ring rot resistance	Rokka *et al.*, 2005
	Glycolkaloid composition	Kozukue *et al.*, 1999
S. berthaultii	Salinity tolerance	Bidani *et al.*, 2007
S. bulbocastanum	Late blight resistance	Bo³towicz *et al.*, 2005
	Meloidogyne chitwoodi resistance	Mojtahedi *et al.*, 1995
	Bacterial stem rot resistance	Rokka *et al.*, 1994
S. brevidense	Tuber characteristics and insect resistance	Serraf *et al.*, 1991
	Tuber soft rot resistance	Polgar *et al.*, 1999
	Early blight resistance	Tek *et al.*, 2004
S. cardiophyllum	Late blight, PVY, Colorado potato beetle resistances	Thieme *et al.*, 2010
S. chcoense	Colorado Potato Beetle resistance	Cheng *et al.*, 1995
S. circaefolium	Late blight and nematode resistances	Oberwalder *et al.*, 2000
S. commersonii	Frost resistance	Nyman and Waara 1997
	Bacterial wilt resistance	Kim-Lee *et al.*, 2005
	Tuber traits	Caruso *et al.*, 2008
S. etuberosum	Tuber characteristics	Novy and Helgeson 1994
	PVY resistance	Novy *et al.*, 2007
	PLRV resistance	Novy *et al.*, 2007
S. x michoacanum	Late blight resistance	Szczerbakowa *et al.*, 2010
S. nigrum	Late blight resistance	Szczerbakowa *et al.*, 2003
S. phureja	Bacterial wilt resistance	Fock *et al.*, 2000
S. pinnatisectum	Late blight resistance	Polzerová *et al.*, 2011
S. stenotomum	Bacterial wilt resistance	Fock *et al.*, 2001
S. tarnii	Late blight, Colorado Potato Beetle and PVY resistance	Thieme *et al.*, 2008
S. torvum	Resistance to *Verticillium dahliae*	Jadari *et al.*, 1992
S. tuberosum	Late blight resistance	Rasmussen *et al.*, 1998

resistance for late blight resistance, whereas *S. etuberosum* (+) *S. tuberosum* have resistance for potato virus Y. These potato somatic hybrids have been confirmed for the hybridity through molecular (RAPD and SSR markers), phenotypic assessments. Hybrids were also evaluated for the disease resistance. Ploidy level (tetraploid) of somatic hybrids has been examined through flow cytometry and guard cell count. At present, work is under progress on the development of more interspecific potato hybrids involving diverse 1 EBN wild species obtained from foreign gene banks. Molecular analyses of the potato somatic hybrids produced at CPRI, Shimla are shown in Figures 8.1 and 8.2.

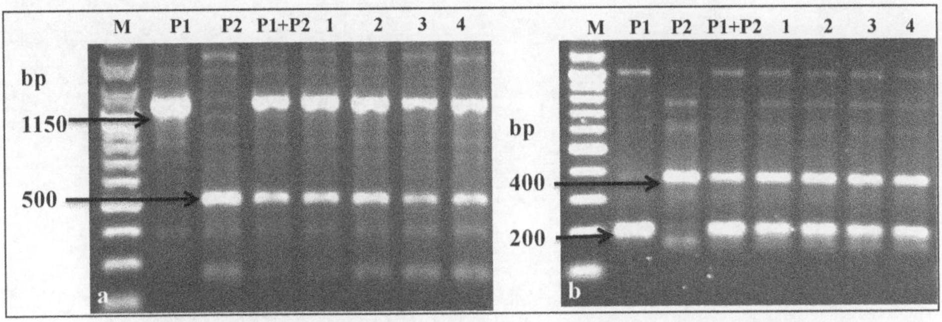

Figure 8.1: Potato Somatic Hybrids between C-13 and *S. cardiophyllum* Confirmed with Possessing Parental Bands by (a) RAPD marker OPAC06, and b) ISSR marker ISSR10. P_1= C-13, P_2= *S. cardiophyllum*, P_1+P_2= pooled parental DNA before PCR.

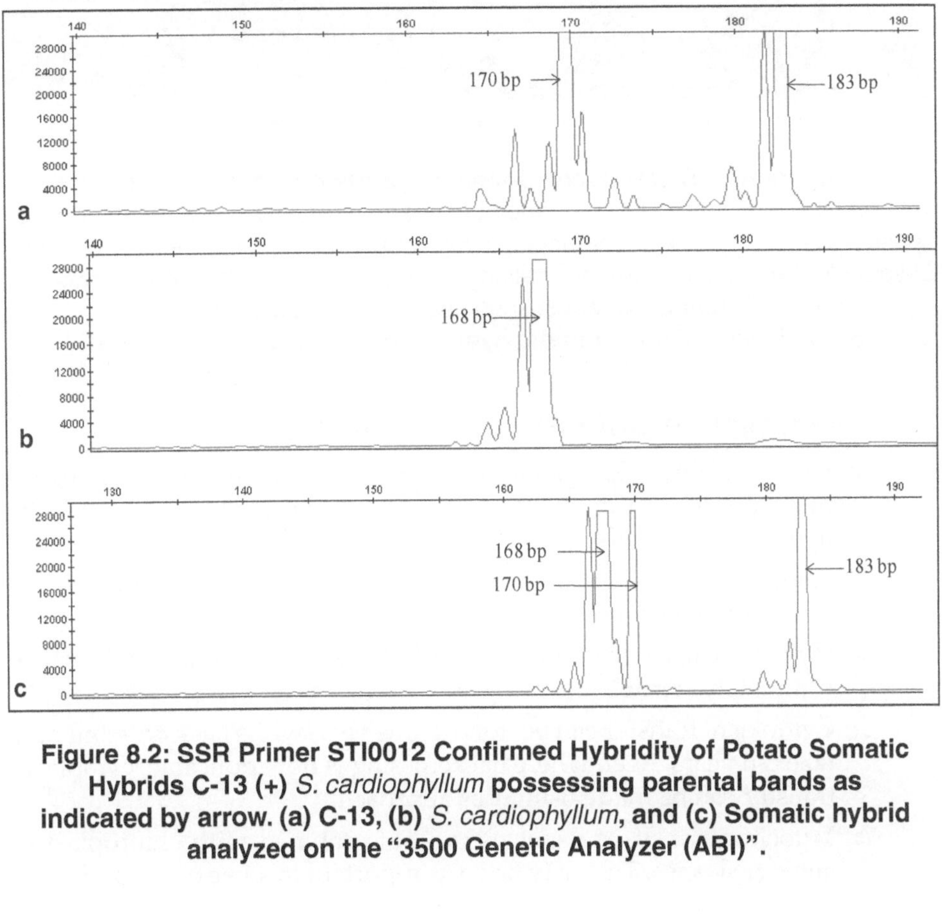

Figure 8.2: SSR Primer STI0012 Confirmed Hybridity of Potato Somatic Hybrids C-13 (+) *S. cardiophyllum* possessing parental bands as indicated by arrow. (a) C-13, (b) *S. cardiophyllum*, and (c) Somatic hybrid analyzed on the "3500 Genetic Analyzer (ABI)".

S. cardiophyllum C-13 Somatic hybrid Somatic hybrid C- 13 S. cardiophyllum

Figure 8.3. (i) Left panel: Somatic hybrids between C-13 and *S. cardiophyllum* confirmed by phenotypes (a) leaf, (b) flower and (c) tuber traits possessing their parental characters; (ii) Right panel: late blight resistance test of somatic hybrids confirmed by challenge inoculation of *Phytophthora infestans*. Parent C-13 was susceptible (S), whereas parent *S. cardiophyllum* **and somatic hybrids were highly resistant (HR).**

4. Advantages of Somatic hybridization

- ☆ Somatic hybrids The potato cyst nematodes arc the most successful plant parasi can be produced between species, which cannot be hybridized sexually.
- ☆ Somatic hybrids can be readily used in breeding programme for transfer of resistance genes
- ☆ Hybrids can be produced even between such clones, which are completely sterile.
- ☆ Cytoplasm transfer can be done in one year, while back crossing may take 5-6 years. Even where backcrossing is not applicable, cytoplasm transfer can be made using this approach.
- ☆ Mitochondria of one species can be combined with chloroplast of another species. This may be very important in some cases, and is not

achievable by sexual means even between easily crossable species.

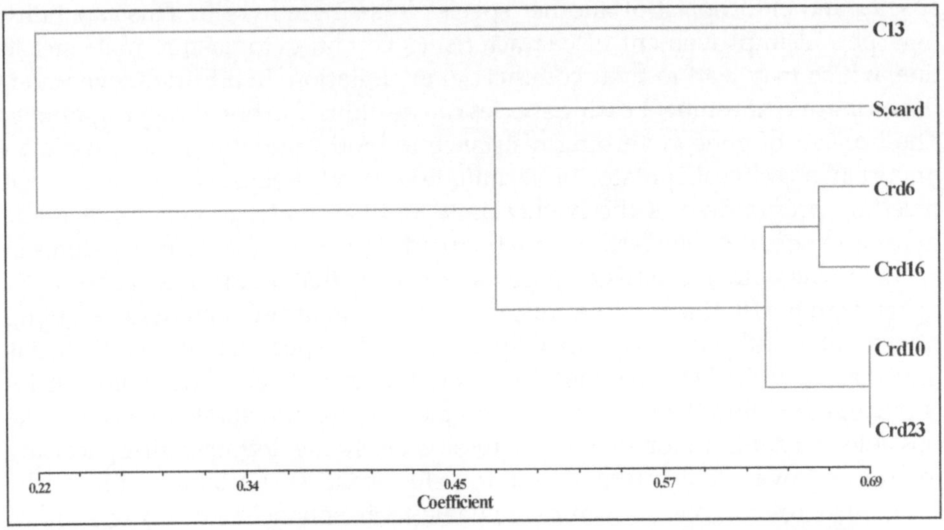

Figure 8.4: Cluster Analysis Based on the Jaccard's Coefficient of Molecular Profiles Generated by RAPD, ISSR, SSR, AFLP and Cytoplasmic Markers Showing Genetic Distinctness among Somatic Hybrid Clones and their Parents.

- Recombinant organelle genomes, especially of mitochondria, are generated in somatic hybrids. Some of these recombinant genomes may possess useful features.

5. Limitations of Somatic Hybridization

- Techniques for protoplast isolation, culture and fusion are very complicated.
- In many cases, chromosome elimination occurs from somatic hybrids leading to asymmetric hybrids. Such hybrids may be useful, but there is no control on chromosome elimination.
- Many somatic hybrids show genetic instability, which may be an inherent feature of some species combinations.
- Many somatic hybrids either do not regenerate or give rise to sterile regenerants. Such hybrids are useful for crop improvement. All interfamily somatic hybrids are genetically unstable and/or morphologically abnormal, while intergeneric and intertribal hybrids are genetically stable, but produce abnormal and/or sterile plants.

6. Conclusion

Somatic hybridization allows transfer of cytoplasmic organelle in a single generation and offer unique opportunities for combining mitochondria of one species and chloroplast of another species in a single hybrid. This capability may permit improvement of characteristics certain cytoplasmic male sterile line, which may lead to their commercial exploitation. In addition, even non-flowering and non-tuber bearing species can be utilized in breeding programme. The transfer of gene governing resistance to biotic and abiotic stresses is an important objective. In potato, this technique is already being used in commercial breeding programme of the Netherlands and Germany. In general, somatic hybrids have low pollen fertility, but they can often be used as female parents in backcrossing with one of fusion parents. It is likely that the approach of somatic hybridization will find grater applications in potato improvement for enabling transfer of useful genes from sexually incompatible species. In this context, it is important that the DNA segment carrying the desired gene from wild species is introgressed into the genome of cultivated potato and stable inherited. The possible mechanissm for such introgression are homeologous pairing leading to crossing over, and intergenomic translocation. An understanding of the gene introgression mechanism may enable their enhancement using suitable treatment; this, in turn will enhance the opportunities for utilization of somatic hybrids in potato improvement.

Chapter 9

Diploid F_1 Hybrid TPS Potato Breeding

Salej Sood, Vinay Bhardwaj, Hemant Kardile, Vikas Mangal and Vinod Kumar

ICAR-Central Potato Research Institute, Shimla, Himachal Pradesh

1. Introduction

Potato is the major staple food in many countries of the world. It has a long history of cultivation in its area of origin *i.e.* Andean region of Peru. It is propagated through tubers to maintain the purity of cultivars due to high heterozygosity and polyploid genome. The cultivated potatoes are auto-tetraploid and self-compatible. But upon selfing they show high inbreeding depression due to high heterozygosity in parental lines and may be expression of lethal alleles on selfing. The homozygous lines therefore could not be developed in tetraploid potato and the varieties as well as advanced breeding material are maintained in the form of tubers only. Maintenance of varieties and other breeding lines through clonal propagation in the form of tubers lead to accumulation of pests and diseases particularly viruses, which keep on multiplying in each cycle of clonal propagation. This leads to diminution of varieties productivity and acceptance. Moreover, the potato breeding programme takes more than 12 years to develop a new variety and is based on clonal propagation of tubers. The seed of the variety is also multiplied clonally through tubers with a seed multiplication rate of approximately 1: 8 tubers per year. Therefore hybrid potatoes at diploid level have been envisioned at the global level for using TPS as the propagation material in potato (Jansky *et al.*, 2016).

The true potato seeds (TPS) *i.e.* botanical seed could not be used for propagation of tetraploid varieties due to genetic non-uniformity *i.e.* each TPS is genetically different. TPS however is easy to maintain, easy to transport, easy to store, amenable to genetic manipulations and is free from diseases and pests inoculums (Sood *et al.*, 2020). Potato breeding through inbred diploid hybrid TPS is a completely new breeding method (Lindhout *et al.*, 2011), being adopted in all potato growing countries including India (Sood *et al.*, 2021). The new diploid hybrid breeding technology works by sexual propagation in each generation like cereal crops. The method is based on identification, evaluation of diploid cultivated species and di-haploids of *Solanum tuberosum* for selection of desired clones with acceptable tuber and other plant traits. Diploid potatoes are naturally self-incompatible. *Sli* is the key gene for self-compatibility in diploid potatoes and now offers a path forward for the inbred-based diploid F1 hybrid potato breeding program (Ma *et al.*, 2021). First identified in *Solanum chacoense* clones 'chc 525-3' by Hosaka and Hanneman (1998), the gene has been recently fine mapped and introduced in diploid tuberosum lines for inbreeding and generation of inbred lines (Eggers *et al.*, 2021). Different sources of self-incompatibility identified in other species also have been found to be regulated by the *Sli* gene (Clot *et al.*, 2020). Recent studies have unraveled the structure and function of the *Sli* gene which has been narrowed down to a region of 12kb on chromosome 12 (Eggers *et al.*, 2021). It was further found that the *Sli* gene encodes an F-box protein which is expressed in the pollen of self-compatible clones only.

The introgression of the *Sli* gene in self-incompatible diploid clones made inbreeding possible in hitherto self-incompatible lines. However, the accumulation of deleterious genes in the recessive phase over a hundred years of clonal selection in cultivated tetraploid potatoes is another major hindrance in inbred lines development. High inbreeding depression is evident from decreased fertility and vigour in selfing generations. To overcome the problem of deleterious genes, Zhang *et al.* (2021) suggested the genome design pipeline for the development of inbred lines where they emphasized to use of the starting material with low heterozygosity and few deleterious mutations. To identify inbred lines with beneficial alleles, the genetic analysis of S_1 inbreeding generation was carried out to identify beneficial alleles and purge large effect deleterious mutations. The diploid inbred lines selection following this breeding pipeline would be highly homozygous and remain true to type on selfing. The inbred lines originating from different lineage could be crossed to develop F_1 hybrids TPS for evaluation. This sexual crossing of the two inbred parent lines results in thousands of true potato seeds per plant instead of only a few potato tubers. So multiplication of a new variety can occur much faster. Since the offspring of a sexual cross is a pristine true seed, these are completely free of diseases and therefore make excellent seed material for potato growers around the world.

2. F_1 Diploid Hybrid Methodology

The basics workflow of the development of diploid inbred lines and hybrid TPS is explained in Figure 9.1.

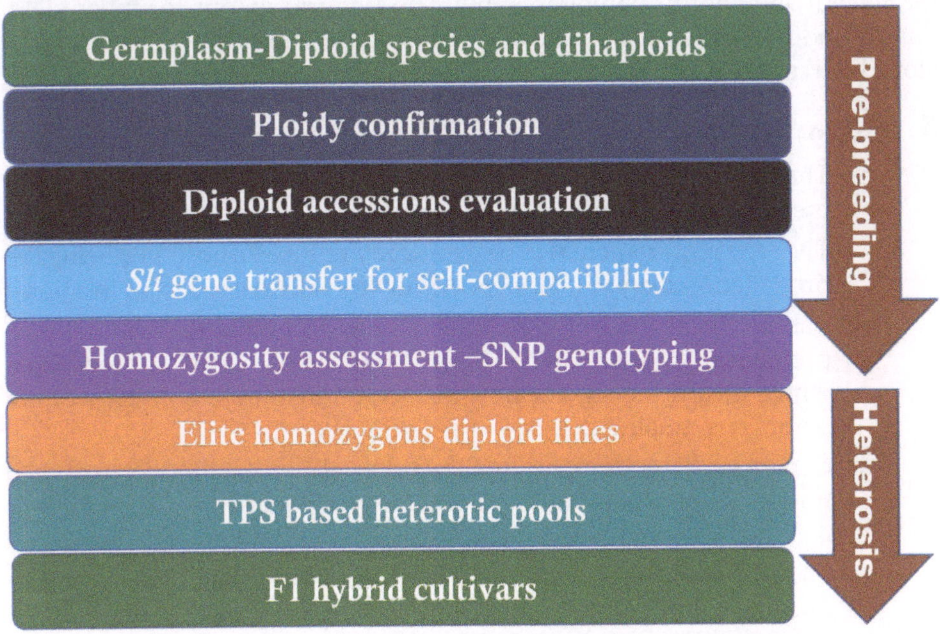

Figure 9.1: Inbred Based Diploid F_1 Hyzrid Potato Breeding.

2.1 Diploid Elite Germplasm Lines

The first and foremost requirement of diploid F_1 hybrid potato breeding methodology is elite diploid germplasm lines in the cultivated background. To do so the tetraploid cultivated potatoes are crossed with haploid inducer lines (HI) and di-haploids are developed, which originate from the female parent only *i.e.* we get exactly half the genetic constitution of tetraploid variety. The di-haploids are weak and mostly self-incompatible. The weak plants with deleterious alleles are removed and strong plants will be selected and maintained.

2.2 Self-compatibility and Inbreeding

The healthy plants will be made self-compatible by introgressing the *Sli* gene from elite *Sli* donor lines to fix the genetic constitution upon selfing in inbred lines. The diploid lines will be selfed for few generations to achieve >95 per cent homozygosity. Robust, vigorous and fertile plants are selected in each selfing generation for advancement to the next generation to accumulate beneficial alleles in inbred lines.

2.3 Heterosis

Selfing will generate many inbred lines originating from different di-haploid parents *i.e.* lineage. The inbred lines from different lineages will be evaluated for general and specific combining ability and hybrid vigour for tuber yield, quality and processing traits. The best hybrid combinations will be evaluated along with commercial tuber-based tetraploid varieties for comparative analysis.

3. Advantages

- ☆ The technology offers numerous advantages over tuber-based potato breeding and cultivation methodology.
- ☆ The seed requirement will be drastically reduced due to TPS being the propagation material. Transport will be very easy to far off locations.
- ☆ The degeneration of tubers due to virus accumulation will not be a problem anymore. Less than 5 per cent of the pathogen are seed-borne which means the significantly reduced transmission of many of the pathogens causing diseases in potato seeds.
- ☆ Genetics will be simple at the diploid level and it will become easy to track and map genes for important traits in potatoes, which at present is difficult to do due to tetraploid inheritance and heterozygosity.
- ☆ The trait stacking will be very easy as backcrossing can be used to introduce the desired trait in true breeding lines without altering the genome of the pure lines.
- ☆ The genetic gain will be higher over time due to the sexual breeding cycle and accumulation of desired alleles in inbred lines.

4. Challenges

- ☆ The inbred lines produced to date in diploid potatoes are not completely homozygous for all the loci in the genome, which means the F_1 hybrids may lack uniformity for some traits governed by these loci. This could be a major deterrent to the acceptance of technology.
- ☆ The approach involves a complete transformation of potato breeding methodology, hence it will require rigorous evaluation in various agro-ecologies to convince the stakeholders on the adoption of this technology. The trials in African countries have shown at par or less tuber yield of F_1 TPS in comparison to tuber raised tetraploid varieties. At par tuber yield and quality to tetraploid varieties is a must for the adoption of diploid F_1 hybrid TPS technology.
- ☆ The TPS is the starting material for growing potatoes, which will require additional time and resources to raise the crop in the nursery and transplanting seedlings in the field. The crop duration of TPS based

approach needs to match with tuber raised crop to fit in the cropping system in India.

- Both scientists and farmers are accustomed to growing potatoes from tubers *i.e.* conducting trials or raising crops. It will be important to standardize the agronomy of TPS-based crop as a complete package and practice of the diploid F_1 hybrid TPS.

5. Conclusion

Potato propagation through diploid hybrid TPS is a completely new breeding method and is being adopted in all potato growing countries including India. The new diploid hybrid breeding technology works by sexual propagation. The method is based on identification, evaluation of diploid cultivated species and dihaploids of *Solanum tuberosum* for selection of desired clones with acceptable tuber and other plant traits. Diploid potatoes are naturally self-incompatible. The *Sli* gene for inducing self-compatibility is introgressed in diploid clones and selfing is done for few generations to get homozygous diploid lines. The diploid homozygous inbred lines are evaluated for various tuber, plant and adaptation traits and potential lines are selected for hybrid combinations to accumulate desirables genes from both the parental lines in F_1 hybrid. The technology provides the opportunity to quickly develop new varieties to meet specific requirements. By improving the elite parent lines we can improve their offspring variety for the target trait more easily, quickly and in a targeted way. The genetic gains in potato breeding will be much higher using TPS based diploid hybrid breeding in potato.

Chapter 10

Potato Nematodes Management

Aarti Bairwa, H.M. Priyank, E.P. Venkatasalam and Anuj Bhatnagar

ICAR-Central Potato Research Institute, Shimla, Himachal Pradesh

1. Introduction

Nematodes are microscopic roundworms that live in many habitats. At least 2500 species of plant-parasitic nematodes have been described, characterized by the presence of a stylet, which is used for penetration of host plant tissue. They mostly attack on roots and underground parts of plants, but some are able to feed on leaves and flowers. Generally, nematode infestations in plants are not easily recognized since the external foliage symptoms are often mistaken to nutrient deficiency symptoms. These nematodes are also enhance the severity of other diseases caused by fungi, bacteria, several other nematodes act as carriers of viral diseases and form disease complexes. The earliest record of parasitic nematode infestation in potato crop was of the cyst nematode recorded in the year 1881 by Julius Kuhn from Germany. This cyst nematode was established as one of the major plant protection globally accounting for an average tuber yield loss of about nine per cent, which is about 43 million tons (Krishna prasad 2006). In India, the root knot nematode infection in potato tubers was first noticed by Dr. Thirumalachar in 1951. Later, the report of cyst nematode occurrence in potato crop from the Nilgiri hills in South India by Dr. Jones in 1961 activated the nematological research not only in potato but also in other crops. About 90 species of nematodes belonging to 38 genera have been reported to be associated with potatoes in India alone. But among these, the potato cyst nematodes (PCN) and root-knot nematodes have been reported as the major pests of potato.

2. Potato Cyst Nematodes

Potato cyst nematodes *viz., Globodera rostochiensis* (Woll) and *G. pallida* (Stone), also popularly called the Golden nematodes, are one of the dangerous pests hindering the sustainable production of potato in any countries including India. Their small size, intimate association with their host and adaptation for long term survival in the soil in the absence of a suitable host present them challenging to the farmers, scientists and policy makers. Potato cyst nematodes are one of the important biotic constraints in the sustainable production of potato in many countries worldwide including India. They are subjected to stringent quarantine and/or regulatory procedures, wherever they occur.

2.1 Origin and Distribution

Potato cyst nematodes originated in the Andean mountains of South America along with its principal host, potato. They were then introduced into Europe in the 1850's and later spread throughout the world through the introduction of improved varieties. Presently, Potato cyst nematodes have been reported from 65 countries with *Globodera rostochiensis* in all countries and *G. pallida* in 47 countries. In India, Dr. F.G.W. Jones first detected the PCN during 1961 from Vijayanagaram farm in Udhagamandalam, The Nilgiris district, Tamil Nadu. Later on, their occurrence was reported from other parts of Nilgiri, Kodaikanal hills, adjoining hills of Karnataka, Idukki District in Western Ghats of Kerala and recently, in the hilly regions of Himachal Pradesh Jammu and Kashmir and Uttarakhand (Bairwa *et al.*, 2021). Accordingly, Tamil Nadu Government imposed domestic quarantine during 1971 to ensure strict checking of seed potato for marketing from infested field of Udhagamandalam and Government of India during 2018 to restrict the movement of potato seed tubers from the infested areas.

2.2 Species and Pathotypes

There are two species of PCN *viz., Globodera rostochiensis* and *G. pallida*. The differential host reactions of PCN populations from Nilgiris and Kodaikanal hills revealed that the pathotypes Ro1 of *G. rostochiensis* and Pa2 of *G. pallida* are the most prevalent forms. The other prevalent pathotypes are Ro2 and Ro5 of the former and Pa1 and Pa3 of the latter.

2.3 Host Range

Potato cyst nematodes are confined to the family Solanaceae with potato, tomato and brinjal as the most preferred hosts.

2.4 Means of Dispersal

The major routes of dissemination of PCN is through the contaminated seed potatoes, movement of farm workers, livestocks and farm implements from infested land, irrigation water, high winds, soil in gunny bags used for transportation, *etc.* Ensuring strict hygiene practices will help minimize PCN spread.

2.5 Yield Loss

Worldwide PCN alone causes estimated losses up to 30 per cent in potato crop yield. The tuber yield loss estimates vary from 5 to 80 per cent depending on the initial inoculum level. The economic threshold level for crop loss due to PCN is usually around 20 eggs per g of soil, which may vary with the environmental interactions and host tolerance.

3. Symptoms

At low PCN population densities, most plants can tolerate nematode invasion and respond by developing more lateral roots as wound response, without affecting their growth and yield. However, as the degree of invasion increases, the plant is unable to compensate and ultimately exhibits a range of symptoms. The typical above ground symptoms of PCN can be divided into two phases. In the first half of the growing season, PCN cause reduced photosynthetic rate, production of fewer stems and smaller, thicker leaves and overall stunting of the haulms. In the second half of growing season, leaves senesce earlier, water and nutrient uptake are reduced causing incipient wilting during hotter days and the number of the tubers and dry matter production are reduced. The above ground symptom first appears in the field as a small patch of stunted plants with pale yellow coloured leaves. These patches increase in size gradually year after year and cover the entire field if potato is continuously grown without any protective measures. Close examination of the roots of infected plants reveal the presence of small pinhead sized, white or yellow female nematodes sticking to the roots.

4. Biology

The hatching of cysts is stimulated by the chemical substances called hatching factors present in the potato root diffusates (PRD). There are at least 25 hatching factors responsible for hatching of both the species of potato cyst nematode. Maximum activity of PRD is reached three weeks after plant emergence and most actively from the root tips. Further two hatching factors identified for *G. rostochiensis* are the potato glycoalkaloids, solanine and α-chaconine. Compound Solanoeclepin A, which occurs naturally in potatoes, is one of the agents involved in hatching of cysts and determined its structure. Hatching depends on host root diffusate, physical conditions before and during juvenile emergence and hereditary preconditions of the cyst. The second stage juvenile (J2) coming out of the cysts moves actively in soil and invade the roots by rupturing with its stylet. It enters through the epidermal cell walls and eventually settles with its head towards the stele and feeds on cells in pericycle, cortex or endodermis by forming a feeding tube. This induces enlargement of root cells and breakdown of their walls to form a large 'syncytium' or 'transfer cell' with dense granular cytoplasm that provides nourishment for nematode development. The nematode molts and remains in the syncytium until its development is complete. The sex of the nematode is determined during J3 stage. The females become sedentary, swollen and remain attached to the roots and posterior part of the body comes out by rupturing the root cells. Males retain their thread shape and come out of the roots to locate and mate the females. The immature females of *G. rostochiensis* are golden yellow in colour while *G. pallida* are white or cream in colour. After the female dies, the body wall thickens to form a hard brown cyst that is resistant to adverse weather conditions. Each cyst contains 200-500 eggs and are easily dislodged in soil at harvest. The cysts can survive in soil for 20-30 years even in the absence of a suitable host crop. The life cycle is completed in 35-40 days and generally, one generation is completed in one year. However, there are evidences of second generation of *G. rostochiensis* being completed because of its shorter dormancy (45 to 60 days) and long crop duration (120 days) and *G. pallida* has a dormancy of 60 to 75 days. Once J2 have developed inside cysts, they enter an extreme form of dormancy, known as 'diapause' and they cannot be stimulated to hatch until the diapause has ended. Only 60 to 80 per cent of juveniles can be stimulated to hatch in the presence of root diffusates and it never reaches 100 per cent. This is a part of the survival strategy of PCN and some juveniles will remain dormant for several years before hatching.

5. Management

The potato cyst nematodes arc the most successful plant parasitic nematodes exhibiting a highly specialized survival mechanism. They have a limited host range in family Solanaceac. It is an accepted fact that once they have established themselves in a given locality, they can not be completely eradicated and thus they have to be managed by adopting several plant protection strategies. Once established, PCN are very difficult to be eradicated from infested fields. Despite massive chemical control measures taken up to eradicate PCN soon after its appearance in the Nilgiris, the nematode remains a serious endemic pest of potato in this region due to intensive cultivation of potato and favourable climatic conditions. PCN have to be managed by a combination of preventive and management methods as follows.

5.1 Preventive Methods

Preventive measures include:

- ✰ Ensuring strict international quarantine regulations to prevent introduction of new pathotypes into areas which are now free of these pathotypes.
- ✰ Implementing strict domestic quarantine procedures to prevent further spread of this pest to new areas that are free from this pest.
- ✰ Adopting good sanitation at field level to prevent the spread of nematodes from infested areas through the farm implements, irrigation water, gunny bags, *etc.*

5.2 Integrated Management

As no single method of control is fully effective in giving desirable level of nematode suppression, an integrated nematode management package incorporating judicious blend of various management options such as chemical, biological and cultural methods is being advocated in the Nilgiris to bring down the PCN population to levels that permit profitable cultivation of potato.

5.2.1 Chemical Control

Application of Carbofuran 3G @2 kg a.i./ha at the time of planting is being recommended as a part of package of practices for potato in the Nilgiris to bring down the PCN population but in some areas use of this chemical is banned. Recently, new fumigant molecule Dazomet (Basamid 90G) @ 40-50 g/m^2 also found to be effective in bringing down the PCN population but after application the soil needs to be covered with polythene sheet which may not be possible in larger areas. However, repeated use of nematicides is not only expensive but also hazardous to the environment. Potato seed tubers from the PCN infested area can be disinfected with Sodium hypochlorite (NaOCl) @2.0 per cent for 30 minutes before transportation to the new potato growing areas.

5.2.2 Resistant Varieties

In India, first cyst nematode resistant cultivar Kufri Swarna was released during 1985, subsequently "Kufri Neelima" during 2012, recently Kufri Shayadri during 2019 and Kufri Karan during 2020 by ICAR-CPRI, Shimla suitable for Nilgiri and Himalyan hills.

5.2.3 Crop Rotation

Since the host range of PCN is restricted to few plants in Solanaceae, crop rotation with non- solanaceous crops is widely recommended for management of PCN. A three to four year rotation with crops like radish, cabbage, cauliflower, turnip, garlic, carrot, green manure crop like lupin *etc.* brings down the cyst population by more than 50 per cent. Growing potato in summer and rotational crop in autumn and spring was found to be successful in managing PCN in the Nilgiris.

5.2.4 Intercropping

Intercropping with French Beans (75:50) was profitable as it recorded higher potato equivalent yield and was effective in bringing down the cyst population.

5.2.5 Integrated Management Practices

For eradication of PCN integration of trap crop (45 days), soil solarization (4 weeks), furadon (3G @ 2kg a.i./ha) application followed by organic amendment and bio- control agent recorded the maximum yield and minimum reproduction factor in the resistant variety Kufri Neelima. Soil solarisation (4 weeks) combined with neem cake (5t/ha) and *Trichoderma viride* (5kg/ha) recorded less nematode population.

5.2.6 Biological Control and Organic Amendments

Recent studies revealed that application of biological control agents *viz., Pseudomonas fluorescens* and *Paecilomyces lilacinus* and organic amendments like neem cake blended with *Trichoderma viride* show promise in suppressing PCN population. Efforts are being made to incorporate these eco-friendly components into the integrated PCN management package.

6. Root Knot Nematodes

The root knot nematodes (*Meloidogyne* spp.) are amongst the most economically important nematode pests of potato. They are prevalent all over the world and can cause significant crop loss in both warm and cool climatic conditions, depending on their species. They have a wide host range and many important cereals, pulses, vegetables, fruits and ornamental crops are good hosts including potato. In India, the root knot nematode infection in potato tubers was first noticed by Dr. Thirumalachar in 1951. Later on, it was reported from almost all potato growing regions of our country.

6.1 Species

Although many species of Meloidogyne are known to infect potato, only four species are reported from India *viz., M. incognita; M. javanica; M. hapla* and *M. arenaria.*

6.2 Distribution

Root knot nematodes are recorded from all potato growing countries of the world. In India, the dominant root knot species affecting potato both in hills and plains has been *M. incognita* while *M. javanica* infestation is in mid hills and plains of northern India. *M. hapla* is confined to hilly tracts of Uttar Pradesh, Himachal Pradesh, Jammu and Kashmir and Tamil Nadu. *M. arenaria* is reported form the plains of Uttar Pradesh.

6.3 Host Range

Root knot nematodes have a very wide host range and have been found to infect more than 2000 plants worldwide that includes major field and horticultural crops. This makesthem very difficult to control because they can always survive and reproduce on other host crops including weeds.

6.4 Yield Loss

Upon severe infection, the root knot nematodes drastically affected the potato yield which recorded even lesser than the quantity planted in Himachal Pradesh and Karnataka, during 1960s. An initial inoculum of 200 juveniles per 100 ml soil resulted in an overall yield reduction of 40 per cent with 100 per cent tuber infestation. Besides the direct yield loss, root knot nematodes also cause indirect damage in potato in the form of blemishes and deformations in tubers which make it unmarketable. Thus the nematodes cause both quantitative and qualitative loss in potato.

6.5 Interaction with other Micro-organisms

Root knot nematode infection is found to predispose potato plants for early blight fungus *Alternaria solani* and brown rot bacterium *Pseudomonas solanacearum*. It is suspected that the nematode favours the spread of brown rot disease in Himachal Pradesh, Uttar Pradesh and Karnataka.

7. Life Cycle

The vermiform second stage juveniles hatch out from the egg masses, penetrate the young potato roots and starts feeding. This leads to the formation of specialized, enlarged cells called 'giant cells' which provide nourishment to the nematodes throughout its development. Hence knots or galls appear in the roots because of the specific host parasite interaction. The second stage juveniles (J2) undergo molting and pass through J3 and J4 stages and finally become adult females or males. The adult females are sedentary and pear shaped while males are migratory and vermiform or thread shaped. Males move out of the root to locate and mate the females. The females lay about 300 to 400 eggs in gelatinous matrix, usually adhering to the root galls. These eggs readily hatch and infect fresh roots. At the time of tuber formation, the juveniles usually enter the tubers since the root system would have started decaying. Life cycle is completed in 25-30 days during summer and 65-100 days in winter. The eggs and juveniles survived for more than 100 days in summer months in Shimla hills. In the hilly regions, two generations are completed by the time of tuberization and hence tuber infestation is invariably observed even under low initial inoculum levels, whereas, in plains, tuber infestation could be low because of the shorter crop duration and preference of the fresh roots by the newly hatched juveniles.

8. Symptoms

The above ground symptoms due to root knot nematode infection include stunting and yellowing of plants with chlorotic leaves due to the hindrance in water and nutrient uptake in roots by nematode infection. In roots, the characteristic swellings called 'galls' are formed. The galls on the root are small and often unnoticed but the warty 'pimple-like' blemishes on the tubers due to nematode infection reduce the commercial value and keeping quality of potato. Brown spots are evident in the flesh of the cut tubers due to the presence of nematodes.

9. Management

Because of the wide host range and semi-endoparasitic adaptation of the root knot nematodes, complete eradication is not possible from an infected area, if once established. The problem may be managed by adopting the following strategies under our conditions.

9.1 Cultural Practice

Avoiding seed tubers from infected areas and planting of nematode-free tubers. Deep ploughing and sun drying during summer months exposes the infective juvenile stages of the nematode to direct sun, thereby killing them. Adopting good sanitation and keeping the field free of weeds which otherwise would serve as alternate hosts for nematodes. Crop rotation with non-host crops like maize or wheat helps in minimizing the nematode damage. Early planting of spring crop in first week of January and late planting of autumn crop in second and third week of October reduce nematode infection in potato due to the lower temperature prevailing during crop period. Growing trap crops such as marigold, *Tagetes patula* in alternate rows with potato reduce nematode population in soil.

9.2 Chemical Control

Application of carbofuran at 1-2 kg a.i./ha reduce nematode infestation and increase yield. The efficiency of these pesticides increase when applied in two split doses once at planting and then at the time of earthing up. This helps keep juveniles of RKN at less than the threshold level of 10 to 20 larvae/100 g of soil sample.

9.3 Breeding for Resistance

The post-inoculation development and reproduction of root knot nematode was hindered in Kufri Dewa. The availability of resistant materials needs to be exploited further to breed nematode resistant variety as this is the only economic means to manage nematode under field conditions.

9.4 Integrated Approach

Since a single method of control is uneconomical and not adequate for better nematode management, a judicious blend of the above methods is always advisable for achieving higher crop production. The false root-knot nematode Nacobbus aberrans (Thorne), the potato rot nematode Ditylenchus destructor (Thorne) and the root lesion nematode Pratylenchus spp. can also cause significant yield losses in potato. Further some nematode species can cause minor problem in potato, including the stubby-root nematodes (*Trichodorus* spp. and *Paratrichodorus* spp.), the lance nematode (*Hoplolaimus galeatus* cobb) and the dagger nematode (*Xiphinema* spp.). Overall, these nematode problems in potato are better managed when integrated management practices are used, *i.e.* exclusion (quarantine regulations, certified plant material, use of clean equipment and machineries), cultural practices (crop rotation, succession, cover crops), genetic control, and ultimately by the use of new effective nematicides.

Figure 10.1: PCN on Potato Root-Ball; PCN Infected Potato Field; Root Knot Nematode Infected Potato Roots, Root Knot Nematode Infected Potato Tubers Showing Warty Pimples, Brown Spots Indicating the Presence of Root Knot Nematodes Inside the Tuber.

10. Extraction of Potato Cyst Nematode and Analysis of Egg Viability

10.1 Extraction of Cysts

Fenwick can method (Fenwick 1940) is a standard method for extraction of cysts from soil sample. In this method extraction is based on the floating properties of dried cysts. The Fenwick can is a metal can tapering at the top with a sloped base. The can has a sloping collar below the rim. In this, 100 CC (cubic centimeter) dry soil is placed in a receptor with a top sieve (1 mm) over the can and wash the air-dried soil sample through the top sieve into the can with a strong jet of water. Water is left running for approximate 5 min. until overflow gets clean. Heavy soil particles fall at the bottom of the can, whereas dried cysts and light soil debris float to the surface and the overflow including cysts are collected on the sieves. An extra sieve of 840 μm is added on top of the 250 μm sieve to remove large debris that moved with the float. The cysts collected on the 250 μm sieve are transferred on a filter paper or white tray. The cysts count (number of cysts) is recorded under zoom stereo-microscope. The can and sieve is rinsed thoroughly before processing every sample.

Figure 10.2: Fenwick Can.

10.2 Differentiation of Live and Dead Eggs/Juveniles in Cysts using Meldola's Blue Stain

For analyzing the viability of eggs/juveniles in cysts of *Globodera*spp. Randomly 100 cysts are collected from sample and placed in small petridish containing 10 ml of distilled water for hydration at room temperature (RT) for 24 h. Hydrated cysts are soaked in 2ml (0.05 per cent w/v) Meldola's blue

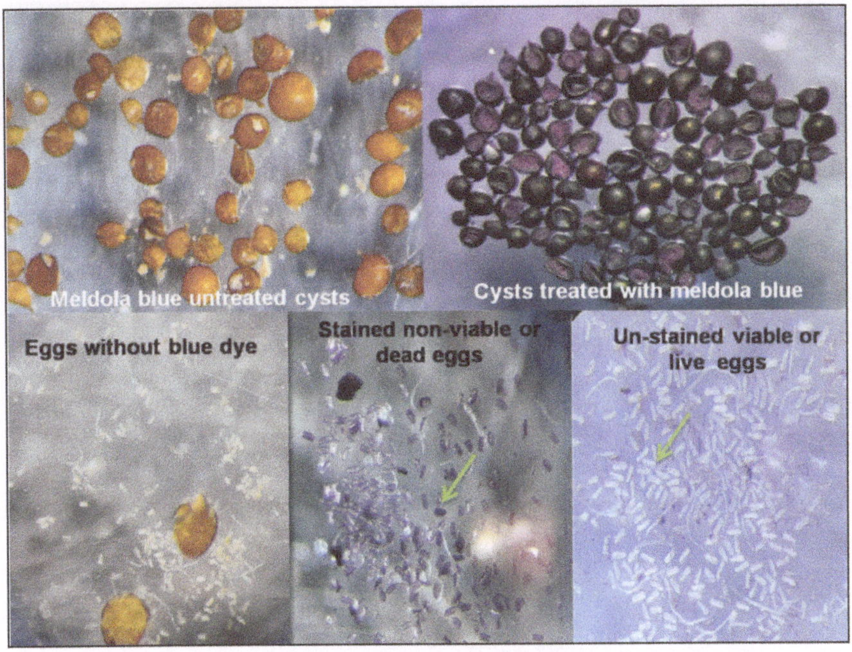

Figure 10.3: Live and Dead Eggs and Juveniles Stained with Meldola's Blue Dye.

stain solution at RT for at least 48 h. After staining, transfer the cysts along with stain solution on filter paper then rinse with water to distain the cysts. Subsequently, 100 washed cysts are placed in a 1 ml eppendorf tube containing 500µl distilled water, disrupted by gently pressing cysts using the tip of the pestle. Due care should be taken not over-crush the cysts/eggs/juveniles. The content from eppendorf tube is collected using 1000 µl pipette and mixed with 50 ml water. From 50 ml stock, 1 ml solution is used for ascertaining the dead/ live, eggs/juveniles under stero-zoom microscope. Live/dead, eggs/juveniles are ascertained based on the colour change eggs/juveniles. Eggs/juveniles stained uniformly dark blue or purple is considered as dead, while unstained one as live (Ogiga and Estey 1974).

Chapter 11

Transgenics and Genome Editing in Potato Improvement

S. Sundaresha

ICAR-Central Potato Research Institute, Shimla, Himachal Pradesh

1. Introduction

Given the significance of the potato, research on the genetic improvement of this crop is important. Potato breeders are challenged by an auto-tetraploid genome, asexual propagation, and breeding principles and practices that are quite different from those employed for the majority of diploid (or allopolyploid), seed-propagated crops and numerous market limiting traits. Moreover, the genetic base of the cultivated potato is considered to be narrow and yield stasis exists within the potato germplasm. However, the potato has an extremely rich gene pool with seven cultivated and 199 wild species. To broaden the genetic base of the cultivated potato and introduce new traits, potato species have been used as a source of new genes. Potato has always been a close companion to biotechnology. Potato, being vegetatively propagated crop, is highly amenable to asexual clonal propagation techniques *in vitro* and consequently genetic engineering. Besides, potato has the distinct advantage of possessing a commercially viable carbon sink in the form of tuber. Therefore, it has also been looked upon as a potential bioreactor for the production of novel compounds of therapeutic and industrial values. A brief account of successful application of biotechnology in potato production and improvement is discussed here.

2. *In vitro* Culture

Potato is perhaps the premier example of a crop plant to which *in vitro* technology has been most extensively applied in all aspects of production, improvement and germplasm handling. This *in vitro* propagation capacity of potato has helped tremendously to overcome genetic nature of this crop imposing several limitations on seed multiplication, conservation of genetic resources and genetic improvement. The first successful establishment of tissue cultures from potato tubers was reported as early as 1951. Since then, the *in vitro* produced disease-free plants, somaclones, haploids and somatic hybrids, plant resistant to diseases and microtubers produced in test tubes have been moved from the laboratory to the field and propagated on a large scale in various countries. Nodal segment culture in which axillary buds and terminal buds grow into new plants is predominantly used for the initial *in vitro* shoot multiplication in potato. *In vitro*-derived micro-plants are used either as direct transplants in the greenhouse/field for the production of mini-tuber/tubers or as explant sources for the production of micro-tubers *in vitro*. Micropropagation revolutionizes potato seed production system by supplying disease-free transplants without any seasonal barrier. *In vitro* potato cultures have also been used to conserve and distribute valuable genetic resources, which are otherwise difficult or impossible using the conventional approaches.

Microtuber has subsequently augmented potato seed production. Micro-tubers are miniature tubers developed under tuber-inducing conditions *in vitro*. These small dormant tubers are particularly convenient for handling, storage and distribution. Unlike micro-propagated plantlets, they do not need the time-consuming hardening period in a greenhouse, and can be adapted easily to large-scale mechanized planting in the field. The ease, with which micro-tubers can be produced, handled and exported, bolsters potato seed production schemes.

3. Transgenic Potatoes

3.1 Potato Transgenics for Viral Resistance

Genetic engineering by employing different strategies for virus resistance is one of the major success stories in potato transgenic biology. The best-documented approach for generating virus resistant transgenic potato is coat protein-mediated resistance, now widely effective against PVX, PVY and PLRV. A novel strategy has been employed to control viroids, which encode none of the proteins necessary to support their own replication but rely on host components for survival. Resistance against potato spindle tuber viroid (PSTVd) was achieved by expressing a double stranded RNA-specific ribonuclease from yeast. Since the first field test of transgenic potato expressing the coat protein gene, there have been many large-scale field trials of transgenic potato, which confirmed durability of the trait. CPRI has pioneered in exploiting the all possible ways of developing viral resistant potato lines.

3.2 Potato Transgenics for Fungal Disease Resistance

Fungal pathogens cause several important diseases in potato. Among these, late blight (caused by *Phytophthora infestans*) was responsible for the infamous Irish potato famine of 1845 and has become a global threat to potato production, especially during the last decade because of emergence of complex races of the pathogen. Recently, a dominant resistant gene (*RB* gene) has been cloned from wild potato species *S. bulbocastenum*. Introduction of RB gene into susceptible cultivars conferred late blight resistance. Limited field trials of RB transgenic potato lines conducted by Central Potato Research Institute demonstrated durable resistance under Indian condition. Another strategy used for developing late blight resistance in potato is RNAi technology. Using Host delivered (HD) RNAi-mediated gene silencing technology successful late blight resistant transgenic potato lines have been developed by silencing the disease causing pathogen *avr* genes.

3.3 Potato Transgenics for Insect Pest Resistance

Potato tuber moth (*Phtorimaea operculella*) is one of the most important pests in tropical and subtropical regions. All attempts to improve potato plants for resistance to Potato tuber moth using conventional breeding methods have been failed. One of the best and common methods for biological control of the pests is the *cry1Ab* gene. This gene encoding a crystal protein that causes perturbance of the digestive apparatus of Lepidopteran insects and insect death follows. Cry1Ab protein has high mortality effects on potato tuber moth. CPRI, Shimla has developed transgenic potato lines carrying *CryIAb* gene from *Bacillus thurengenisis* against potato tuber moth (PTM). Transgenic tubers stayed free from PMT damage in both *In vitro* and field assays.

3.4 Potato Transgenics for Improved Quality Traits

Potato protein suffers from deficiencies in sulfur containing amino acids. Plant breeding has had limited success in improving the nutritional quality of plant proteins. Introduction of gene encoding seed albumin protein from amaranth (*AmA I*), which is rich in cysteine and methionine, has elevated the level of sulfur containing amino acids of potato proteins. CPRI, Shimla in collaboration with National Centre for Plant Genome Research (NCPGR), New Delhi has already introduced this gene into several Indian potato cultivars. This may partly help addressing protein-calorie malnutrition in the country. Transgenic potato lines to improve cold-chipping attributes by RNAi-mediated silencing of vacuolar invertase gene (*Va-INV*) and UDP Glucose Phosphatase (*UGPase*) gene are in different stages of development and evaluation.

4. Genome Editing

Genome editing is a versatile tool which can be employed for crop improvement programme by removing unwanted sequences and make it desirable character of interest by gene knockout, knock in and overexpression studies. Sequence editing either by deleting or modifying the genes individually and then studying the subsequent mutant phenotypes can address the challenges of understanding the function of genes. For precise DNA manipulations, new genome editing systems which are able to induce double-stranded breaks (DSBs) at specific sites in the genome and repaired naturally using DNA repair mechanism (non-homologous end-joining or homologous recombination mechanism) thereby, ensuring the gene mutation at target site. This system is facilitated by protein-guided nucleases, such as zinc finger nucleases (ZFNs), transcription activator-like effector nucleases (TALENs), or special RNA/DNA-guided nucleases, including RNA-dependent DNA cleavage system like clustered regularly interspaced short palindromic repeats (CRISPR)–associated protein (Cas) system.

5. CRISPR

CRISPR is an advanced and improved molecular breeding technique in terms of creation of mutagenesis for better improvement programme. CRISPR (clustered regularly interspaced short palindromic repeats)/Cas (CRISPR-associated) system, has been developed as an easy, specific targeted genome modifications and has emerged as the most powerful method due to ease in designing and construction of gene-specific single guide RNA (sgRNA) vectors. These sgRNA vectors are easily reprogrammable to direct *Streptococcus pyogenes* Cas9 (SpCas9) to generate double stranded breaks (DSBs) in the target genomes that are then repaired by the cell via the error-prone non-homologous end-joining (NHEJ) pathway or by precise homologous recombination (HR) pathway. A recent article by Nadakuduti *et al.* (2018) reviewed the developments in gene editing technologies and their role in trait improvement in different crops including potato. CRISPR/Cas9 technology has been successfully implemented in potato (MSU group USA) for targeted editing to generate mutations (by means of NHEJ) as well as gene targeting to edit an endogenous gene (by HR). In this bulletin, we describe procedures for designing sgRNAs, protocols to clone sgRNAs for CRISPR/Cas9 constructs to generate knockouts, design of donor repair templates. Accordingly, CPRI also initiated CRISPR based targeted editing of potato genome for generating variety specific potato seeds and late blight resistance lines.

6. Conclusion

From the 16th century Andean highlands to the 21st century cultivated potato species worldwide, biotechnology has transformed potato from test tube to the field in such a way that its production, especially in developing countries, has outpaced all other crops. These developments have far-reaching implications not only in the production and improvement of present day potato but also for the induction of genetic variability, which would enable the synthesis of novel future potatoes.

Chapter 12

Status of Adoption of Potato Technologies and Varieties in India

Pynbianglang Kharumnuid and N.K. Pandey

ICAR-Central Potato Research Institute, Shimla, Himachal Pradesh

1. Introduction

India is predominantly an agriculture based country in which agriculture and allied activities contributes about 17 percent to the country's Gross Value Added and provides employment to about 54.6 percent of the population. Potato plays a very important role in Indian agriculture as potato alone contributes about 21 percent of the total vegetable area and 26 percent of total vegetable production of India. Potato is a nutrient-rich crop which provides more calories, vitamins and nutrients per unit area than any other staple crops. FAO declared potato as the crop to address future global food security and poverty alleviation during 2008. As per FAOSTAT data for the year 2017, India with 48.6 million t is ranked second in potato production in the world, only behind China with 99.2 million t. The productivity in India is higher than in China and Russia, the third largest potato producer. However, the productivity is lower than most of the developed European countries. One of the major reasons is that, in Europe, potato crop is grown in summer having long photoperiod of up to 14 hours and the crop duration of 140-180 days. The potato in Indian plains is, however, grown in completely contrasting situations. Nearly 85 per cent of the crop is grown during winters having short photoperiod (with about 10-11 hours sunshine) and the crop duration is also limited to 90-100 days because of short and mild winter.

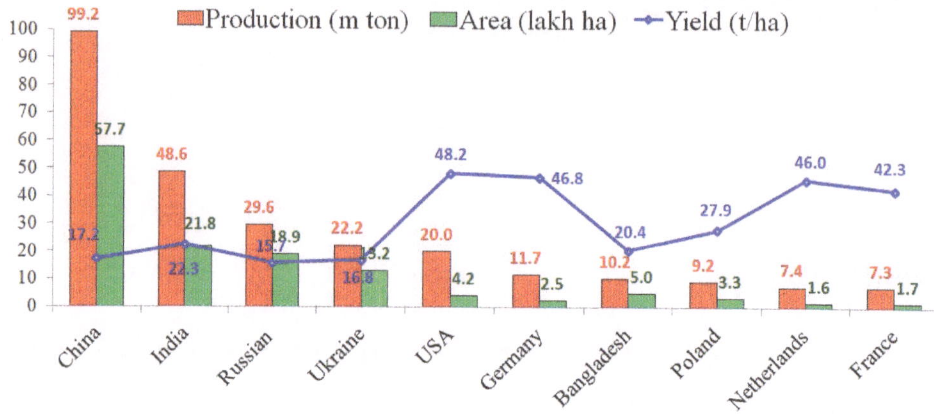

FAOSTAT, 2017

Figure 12.1: Top 10 Potato Producers in the World.

2. Trends in Area, Production and Productivity of Potato in India

At the time of inception of CPRI, in the year 1949, India produced only 1.54 million t potatoes from 0.234 million ha area at an average productivity level of 6.58 t/ha. As per the 1st advance estimatesby National Horticulture Board, the potato production in India during 2018-19 was 52.6 million t from 2.18 million ha area with a productivity of 24.13 t/ha. Over the period of 1949-50 to 2018-19, the compound annual growth rate (CAGR) in potato area, production and yield were 3.3 percent, 5.4 per cent and 2.0 per cent per annum, respectively. It is the hard work of potato farmers, scientists and policy makers that potato area, production and productivity increased over 7 decades by 8.5, 34.06 and 3.65 times, respectively. Contribution of CPRI and AICRP on Potato has been adequately recognised by the nation on several occasions.

Table 12.1: Compound Annual Growth Rate (per cent) of Area, Production and Yield of Potato in India

Year	*Area*	*Production*	*Yield*
2009-10 to 2018-19	2.1	3.2	1.1
1949-50 to 2018-19	3.3	5.4	2.0

3. Adoption of Improved Potato Varieties in India

The adoption of improved potato varieties is critical for achieving sustainable and higher productivity and production. So far, Central Potato Research Institute, Shimla has developed 66 potato varieties for different agro-climatic regions of the country and majority of them are for north Indian plains. Potato varieties developed by CPRI are very popular among farmers and cover

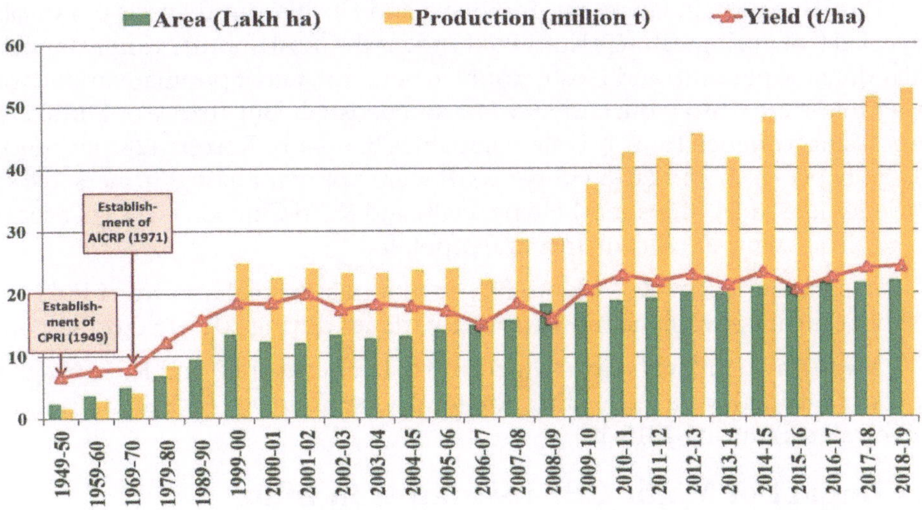

Figure 12.2: Trends in Area, Production and Yield of Potato in India.

nearly 95 per cent of total area under potato. Four varieties, *viz.*, Kufri Jyoti, Kufri Bahar, Kufri Pukhraj, and Kufri Chipsona1 together contributed around 75 per cent of total area under potato. The varietal adoption patterns of potato varieties in major producing states are as follows:

Uttar Pradesh: The study conducted during 2016-17 crop season showed that overall, varieties like KufriBahar (38 per cent of potato area), KufriPukhraj (21 per cent), KufriKhyati (12 per cent) and Kufri Chipsona-1 (9 per cent) were the leading potato varieties in Bulanshahr and Meerut district of Uttar Pradesh.

West Bengal: The survey conducted in Bardhaman and Hooghly districts of West Bengal for 2017-18 crop season revealed that overall, Kufri Jyoti (52 per cent area) was the most popular variety, which was followed by KufriChandramukhi (18 per cent) and Kufri Pukhraj (16.2 per cent).

Bihar: A survey for assessing varietal adoption patterns in two major potato growing districts, *viz.*, Nalanda and Patna of Bihar during 2016-17 crop season revealed that Kufri Pukhraj (69.53 per cent potato area) was the most popular variety, followed by Kufri Sindhuri (6.8 per cent) and Kufri Khyati (5 per cent).

Madhya Pradesh: In Madhya Pradesh, Kufri Chipsona-3 (26.5 per cent area) and KufriChipsona-I (22.3 per cent area) were the leading potato variety closely followed by Kufri Jyoti (22.07 per cent area) during 2014-15 crop season.

Gujarat: The survey conducted in Banaskantha and Aravalli district of Gujaratduring 2017-18 crop season revealed that the most popular varieties were Lady Rosetta (40.5 per cent) and Kufri Pukhraj (33.3 per cent). Other important varieties were Kufri Lauvkar (12.6) and Kufri Badshah (6.3 per cent).

Punjab: Study in Jalandhar, Ludhiana and Hoshiarpur districts of Punjab revealed that during 2016-17, Kufri Pukhraj was the most popular variety in both Jalandhar (50 per cent) and Hoshiarpur (76 per cent). Other popular varieties of Jalandhar were Kufri Jyoti, Lady Rosetta and Badshah. But, in case of Ludhiana processing varieties from private companies,like Lady Rosetta (28 per cent), FC-3 (25 per cent) and FC-5 (15 per cent) were popular among farmers. CPRI varieties like Kufri Chipsona-1 (16 per cent) and Kufri Chipsona-3 (13 per cent) were also having significant area under potato.

Assam: In Assam, Kufri Jyoti (50 per cent potato area) and Kufri Pukhraj (33 per cent area) were the most popular varieties during 2015-16 crop season.

Meghalaya: In Meghalaya, Kufri Jyoti (34 per cent area) and Kufri Megha (14 per cent area) were the leading potato varieties followed by Kufri Giriraj 10 per cent area) during 2015-16.

4. Impact of Major CPRI Varieties in India

Potato varieties developed by ICAR- Central Potato Research Institute cover nearly 95 per cent of total area under potato. India produced about 45.87 million tonnes of potato annually during the triennium 2014-17 and contributed Rs. 57,512 crore annually to the Gross Value Added (GVA) at current price. The varieties developed by ICAR-CPRI contributed Rs. 54,636 crore annually during this period.

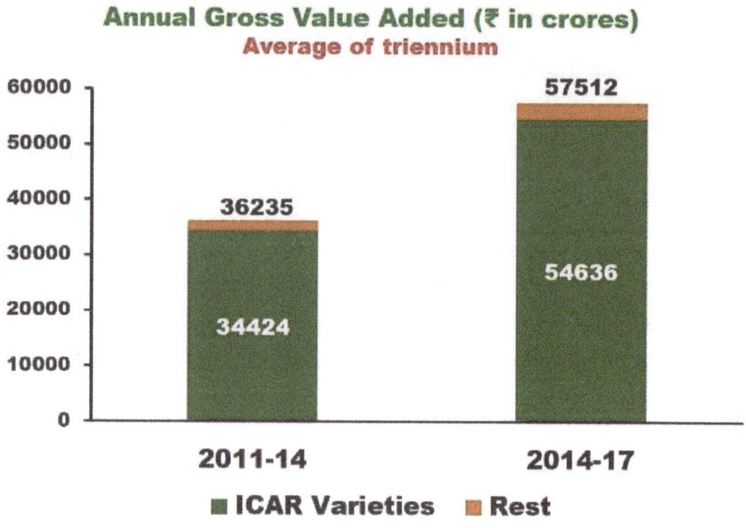

Kufri Jyoti, a day-neutral variety released in the year 1968 played a lead role in popularizing potato throughout the country. Because of wide adaptability of this variety, potato emerged as a principle component of the profitable maize-potato-wheat sequence in the vast North-western plains. Presently it occupies about 21 per cent area in the country and thereby contributing about Rs. 11,800/- crore to the agricultural economy.

Kufri Bahar released in 1980 became the principal potato variety of Uttar Pradesh, the highest potato producing state of the country. It occupies about 17 per cent area in the country and contributes about Rs. 9,551/- crore to the agricultural economy.

An early bulking variety (60 days), Kufri Pukhraj released in1998 is the leading variety today in India occupying >33 per cent area. It contributes about Rs. 18,541 crore in the agricultural economy of the country. It is most suitable forrice-wheat basedsequencein the northern plains.

Kufri Chipsona1 released in 1998 is 4th leading variety of the country today occupying about 4 per cent area under potato and contributing about Rs. 2,247/- crore to the agricultural economy. Besides, five more processing varieties have been released by the Institute catering to almost 35 per cent of raw material for the processing industries.Potatoprocessing companies like M/S Frito Lay India (Pepsico), ITC Ltd., Merino Industries, Mc Cains India, PRANFood Ltd., Haldiram, Pailan Food Ltd., Balaji Wafers, Golden Fries, Bikaji Foods International Ltd. *etc.* use Indian processing varieties and sell their products as >100 brand names. This has resulted in tremendous growth of processing sector from mere <1 per cent during 1990s to > 8 per cent now.

5. Impact Assessment of Potato Varieties using Economic Surplus Model

Kufri Pukhraj: The Economic Surplus Method with closed economy was used to measure the benefits of and returns to research and development (R and D) of an early bulker and drought-tolerant potato cv. Kufri Pukhraj in India. A period of 40 years from 1978 (the year in which breeding of Kufri Pukhraj started) to 2018 was considered for developing the cost–benefit stream. All these estimates of costs and benefits were expressed in real terms by adjusting to Wholesale Price Index with 2018 as the base year. The results of the model revealed that the total cost of R and D of Kufri Pukhraj was Rs.12.1 crore. The Net Present Value (NPV) of the benefits of Kufri Pukhraj R and D was Rs. 10974.5 crore and 4215.5 crore at 5 and 8 per cent discount rate respectively and the internal rate of return (IRR) of the net benefit was 43 per cent.

Kufri Chipsona-3: Economic Surplus Model was used for assessing the impact of returns to research investment for developing processing potato cv., Kufri Chipsona-3. Research cost computed for the period 1990-91 to 2006-07 while the adoption cost was estimated during 2007-08 and 2012-13. All estimated expenses were adjusted to consumer price index of industrial workers with the 2012-13 as base year. Returns to investment for developing Kufri Chipsona-3 were evaluated during different scenario in order to articulate different funds availability situations. It was found that the IRR was 58.90 per cent. The net present worth was found to be Rs. 5871 million and 2726 million at 5 per cent and 10 per cent interest rate.

Chapter 13

Impact of Climate Change on Potato: Current Scenario and Adaptation Strategies

V.K. Dua

ICAR-Central Potato Research Institute, Shimla, Himachal Pradesh

1. Introduction

Climate change is posing serious threat to human survival on the earth in one or the other way. The disturbances in earth's climate have been greatly accelerated during the past century. It's a well-established fact that greenhouse gas (GHG) emission is responsible for climate change and is increasing rapidly at an alarming rate. Potato is the third major food commodity of India after rice and wheat in terms of production. India has made tremendous progress in potato production and the per capita availability of potato has increased from 4.37 kg in 1950 to about 30 kg in 2019. However, the climate change is likely to have a negative effect on potato growth in India, including productivity, production and profitability. The increase in CO_2 is expected to bring on increase in productivity of potato as reported by many workers. However, increase in temperature and atmospheric CO_2, both are interlinked and occur simultaneously and the CO_2 enrichment does not appear to compensate for the detrimental effects of higher temperature on tuber yield. Worldwide potato crop losses due to late blight have been estimated at €12 billion (Haverkort *et al.*, 2009). In India, severity of late blight varies from region to region, being more severe in temperate highlands than in the sub-tropical plains with an average of 15 per cent crop losses annually (Collins 2000). In case of late blight,the effect of climate change on the most dreaded disease of potato in India also, could be mixed. Late blight caused by *Phytophthora infestans* is often regarded as the most important disease of potato globally. In potato cultivation, potato seed is most expensive input accounting for 40 to 50 per cent of the production cost. Since the eastern, northeastern, Deccan and south western parts of the country are not suitable for quality seed production barring few locations, the farmers of these areas have no option but to buy the seed potato from northern India. The high hills of Himachal Pradesh, Indo-Gangetic plains of Punjab, Haryana, north-western part of Uttar Pradesh, and Bihar are suitable for nucleus and breeder seed production. However, Punjab and western Uttar Pradesh are two major states which supply quality potato seed to rest of the country. In case of climate change scenario, the temperature change is likely to affect the late blight outbreak in Punjab and western Uttar Pradesh, thus affecting the potato seed production.

Bacterial wilt is causing huge losses in warmer area of Madhya Pradesh and plateau region. Presently, in North Western Indo-Gangetic region, the disease severity is very less but due to warming, the disease may enter the new areas where it is not present. The present studies were conducted to assess the yield losses due to climate change, to develop suitable adaptation strategies and to assess effect of climate change on late blight and bacterial wilt prevalence. Work of ICAR-CPRI, Shimla on climate change impact on potato production.

2. Impact on Potato Productivity

For impact on potato productivity, we used WOFOST (WOrld FOod STudies) crop growth simulation. WOFOST, a mechanistic model which simulates the growth of a crop based upon eco-physiological processes is widely used to assess the effect of climate change on the growth and yield of many crops which includes wheat, rice, maize potato, barley, soybean, sugar beet throughout the world (Boogard *et al.*, 1998, Wolf *et al.*, 2000) and is one of three most widely used crop growth model for climate change studies (Diepen *et al.*, 1987). WOFOST has been calibrated for Indian potato cultivars using the time course data on potato growth and development (Dua *et al.*, 2014). Hence, WOFOST model was used for impact assessment of climate change on potato productivity and scheduling planting date and selection of suitable cultivar to minimize climate impact in Uttar Pradesh, Punjab, Madhya Pradesh, Gujarat, Bihar and West Bengal. The simulation studies were carried out for four potato cultivars, Kufri Badshah, Kufri Bahar, Kufri Pukhraj and Kufri Jyoti. The model was run for different planting dates depending upon the locations and the cultivar selected.

JHULSACAST model was used to work out the effect of climate change on potato late blight outbreak, duration of favourable period and disease severity in Uttar Pradesh, Punjab and West Bengal. The model requires hourly temperature data (°C), relative humidity (RH per cent) and daily rain (mm) data as input. The model was run for baseline scenario (year 2000) and future

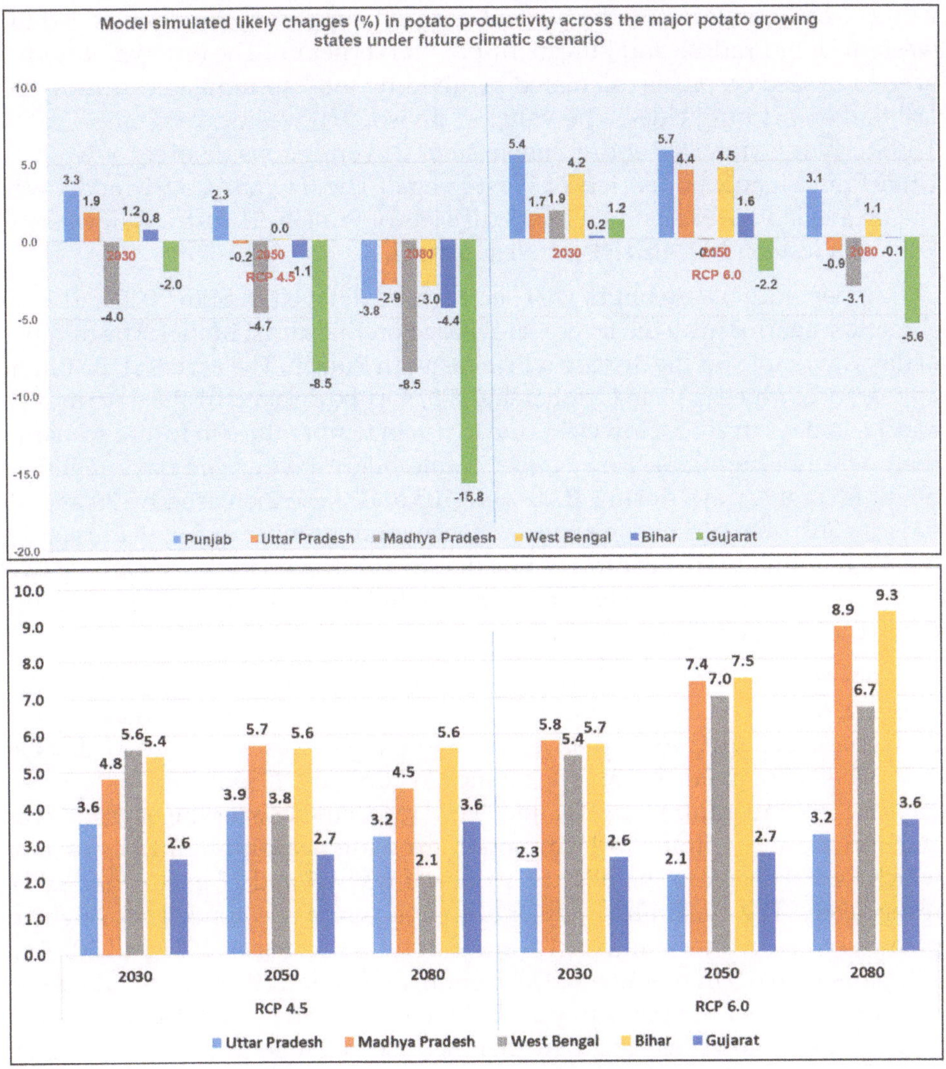

Figure 13.1: Impact on Late Blight.

climate scenarios (years 2020 and 2055). IMD district normal of 1971-2000 were used for baseline scenario (year 2000) and for future climate scenarios of the years 2020 and 2055 A1FI scenario of temperature (SRES A1FI pathway) was used. It is not possible to predict the daily humidity in years 2020 and 2055 during cropping period, therefore, we assumed that favourable RH would start from a week of emergence of the crop in western Uttar Pradesh and Punjab respectively and proceed with 10 days intervals after the setting of required RH. Thus, the model results are the outcome of the effect of temperature input. To estimate the number of sprays required in seed crop, the daily severity values were calculated on the basis of decision support system developed for western Uttar Pradesh and Punjab (http://cpri.ernet.in). The number of sprays needed in seed crop was calculated by dividing total accumulated severity by 180 for contact fungicides. The value of divisor 180 was derived empirically. The study was done for entire Punjab state, 23 contiguous districts of western Uttar Pradesh and 11 districts of West Bengal. The JHULSACAST model was run on available weather data for 13 districts of Punjab, 21 districts of western Uttar Pradesh and 11 districts of West Bengal.

Appearance of late blight is expected to be delayed in both 2020 and 2055 climatic scenarios in all the three states based on Jhulsacast Model. The number of days favorable for the disease will increase in Punjab. The earliest Late Blight appearance, which is 13-15 October, in 2000, may be delayed by 0-6 days in 2020 and 12-14 days in 2055. However, due to rise in temperature in future scenario, there will not be mid-season breaks in late blight favourable days. The late blight favorable days during 2000 were 105 and these increased by 30 and 35 days by 2020 and 2055, respectively. Hence more number of fungicide sprays will be required for the control of this disease. However, in the remaining two states the number of favorable days for late blight may decrease and hence less number of fungicides sprays will be required.

In western Uttar Pradesh, earliest late blight appearance during the potato crop season is predicted during 13th October to 1st November in baseline year 2000 and is expected to be delayed by 0 to 8 days in 2020 and 10-21 days in 2055. The delay in late blight appearance in 2020 and 2055 is due to expected increase in temperature by 1.08 and 2.98 °C respectively over the baseline year 2000. In western Uttar Pradesh, potato growing season would be warmer which would decrease late blight favorable days by 7 and 27 in 2020 and 2055, respectively. The maximum number of sprays required to control late blight in potato seed crop thus would be 7.3 and 8 in future scenario (2020 and 2055) in comparison to 6.5 in baseline (2000) in Punjab. In contrast, there would be no change in number of sprays in year 2020 (over baseline year 2000) in western Uttar Pradesh, however, due to further increase in temperature in year 2055, it could be reduced by 2.

In West Bengal, earliest late blight appearance during the potato crop season was predicted during 4th to 11th October in baseline year 2000 and is expected to be delayed by 7.6 days in 2020. The delay in late blight appearance in 2020 is expected due to increase in temperature by 1.08 over the baseline scenario. However, due to delay in appearance of late blight in the season, the total late blight favorable period is likely to be reduced by 17.7 days in 2020, respectively. The number of sprays required to control late blight in potato seed crop would be 4.2 in future scenario of 2020 in comparison to 5.5.5 in baseline (2000) in West Bengal. Hence less number of fungicides sprays will be required.

3. Impact on Survival of *R. solanacearum* in Potato Growing Districts of Plateau Region

For studying the survival of *Ralstonia solanacearum* in soils of plateau region of India, results of the experiment conducted under controlled condition of various temperatures on the survival of *R. solanacearum* in soil were used to develop. A logical mathematical model which give the time (in days) in which the bacterium will not survive in the soil. The model gives the results on the basis of daily mean soil temperature. For baseline survival, IMD normal were used and for future climate, temperature rises for A1F1 scenario of the climate change for 2020 and 2055 were added on the baseline temperature. The presence of *R. solani* in the soil was presumed on 1st of January and the days it will survive in the soil of 60 potato growing districts of eight states in plateau region of India were worked out, based upon soil temperature.

- *Tamil Nadu and Kerala:* Under the baseline climate scenario (2000), the bacterium would survive till 18th March in soils of Idukki, Kannur, Malappuram, Pathanammitha, Wayǎnad districts of Kerala and Dharamapuri, and Virudinagar districts of Tamil Nadu and no changes are expected in 2020. However, in 2055, the bacterium is likely to survive in soil of these districts till 03 March only. The bacterium is likely to disappear 0-05 days earlier in 2020 and 05 to 15 days in 2055.
- *Karnataka:* The bacterium could survive in soil till 18 March in Gadag, 7 April in Dharwad, Bidar, Chitradurga, Karwar, Udupi, and 27 April in Rural and Urban Bangalore districts of Karnataka in baseline scenario. In 2020, the bacterium is likely to survival in soil till 18th march in some areas of Dharwad and Chitradurg while in rest of districts, no change is expected. The bacterium is likely to disappear 0-08 day earlier in 2020 and 11-15 days in 2055.
- *Andhra Pradesh, Maharashtra, Odisha and Chhatisgarh:* The bacterium may survive in soil from 7 to 27 April in different districts and is likely to disappear 3-8 and 11-25 days earlier in Andhra Pradesh and Maharashtra, 3-8 and 11-20 days in Odisha and 3-5 and 11-15 days in Jharkhand in 2020 and 2055, respectively.
- *Jharkhand:* Under the baseline (2000), the bacterium could survive in soil maximum till 27 April in Gumla and 17 May in Ranchi and Singhbhum. No change is expected in Gumla in 2020, however, in Ranchi and Singhbhum, the survival is likely to be up to 6th May only.
- *Madhya Pradesh and Gujarat:* The bacterium may survive in soil from 27 April to 17 May in different plateau districts of Madhya Pradesh and Gujarat in 2000. The bacterium is likely to disappear 3-8 days earlier in 2020 and 5-15 days in 2055 in Madhya Pradesh and 3-5 days earlier in 2020 and 11-26 days in 2055 in Gujarat.

- *Rajasthan:* Under the baseline climate scenario (2000), the bacterium survived in soil till 17 May. The period is likely to reduce by 0-5 days in 2020 and 5-15 days in 2055.

4. Adaptation and Input Strategies

The various adaptation strategies to combat the impact of climate change on potato productivity may include breeding short duration and heat tolerant cultivars, developing potato cultivars that tuberises at higher night temperatures. Mining for biodiversity to heat tolerance should be given priority. Breeding drought and salinity tolerant cultivars would be effective to face the future challenges of climate change. Use of wind breaks around fields and crop residue mulches for some period after planting, using drip and sprinkler irrigation in place of furrow and basin methods and altering cultural management in potato based cropping systems are few examples of agronomic management practices to reduce the impact of climate change. Besides, conservation tillage and on farm crop residue management are required to increase input use efficiency. Advance planning for possible relocation and identifying new areas for potato cultivation is needed. Improvement and augmentation of cold storage facilities and air conditioned transportation from producing to consumption centers will be required for storage and transportation of this semi-perishable commodity. Strengthening education, research and development in warm climate production technology for ware and seed potato crop is also required to meet the production targets in future climates.

Chapter 14

Soil and Plant Tests for Fertilizer Recommendation

Jagdev Sharma and V.K. Dua

ICAR-Central Potato Research Institute, Shimla, Himachal Pradesh

1. Introduction

Potato (*Solanum tuberosum* L.) is a highly attractive crop in agricultural production systems. However, while countries with a high-input agriculture, such as the USA, France and Germany, Netherlands can reach average potato yields of more than 45 t ha^{-1} the average yields of most other countries including India are considerably lower leading to average potato yields of about 20 t ha^{-1} worldwide. Improper use of fertilizers besides irrigation and other factors is one of the reasons for low productivity. The primary aim of agriculture in twenty-first century is to maximize food production while maintaining soil health and improving environmental quality. Manures and fertilizers have played most important role in increased food grain productivity and will continue to do so. Indiscriminate and non-judicious use of chemical fertilizers, escalating fertilizer prices, and constraints on the availability of fertilizers as well as the hazardous effect of underuse or overuse of fertilizers on environmental health, especially soil health. As soil health is directly related to human health, soil quality could be maintained through balanced application of manures and fertilizers and is possible through evaluation of soil fertility and assessment of crop demand for nutrients. Soil fertility is the key component of soil quality, which can be assessed through soil testing. Besides maintaining soil quality a lot of attention is being used to enhance the NUE as the cost of fertilizers is on rise and is reducing the profits. Fertilizer recommendations based on soil nutrient availability (soil tests) and crop needs enhance nutrient-use efficiency.

At the end of the nineteenth century and at the beginning of the twentieth century, a lot of effort was put into investigating the relationships between production factors, such as the nutrient supply of the plant, and the yield of crops. During this period the most famous yield-response law originating from this time is the law of the minimum was developed by Carl Sprengel which was later spread by Justus von Liebig in the early nineteenth century. According to the law of the minimum, optimal crop growth can take place only if all required nutrients are at the optimum level. In other words it states that plant growth is controlled not by the total amount of nutrients available but by the amount of the scarcest nutrient and points to the importance of balanced nutrition for optimal plant growth. The law of diminishing yield increase states that the higher the nutrient supply, the lower the yield increase obtained from incremental feriliser doses *i.e.* which means that yield response to fertilisation follows a saturation curve. These laws are the basis for modern approaches to develop strategies for efficient resource use in plant production. Modern approaches uses both soil and plant tests increasing the production of crops and maintaining soil quality. Soil testing dates back to 1956-57 when 24 soil testing laboratories were established under Indo-US collaboration. A chemical soil test is a method for estimating nutrient supplying power of a soil by using different extractants, estimating organic components (organic carbon per cent) and salinity/sodicity *etc.*, while physical soil tests (soil texture (proportion of sand silt and clay),water holding capacity, bulk density and particle density *etc.* provide information regarding suitability of the soil for potato crop growing. Compared to plant analysis, the primary advantage of soil testing is its ability to determine the nutrient status of the soil before the crop is planted. The result of a soil test is called as soil test value which measures a measures a part of the total nutrient supply in the soil. Soil test do not measure the exact quantity of a nutrient potentially taken up by a crop. Soil test values determine the degree of yield response. The fixed value of a soil test called as critical value. If the soil test value is below critical value the crop will respond to applied fertilizers. To predict nutrient needs of crops a soil test must be calibrated against the response of crops in nutrient rate experiments. Thereafter, interpretation and evaluation of the soil test values primarily form the basis for fertilizer recommendation.

Soil test programmes for potato are generally have the following objectives: i) To predict the probability of profitable response to fertilizer. ii) To evaluate the soil fertility status of the soil to provide a basis for fertilizer recommendation. iii) Assessing suitability of different soils for potato cultivation. iv) Need for soil ammendments. Phases of soil testing programme includes: i) Collecting the soil samples ii). Extraction and determining the available nutrients iii) Calibrating and interpreting the analytical results iv). Making the fertilizer recommendations. The success of a soil testing programme depends upon the operative precision in all the four stages. Soil test interpretations involves an economic evaluation of relationship between soil test and crop response to applied fertilizer.

2. Soil Sampling

It is the most critical aspect of soil testing is obtaining a soil sample that is representative of the field. The soils are normally heterogeneous, and wide variability can occur even in a uniform fields. There is always a considerable opportunity for sampling error. The sampling error in a field is generally greater than the error in the laboratory analysis. Reliable fertilizer recommendations can only be provided based on a representative soil sample. Intensive soil sampling is the most efficient way to evaluate variability. Soil sampling unit is an area of soil which is to be represented by a composite sample. It should be an homogeneous area. Soil samples are collected throughout the soil unit area. Variations in colour, texture, slope and management should be considered while selecting an unit area. Ten to fifteen subsamples are collected in a zigzag pattern from an apparently homogeneous area with the help of stainless steel augers/spades/khurpi after scrapping off surface litter. If a spade of Khurpi is to be used a V shaped pit is dug in the plough layer an uniform cm thick slice of soil is from both sides is collected. The subsamples should be mixed well in stainless steel, plastic container or on polythene sheet and the bulk is reduced by quartering method to about 500 g of the composite sample is taken for analysis. The samples are dried in shade, ground with wooden pestle and mortar gently, sieved through 2 mm stainless steel sieve and stored in plastic or polythene bags with proper labelling and description. One half is moved to the side quantitatively. Then each half is further divided into half, the four quarters being separated into separate 'quarters'. Two diagonally 'opposite quarters' are discarded quantitatively. Soil physical tests include determination of soil texture (proportion of sand silt and clay),water holding capacity.

Soil test do not measure the exact quantity of a nutrient potentially taken up by a crop. Soil test values determine the degree of yield response. The fixed value of a soil test called as critical value. If the soil test value is below critical value the crop will respond to applied fertilizers. To predict nutrient needs of crops a soil test must be calibrated against the response of crops in nutrient rate experiments. Thereafter, interpretation and evaluation of the soil test values primarily form the basis for fertilizer recommendation.

2.1 Soil Testing and Target Yield Approach

For better utilization of nutrients from fertilizers and soils its better to have appropriate achievable yield targets and make fertilizer recommendations accordingly. This approach uses soil test value, efficiency factor for soil and fertilizers and crop nutrient requirements to formulate a model equation and workout fertilizer dose for specific yield. This approach at Jalandhar helped in increasing fertilizer use efficiency by 109 per cent, decreasing input costs by 50 per cent and increasing net returns by 5 per cent.

2.2 Sensors and Soil Testing

Sensors that measure a variety of essential soil properties on the go are being developed. These sensors can be used either to control variable rate application equipment in real-time or in conjunction with a Global Positioning System (GPS) to generate field maps of particular soil properties. Depending on the spacing between passes, travel speed, and sampling and/or measurement frequency, the number of measurement points per acre varies; however, in most cases, it is much greater than the density of manual grid sampling. The cost of mapping usually is reduced as well. Optical Sensors use light to measure soil properties. The sensors measure different frequencies of light reflectance in near-infrared, mid-infrared, and polarized light spectrums can be placed on vehicles or aerial platforms such as drones or even satellites. Soil reflectance and plant color data are t two variables from optical sensors that can be aggregated and processed. Optical sensors have been developed to determine clay, organic matter and moisture content of the soil. Electrochemical Sensors could provide the most important information needed for precision agriculture. Several researchers are trying to adapt existing soil preparation and measurement procedures to essentially conduct a laboratory test on the go. The values obtained may not be as accurate as a laboratory test, but the high sampling density may increase the overall accuracy of the resulting soil nutrient or pH maps. Airflow sensors were used to measure soil air permeability on the go. The pressure required to squeeze a given volume of air into the soil at fixed depth was compared to several soil properties. Experiments showed potential for distinguishing between various soil types, moisture levels, and soil structure/compaction. Farmer's acceptance of soil testing is strongly dependent on the extent and severity to which nitrogen, phosphorus, potassium and pH are problems for crop production in the area, and the accuracy with which the tests can be used to predict crop responses and fertilizer needs.

2.3 Tissue Analysis

Tissue analysis has been used for many years as an additional nutrient management tool. The basis behind tissue analysis is that maximum yield and quality are associated with an optimum range of nutrient in the tissue sampled. If the level of nutrient falls outside this range then, then corrective measures should be taken. The deficiency/toxicity of nutrients is reflected in the form of peculiar symptoms on plant parts such as leaves. However, in many cases the symptoms are hidden and by the time they appear considerable damage has already been done. So quantitative analysis can help in identifying the limiting or toxic nutrients in advance and thus help in taking corrective measures well in advance as it will i) diagnose a nutrient deficiency or toxicity, ii) to help predict the need for additional nutrients and iii) monitor the effectiveness of a fertilizer program. In case of potato stolon formation, tuber initiation and flowering/tuber bulking are three most critical stages for realizing optimal yields potential. Stolon formation and tuber initiation stages are ideal for tissue testing as testing at these stages allows for taking correction measures well in time. For destructive analysis 4th leaf/petiole from fully expanded leaf from top is sampled. The plants are sampled before noon since in morning hours plants are not under stress. It is critical that this tissue stage is collected because younger or older tissue will have different nutrient concentrations and can lead to erroneous interpretations. For sampling, approximately 30 to 40 leaves should be collected. Non-destructive technique used for judging nutrient status of plants involves use of sensors like chlorophyll meters. An overfertilized plot is taken as reference plot and chlorophyll content of the plots to be fertilized is measured periodically with the help of chlorophyll meter and compared with overfertilized plots. This technique has been able to save 10-20 per cent of N dose in potatoes.

These sensors use spectral reflectance data which is inversely related to leaf chlorophyll level and relies on the interaction that occurs when light penetrates plant tissue, where it will be absorbed, reflected from the surface or transmitted through the leaf. These optical spectra are dependent on the leaf pigment content of different absorption wavelengths. Chlorophyll has absorbance peaks in two distinct regions: the blue region (400 nm to 500 nm) and the red region (600 nm to 700 nm), with no transmission in the near infra-red (NIR) region. Taking advantage of this fact, scientists designed sensors that emit light in the red region and the NIR region. By comparing the reflectance or the absorbance of these transmittances at the two wavelengths, a value is generated that represents green vegetation of the sample.

Chapter 15

Post-Harvest Management of Potatoes

Brajesh Singh

ICAR-Central Potato Research Institute, Shimla, Himachal Pradesh

1. Introduction

Potato contains approximately 80 per cent water and is a semi-perishable commodity. About 90 per cent of the potatoes produced in our country are grown in the Indo-Gangetic plains and are harvested during January to March, the beginning of the long hot summer. This makes it very difficult to store the harvested potatoes, as under these conditions potatoes cannot be stored without refrigeration for more than 3-4 months after harvest because of enormous losses due to rottage, shrinkage, sprouting and attack of different microbes and pests. According to an estimate, a higher proportion of potatoes (16 per cent) in the country is wasted as post-harvest losses (PHLs) than that used as seed (8.5 per cent) or processing (7.5 per cent). This is an unfortunate situation that nearly 1 million t in excess to the total potato production in Canada during the triennium ending 2010 (FAOSTAT) itself was wasted in India. Losses occur at every stage of the supply chain. These losses result due to lack of proper storage facilities, absence of proper handling, transportation, pre- and post-harvest treatment. Besides, the attack of pathogens at conducive temperatures and environment increase the losses tremendously and therefore, proper handling, desired treatments and safe post-harvest environment is required to reduce the losses in potatoes after harvest.

2. Post-harvest Losses

Hot summer temperatures, lack of state of the art cold storage facilities and massive transportation of potatoes from northern to southern states are the causes of the high wastage of potato in absolute terms. Various factors which play key role in post-harvest loss of potatoes are being discussed in subsequent heads.

2.1 Pathogenic Factors

Losses in potatoes caused by pathogens are greater than the losses due to physiological causes. Physical damage to tubers during harvesting and handling aggravates the attack of bacteria and fungi leading to quantitative loss. Generally, the more common storage diseases caused by fungi are late blight, dry rot and pink rot. The most severe bacterial disease that causes rotting is soft rot. When infection occurs in the field, rotting begins in the field and continues during storage. When infection occurs after harvesting, it is generally through mechanical injury as in the case of dry rot. High humidity and condensation of water on tuber surface can lead to infection by soft rot. Among the insect pests, tuber moth causes maximum damage during storage and is common in potatoes stored under higher temperatures, as is the case with non-refrigerated storage.

2.1.1 Fungal Diseases

Among the diseases of fungal origin, the most important ones causing pre and post-harvest losses are the late blight and the dry rot. Late Blight is caused by *Phytophthora infestans* and the disease affects all plant parts, *viz.*, leaves, stems and tubers. Tubers develop reddish brown, shallow to deep, dry rot lesions. The affected tuber flesh becomes 'caramalised' with a sugary texture. Tubers carrying the pathogen are the real carriers and serve as the source of the disease in the subsequent season. When the storage conditions are good, the rot remains dry and doesn't extend. However, when the storage conditions are faulty, the dead tissues open a route to the secondary soft rot. The Dry Rot caused by *Fusarium sp.* on the other hand appears during storage. The skin wrinkles (Figure 15.1.) forming irregular concentric rings, and in general tuber tissue decay forms pockets of pink, red or bluish colour, which contain spore masses in the mycelium. The disease incidence can be reduced with proper curing and avoiding bruising during harvest and transport.

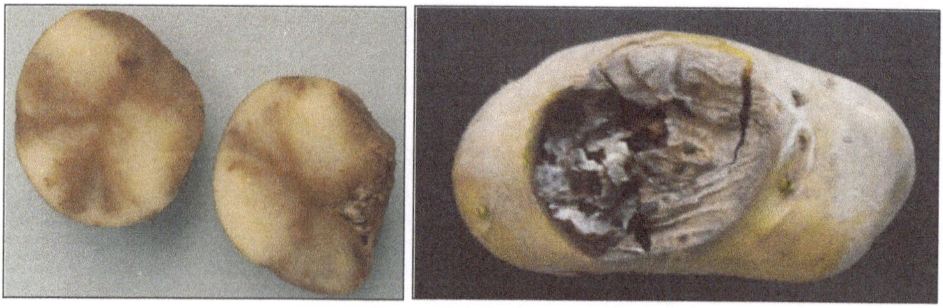

Figure 15.1: Tubers Infected with Late Blight and Dry Rot.

2.1.2 Bacterial Diseases

Among the main diseases of bacterial origin, Bacterial wilt (Ralstonia solanacearum) and soft rot (*Erwinia carotovora var. carotovora* and *Erwinia carotovora var. atroséptica*) are the major threats. Brown Rot is a destructive disease to the plant and causes rotting of the tubers in transit or storage. The Soft Rot causes high amount of losses particularly during storage and every year about 2 per cent losses have been estimated due to rots. Excessive moisture, high temperature, excess nitrogen, tuber injuries and poor ventilation during storage are the important factors helping the disease development and spread. Initially a small area of tuber tissue around lenticel or stolon attachment point becomes water soaked and soft (Figure 15.2). Under low humidity, the initial soft rot lesions become dark and sunken, while under high humidity, the lesion may enlarge and spread to larger area. Tubers in advanced stages of decay are usually invaded by other organisms and the decaying tissue becomes slimy with foul smell and brown liquid ooze. At harvest, many small rotted tubers with intact skin can be seen. Among the control measures, avoiding excess irrigation and nitrogen, providing proper drainage prevents the spread of the disease. Harvesting the crop before soil temperature rises above 28 °C is recommended. The crop should be harvested only when the tuber skin is fully cured. Care should be taken to avoid injury to tubers and bruised injured tubers are sorted out. Treating the tubers with 3 per cent boric acid for 30 min and drying them under shade minimizes infection during storage.

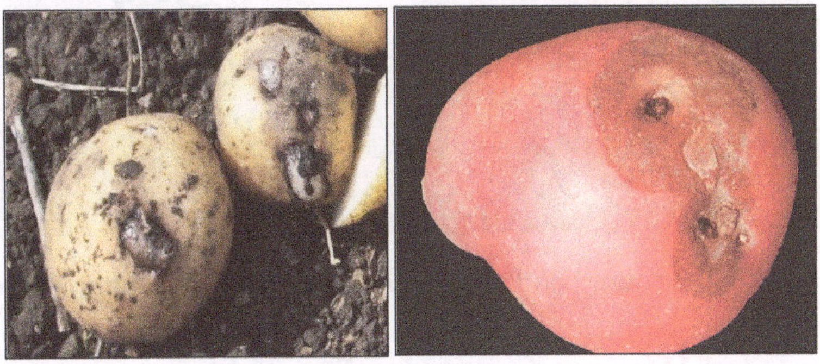

Figure 15.2: Tubers Infected with Brown Rot and Soft Rot.

2.1.3 Potato Tuber Moth (PTM)

PTM (*Pthorimea operculella*) in larvae state damages foliage mining between the inferior epidermis and superior of the leaves. It is also introduced in the tuber mining burrows and filling them with their excrements. The initial invasion happens in the field and continues in storage. The tunnels formed in the tubers make them unfit for any consumption and in conducive conditions the rottage by secondary pathogens is aggravated. Therefore, in the areas of PTM infestation in traditional and room temperature storages proper care like removal of infested tubers, restriction of PTM entry into storage and use of dried leaves of some herbs (*e.g.* Lantana) in storage is recommended.

2.2 Losses Due to Pre-storage Factors

The quality during potato storage depends on the quality of tubers reaching the store and its management starts from the production stage itself. Cultural practices affect the quality of tuber, since potatoes grown in improperly prepared fields are mechanically damaged at harvesting and have reduced storability. Adequate pest and disease management is also necessary for producing potatoes with good keeping quality. Hence, management of pre-storage factors that affect the keeping quality of potato is prime step in good storage management and about 3-8 per cent losses have been attributed to these factors. The potatoes need to be harvested in dry weather and irrigation has to be stopped two weeks before dehaulming. Harvesting is done 10-15 days after haulm cutting for facilitating skin setting. Besides this, curing/suberization (the process by which wounds are healed in potatoes under optimum conditions of <25 °C temperature and 95 per cent relative humidity) is also important. Potatoes harvested under wet soil conditions must be dried before storage because even little moisture on the surface of the tubers could lead to infection and rotting during the storage. Only mature tubers with good skin setting are ideal for storage under any storage system.

Figure 15.3: Uncured and a Cured Potato Tuber and Curing in Field

2.3 Losses at Harvesting and Handling

The mechanical damage during harvest is responsible for loss initiation in potatoes in form of cuts and bruises hence care should be taken to ensure that cuts and bruises are minimized and the exposure to sun should be avoided. Exposure from sun heat causes excessive rotting and damages can take place as a result of direct contact with the sun at the time of harvest. This may also occur before harvesting when the plant is dead or cut. Potatoes harvested in hot weather rot more than those removed in more temperate conditions. When air temperature is 32 °C or higher, it is not advisable to dry potatoes and optimum temperature of air for drying has been recommended as 26.5 °C or less. Proper care at the time of harvesting, curing of potatoes in temporary heaps and careful handling may help in reducing the losses to a great extent.

2.4 Losses during Storage

Losses during storage are affected by factors like physiological tuber condition, mechanical damage suffered during harvest and handling, as well as by storage conditions. Mechanical damage (cuts and bruises) facilitates invasion and development of microorganisms that cause illnesses and rotting. It is appreciated that weight loss shows a lineal relationship with regard to nature or magnitude of physical damage. On the other hand, losses by rottage increase exponentially with regard to magnitude of physical damage. In general, it is necessary to reduce tubers physical damage to minimize losses during storage.

2.5 Losses during Transport, Marketing and Distribution

Losses to potato crop during transport, marketing and distribution have been estimated to be about 3-8 per cent. It is recommended that transport of potatoes should be completed rapidly to avoid sun damage. Transferring tubers in sacks generate a larger percentage of damages unless the handling is done very carefully. Thus crop handling inside the field or transport to exterior with trucks or trailers has great importance. There are mechanical damages to potatoes before leaving the field. These damages become more evident later during storage. To avoid mechanical damages in all the operations, it is necessary to convince the personnel to utilize proper handling like: Potatoes should be placed inside containers and not thrown from distance, truck drivers should not stand on potato sacks but on the platform of the truck and full sacks of potatoes should be placed in position and not thrown at truck loading and discharging. Use of soft linings is recommended in trailers and trucks that transport potatoes. A straw bed should be used in trucks, or pads can be made with sewn sacks half filled with straw and laying potato bags on these shall significantly reduce the bruising. It is also necessary to securely tie the load to avoid movement of sacks, resulting in bruising. Another point is that potatoes should be handled the absolute minimum number of times possible. All labour involved with potato handling should be supervised carefully to guarantee an appropriate operation.

3. Storage Practices to Minimize Losses

In India, potatoes are generally stored in cold stores and used for both seed and table purposes. Seed potatoes are best stored in cold store maintained at 2-4 °C and at about 95 per cent relative humidity. But cold stored potatoes are not suitable for table and processing purposes. Potatoes stored in cold store accumulated sugars and become sweet in taste and are therefore, less suitable for consumption. Because of high accumulation of reducing sugars, these potatoes produce dark coloured chips which are unacceptable both color-wise and taste-wise. Thus, the storage requirements of potatoes should be in accordance with the purpose for which potatoes are stored. It is suggested that cold stores maintained at 2-4 °C should be used exclusively for the storage of seed potatoes only. Table and processing potatoes should be stored at 10-12 °C, after treating the potatoes with a sprout suppressant for long-term storage. For short-term storage of table and processing potatoes, non-refrigerated storage methods like heaps can be used profitably. Handling of potatoes with due care is expected to reduce the post-harvest losses.

3.1 Storage of Seed

Seed potatoes are best stored at 2-4 °C and 90-95 per cent RH in cold stores, because at this temperature, sprouting does not occur and weight loss is minimum. The farmers have to pay the rent for storing potatoes in these cold stores and they take out these potatoes at the time of planting in the next potato season. The only problem regarding the storage of seed potatoes in cold stores is the uneven distribution of these cold stores over locations and non-regulatory storage charges.

Figure 15.4: A Traditional Cold Storage and Stack of Potatoes Inside the Store.

3.2 Storage of Table and Processing Potatoes

The potatoes to be used for table and processing purposes have different storage requirement than the seed potatoes. Essentially storage for table and processing potatoes may be divided into two categories *viz.*, Non-refrigerated storage systems and refrigerated storage systems as discussed below.

3.2.1 Non-refrigerated Storage Systems for Table and Processing Potatoes

These are the storage systems where no artificial refrigeration is used. These storage methods can be used to store potatoes for 3-4 months (*i.e.* up to June) after harvest in February/March. However, once the monsoon begins, potatoes cannot be stored in a non-refrigerated store because it works only under dry conditions. Short-term storage of potatoes for 3-4 months is good enough because it helps avoid distress sale immediately after harvest and the farmer can get better returns by selling the potatoes in May or June, when potato prices start rising. It is relatively easy to store potatoes under non-refrigerated conditions when they are dormant. Indian potato varieties have a dormancy period of 6-8 weeks. Once the dormancy period is over, sprouting begins. Weight loss in potatoes during storage is mainly due to water loss from the tubers and the weight loss is much more in sprouted tubers. The extent of rotting is high under non-refrigerated storage systems as the temperature and humidity conditions are favorable for aggravating the infection.

Figure 15.5: A Modified Potato Heap.

3.2.2 Refrigerated Storage for Table and Processing Potatoes

Fresh potatoes are available for consumption only for a few months, after the harvest. For a major part of the year, potatoes from cold stores are consumed. Though cold storage facility was developed primarily for the storage of seed potatoes, they have been used for the storage of table potatoes as well. But, 2-4 °C is not the ideal storage temperature for table and processing potatoes as at this temperature, potatoes become sweet due to sugar accumulation and are not preferred for consumption. Further, cold stored potatoes do not keep well once they are taken out of the store. Therefore, storage of table and processing potatoes is suggested at 10-12 °C. This method of potato storage is comparatively new to our country, but a good number of cold stores (approximately 700) are now storing potatoes by this technology and these potatoes are being marketed as low sugar potatoes (table) and depending on the variety stored, potatoes are suitable for processing as well up to about 6-7 months. However, at this storage temperature, potatoes sprout. Therefore, it is necessary to treat the stored potatoes with a sprout suppressant like CIPC (Isopropyl N-(3-chlorophenyl) carbamate). The treated potatoes are safe for consumption after 3-4 weeks of treatment.

Figure 15.6: A Commercial Cold Store using 10-12 °C Storage Technology, Potatoes and Chips.

With the use of correct pre and post-harvest practices of harvesting, handling and storage, the losses incurred in potato value chains can be minimized and a substantial volume of potatoes can be saved which will result in additional revenue to the potato grower.

Chapter 16

Application of Proteomics and Metabolomics in Potato

Som Dutt, Sushil S. Changan, Milan K. Lal, Dharmendra Kumar and Brajesh Singh

ICAR-Central Potato Research Institute, Shimla, Himachal Pradesh

1. Introduction

Genome sequencing at large scale has provided a platform to gain better insights into the genome structure and organization to a great extent. However, real challenge still remains in deciphering the function of the encoded proteins, their interaction with other proteins or other biomolecules through which these contribute to the functioning of the organism (including plants) as an entity and at systems level. A systems-level understanding is essentially required if we intend to fully capitalize on plants to address global challenges such as food, and fuel shortages, changing global climate *etc.* The first step in a systems approach is to identify all of the system components (*e.g.*, genome, transcripts, proteins, metabolites). Information regarding the function of the system as a whole is desired and the approached to achieve this, has given rise to new field known as Systems Biology. Systems biology is the study of the interactions and interplay between all properties of biological systems. It involves integrating the existing knowledge about biological components, building a model of the system as a whole, and extracting unifying organizational principles that explain the form and function of living organisms. This lecture will briefly describe the two important components of systems biology. These are: proteomics and metabolomics which hold great importance in characterization and improvement of plants crops including potato.

2. Proteomics

The proteome refers to the protein complement of a particular biological system at any given point of time. The term "proteome" was first used by Wilkins *et al.* (1995) to describe the protein complement to the genome. Though proteomics is theoretically the analysis of the full protein population of complex biological systems, the available technologies are not capable of fulfilling this description experimentally. Conceptual progress in proteomics has made possible the analysis of entire proteomes at previously unprecedented density and accuracy. Various steps/methodology involved in proteomics studies are described below.

2.1 Extraction of Proteins

Sample preparation procedures for proteomic analysis requires the reliable and consistent extraction of proteins from tissue samples. Plant tissues usually contain high levels of proteases and secondary metabolites which cause problems with protein extraction and proteomic procedures. The existence of secondary metabolites displays species/tissue- specificity and varies with age or developmental stage. Secondary metabolites can severely affect the performance of protein extraction and separation. For example, phenolics can build irreversible complexes with proteins, and the oxidation of phenolics by phenoloxidases and peroxidases can cause streaking on 2-DE gels. Pigments, polysaccharides, and lipids can also cause severe disturbances in 2-DE gels. For total protein extraction, an ideal protocol should reproducibly capture all the protein species in a proteome with low contamination of other molecules. In many cases, the extraction procedure must be optimized for plant species, tissue, or cell compartment. The three main extraction protocols commonly employed for plant proteomics are (1) TCA/acetone precipitation, (II) phenol extraction and (III) TCA/acetone/phenol extraction.

2.1.1 TCA/Acetone Extraction

The method is based on protein denaturizing under acidic and/or hydrophobic conditions that help to concentrate proteins and remove contaminants. TCA/acetone extraction involves the cleanup of tissue powder with 10 percent TCA/acetone, which is usually more effective than either TCA or acetone alone. Plant tissues are rich in proteases, which cause the proteolytic degradation of proteins and the loss of high molecular weight proteins. However, homogenizing plant tissue in TCA/acetone almost immediately inactivates proteases and precipitates proteins. Also, TCA/acetone extraction is valuable for removal of interfering compounds and the enrichment of very alkaline proteins such as ribosomal proteins from total cell lysates. *Limitation:* The resulting pellets from TCA/actone precipitation may be difficult to dissolve.

2.1.2 Phenol-based Extraction

In this method, plant tissue powder is extracted in the extraction buffer/ buffered phenol (pH 8.0) and followed by methanol (or acetone) precipitation of phenol phase. Phenol dissolves proteins (including membrane proteins) and lipids leaving water-soluble substances (polysaccharides, nucleic acids, *etc.*) in the aqueous phase, thus proteins in phenol phase are purified and concentrated simultaneously by subsequent methanol (or acetone) precipitation. In the case of tissues containing high levels of phenolic compounds pulverizing plant tissues with PVPP can help to remove phenolic compounds. The important points for successful use of phenol extraction are (i) keeping samples at low temperature during the first extraction step and (ii) careful recovery of the phenolic phase after centrifugation. Phenol can minimize protein degradation resulting from endogenous proteolytic activity. Many studies showed that phenol extraction gave satisfactory results in recalcitrant plant tissues rich in components which inhibit electrophoresis. Overall, phenol protocol is more efficient than TCA/ acetone precipitation. **Limitation**: toxicity nature of phenol and more time-taking.

2.1.3 TCA/Acetone Precipitation/Phenol Extraction

This is integrated phenol-based extraction with TCA/acetone cleanup steps prior to protein extraction. The protocol holds the merits of both TCA/ acetone precipitation (for removal of phenolics, lipids, and pigments) and phenol extraction (for removal of phenolics, lipids and pigments, polysaccharides, nucleic acids, and salts). In the TCA/acetone/phenol extraction protocol, tissue powder purged by TCA/acetone is used as the starting material in phenol extraction of proteins. The protocol has been proved to be very effective to handle with recalcitrant tissues and provides enhanced 2-DE-based proteomic analysis of plant tissues. Although the TCA/acetone/phenol extraction is time consuming and laborious compared to both the TCA/acetone precipitation and the phenol-based extraction, it has the potential to generate high-quality protein samples for some kinds of recalcitrant plant tissues. *Limitations:* Possible protein loss due to multiple cleanup steps.

2.2 Separation and Detection of Proteins

There are two preferred methods for separation of complex protein or peptide samples: (1) two-dimensional (2-D) gel electrophoresis; and (2) liquid chromatography (LC).

2.2.1 Two-Dimensional Gel Electrophoresis

Majority of proteome studies undertaken on plant samples have used two dimensional gel electrophoresis (2D-GE) for separation of proteins. This is mainly because of advances in isoelectric focusing (IEF) which now routinely utilize immobilized pH gradient (IPG) strips in the first dimension. This has significantly enhanced the ease and reproducibility of 2D-GE for proteomic analysis. The basic procedures and techniques for undertaking 2D-GE are available in many biochemistry books/manuals. After the IEF has completed, strips are equilibrated in an SDS loading buffer and placed on top of a precast acrylamide gel. Once the sample has successfully been arrayed in the second dimension acrylamide gel, sample analysis requires a successful digestion, and peptide extraction to be undertaken. The basic procedure has been widely reported and used with many minor modifications. One of the underlying objectives of 2D-PAGE is the reproducibility and consistency of protein extractions. Many attempts have been made to improve extraction reliability and performance utilizing a variety of detergents, chaotropes and mechanical disruption techniques. Such optimization is often required due to the nature of different samples and the lack of completer solublization in non-denaturing homogenization buffers. After 2D-GE the gels need to be stained for detecting proteins. Although several methods for protein detection have been reported, silver and Coomassie Brilliant Blue (CBB) staining methods are still the preferred methods for in-gel protein detection. Silver staining is ~ 10-fold more sensitive than CBB, with a detection range of 0.1–1 ng. However, silver staining has problems such as inferior reproducibility, poor linear dynamic range, and non-quantitative negative-staining of some modified proteins, all of which complicate downstream quantitation and spot matching. Broad dynamic range fluorescent protein stains including SyproRuby™, Deep Purple™, and ruthenium II, which have detection sensitivities around 10–20 ng are promising, but expensive, alternatives to Coomassie and silver staining as general protein stains.

2.3 Identification of Proteins

2.3.1 Protein Identification using Mass Spectrometry

Mass spectrometry is the most preferred technique employed for identification of proteins. The development and commercialization of two different "soft" ionization approaches, electrospray ionization (ESI) and matrix assisted laser desorption ionization (MALDI) enabled large macromolecules such as proteins to be analyzed either in flowing, liquid solution or in a dry, crystalline state, respectively. The importance of these developments was appreciated by the scientific community and in 2002 led to John Fenn and Koichi Tanaka sharing the Nobel prize in Chemistry. The utilization of MS for the identification of proteins has become the primary technique for large scale protein analysis in recent years. The choice of sample delivery methods and subsequent analysis are dictated by availability of hardware, costs and experimental objectives. As stated above the two most popular sample delivery devices currently being used in proteomics are MALDI and ESI. These sources are connected to various types of mass spectrometers, which further influence the type of data and quality produced.

2.3.2 Matrix Assisted Laser Desorption Ionization

The most widely used method of sample delivery and analysis uses the MALDI source attached to time of-flight mass spectrometer (MALDI-TOF). This combination has been successfully employed in the proteomic identification of proteins. This method of analysis is best suited for gel-separated proteins producing samples of relatively low complexity. This approach uses the absolute mass of each peptide to produce a peptide mass fingerprint (PMF) of the protein, which is then used to search a database of theoretical peptide masses. The analysis of samples by MALDI-TOF to produce PMF data is relatively high-throughput technique but produces data of poorer quality when compared to MS/MS –type analysis. This has a direct impact on pattern-matching confidence scores and creates more ambiguity when interrogating a large database for a match. The delivery process relies on aromatic acid matrices (*e.g.* dihydroxybenzoic acid or hydroxycinnamic acid) mixed with a digested sample. The matrix transforms the energy from LASER to the sample leading to a release of peptides from the matrix. The LASER desorbs the peptides from a solid dried sample into the gas phase for analysis by mass spectrometer. No amino acid sequence information is obtained by basic MALDI-TOF analysis. Matching of data is based on the similarity of the masses to the predicted patterns from an *in silico* digestion of the sequence database. In most MALDI-TOF instruments there are now techniques that provide some sequence information by a post-source decay (PSD) process. However, PSD will generally only provide sequence tags for three to four amino acids from 10-15 amino acids peptide. While these can be useful and provide an increase in matching confidences, they do not offer the same matching capabilities as authentic MS/MS spectra from a tandem mass spectrometer. More recently, MS/MS capabilities have been made available with the coupling of the MALDI source to tandem mass spectrometers allowing full-peptide fragmentation. The MALDI source has commonly been attached to a tandem mass spectrometer with a TOF component due to synergistic reasons involving timed points of ionization and the pulsing nature of the TOF (*e.g.* TOF-TOF and Quadrupole-TOF (Q-TOF), but more recently has also been coupled to Ion trap (IT) mass spectrometers. This arrangement of MS components provides the advantages of sample throughput and high-confidence matching of MS/MS spectra into a single system. Moreover, these systems now have capacity to deal with complex lysates with the development of LC-MALDI applications allowing chromatographic separations to be directly spotted onto MALDI plates. One of the considerable advantages of this technique is the ability of lockdown samples providing the capacity to return to a plate for a subsequent detailed analysis of a particular spot of interest. The LC-MADLI setup appears to be providing an excellent means of obtaining proteomic depth and protein coverage when compared to other methods. These systems thus provide significant advances in the large-scale analysis of complex systems with unambiguous matching capabilities, high-throughput applications and the capability to analyse complex samples.

2.3.3 Electrospray Ionization

The ESI source is commonly used to analyse complex mixture and is capable of interfacing with a wide range of mass spectrometers that usually provide MS/MS capabilities. ESI is an excellent method for the ionization of a wide range of polar molecules and relies on the sample of interest being dissolved in a solvent and delivered to the source in a thin capillary. A high voltage is to the tip of the capillary, producing a strong electric current which causes the emerging sample to be dispersed into an aerosol of highly charged droplets. These droplets are vaporised by an inert gas such as nitrogen-releasing charged species for analysis by the mass spectrometer. Whatever the arrangement of MS components that is utilized, this approach relies on the fragmentation of each peptide in some sort of collision cell within the mass spectrometer yielding a fragmentation spectra or MS/MS spectra. Each MS/MS spectrum can then be interpreted individually to suggest a probable amino acid sequence (de novo sequencing) or can be used to search a database containing theoretical MS/MS spectra to suggest both an amino acid sequence and probable identity of the original protein. The analysis of sample using ESI can readily be achieved manually using a syringe apparatus often provided with the system, but by far the most convenient and sensitive procedures involve a high-performance liquid chromatography (HPLC) connected to the ionization source. This provides sample automation and higher throughput but also has an added level of complexity. Samples are usually injected onto a reverse phase HPLC column such as C18 allowing the concentration and separation of peptides prior to analysis by the mass spectrometer. This analysis produces MS/MS data from the fragmentation of a parent ion (peptide) in the process termed collision-induced dissociation (CID). These CID spectra are high –quality data for pattern matching and interpretation of peptides.

3. Data Analysis

3.1 Peptide Mass Fingerprints

The ability to rapidly screen many hundreds of gel-separated samples at relatively low costs still makes PMF-based analyses a practical route. Since no sequence data are obtained (unless using PSD), data matching is heavily reliant upon mass accuracy and sequence availability. Most modern TOF mass spectrometers provide a high mass accuracy (±10-30 ppm) which is a crucial aspect to the analysis of PMF data. Thus the real limitation to using PMF in proteomic studies is sequence availability.

3.2 Peptide Fragmentation Data (MS/MS)

The use of MS/MS data produced by a mass spectrometer capable of targeted peptide fragmentation provides an unsurpassed method of data matching and analysis in species with both a well characterized genome and those lacking genomic data. The data format provides the mass of the analysed peptide as well as some fragmentation data which can either be used directly for interrogation with a database using available software (*e.g.* MASCOT of SEQUEST) or can be deconvoluted to provide some sequence information (*de novo sequencing*), which can then be used to match to the sequence databases using the BLAST. The use of either method provides a far greater confidence in the resulting match, but just as with PMF matching, the final interpretation still requires some level of assessment.

4. Application of Plant Proteomics

Proteomics is a vital component of functional genomics and systems biology. Proteins being the actual functional entities of the biological systems their in-depth analysis using proteomics approach will be helpful in all the research areas aiming to have deeper molecular insights into the complex metabolic processes influencing plant growth, development, biomass potential and their interactions with environment.

5. Metabolomics

Metabolites represent a diverse group of low-molecular-weight structures including lipids, amino acids, peptides, nucleic acids, organic acids, vitamins, thiols and carbohydrates, which make global analysis a difficult challenge. It is estimated that more than 100,000 secondary metabolites are produced by plants, while the total number is estimated to exceed over 500,000. To assess this diversity of structurally complex chemical compounds, various approaches have been initiated largely due to the tremendous advances in the instrumentation and data handling capabilities. The comprehensive assessment of endogenous metabolites and attempts to systematically identify and quantify metabolites from a biological sample is known as metabolomics. Various methodologies for metabolomics are described here.

5.1 Sample Preparation

With the exception of nondestructive nuclear magnetic resonance (NMR)-based, nevertheless still have certain limitations to application with plants, essentially all metabolomics approaches give us what is basically a 'metabolic snapshot' which is specific to the time of sampling. This so-called 'point-in-time-chemistry', provides us with a hugely valuable metabolic insight but it is also important to bear in mind that failing to take full and proper care of the cultivation and sampling of the plants will result in potentially misleading and incorrect conclusions. Full consideration must be given to these temporal and spatial dynamics of plant metabolism when wishing to perform comparative analyses on contrasting samples. Producing and comparing metabolic profiles of any biological materials is always possible but determining the significance and biological relevance of the differences observed is a challenging task.

5.2 Extraction, Separation and Detection

Chemical complexity, metabolic heterogeneity, dynamic range and ease of extraction remain the main challenges in developing an effective metabolomics technology platform. Multiparallel technologies are required to gain the desired broad metabolic picture. Careful selection of suitable combinations of extraction, separation and detection protocols can however, lead to a rapid build-up of complementary biochemical data on the composition of biological samples (LC-MS, GC-MS, high-pressure liquid chromatography (HPLC)-PDA); (GC-MS, LC-MS); (HPLC-PDA, FT-ion cyclotron resonance (ICR)- MS); (CE-MS, CE-PDA). Consequently, both polar/semipolar (*e.g.* methanol (MeOH)/ water) and lipophilic (*e.g.* chloroform) extractions are usually analysed and, especially for plants, an additional analysis of the volatile components via solvent extraction (*e.g.* pentane) or through headspace extraction (*e.g.* solid phase (micro)extraction, SP(M)E) is often desirable. The main detection technologies are described as follows:

5.3 Mass Spectrometry (MS)

MS is the primary detection method of choice for plant metabolomics due to its sensitivity, speed and broad application. Depending upon the type of extract made, GC (gas chromatography) or LC (Liquid Chromatography) are most routinely used for metabolite separation before the samples pass into the mass spectrometer. Capillary electrophoresis separation is, however, gaining interest.

5.4 Gas Chromatography–Mass Spectrometry (GC-MS)

GC-MS is currently proving to be the most popular global analysis method. Popularity stems primarily from the robustness of both the separation and the electron impact spectrometry technique and the availability of some excellent deconvolution and metabolite identification software. Together, these can be used for complete metabolite identification and quantification. Running standards and comparing with commercially available compound databases can be used to confirm metabolite identity. Gas chromatography-MS is the principal technique for separation and detection of metabolites that are naturally volatile at higher temperatures. However, the technology is more broadly applicable to groups of nonvolatile, polar (mainly primary) metabolites, such as amino acids, sugars and organic acids, by converting these into volatile and thermostable compounds through chemical derivatization. These derivatized samples can then be analysed by GC-MS.

5.5 Liquid Chromatography-MS

LC-MS is a particularly important additional, versatile technology for plants as it also provides a means to analyse many large groups of 'secondary' metabolites often present in plant tissues. Advances in chromatographic technologies (*e.g.* the ultraperformance liquid chromatography (UPLC) system from Waters Corporation, Milford, MA, USA) together with advances in column chemistry (*e.g.* hydrophilic interaction chromatography (HILIC) and long monolithic columns) are yielding significantly improved separation potentials. The technology is inherently restricted to molecules which can be ionized, either as positively or negatively charged ions, before moving through the MS. A range of possible techniques is available in order to ensure the ionization of a broad range of (LC-separated) metabolites (*e.g.* electrospray ionization (ESI); atmospheric pressure chemical ionization (APCI); photoionization (PI). The high analytical precision of modern LC techniques combined with the high sensitivity and mass accuracy and resolution of MS systems, in particular timeof- flight (TOF) and Fourier transform ion cyclotron resonance (FT-ICR) instruments, is proving very useful in the analysis of complex metabolite mixtures typified by plant extracts. Working in positive and negative ion modes can provide broader coverage of molecules which more readily either gain or lose a proton. Rather than performing separate analyses, some machines now have the capacity to switch continuously between positive and negative modes during each run. Unlike GC-MS, few mass spectral libraries are available for LC-MS and this is a key topic being given considerable attention at present.

5.6 Capillary Electrophoresis

Capillary Electrophoresis is an alternative separation technology growing in popularity when combined with MS (CE-MS) for extra selectivity and sensitivity. The value of application with the less complex microbial extracts has already been clearly demonstrated and wider use with plants can be expected. High-resolution chromatographic separation and sensitive detection of water soluble extracts make a strong combination suitable for the analysis of a diverse range of primary and secondary metabolites.

5.7 Fourier Transform-ICR-MS (abbreviated as FTMS)

FTMS is a much discussed but, as yet, rarely used technology for plant metabolomic analysis. More recent applications are beginning to emerge. This technique has the advantage of providing more accurate molecular mass estimations through a high resolving power. Recently introduced orbitrap FTMS instruments offers not only faster and more sensitive analyses but also are significantly less expensive than the cyclotron FTMS machines are also worthy of attention. The technology is predicted to have a great future in the field of metabolomics, although for many the inability to separate structural isomers which have identical mono-isomeric masses is still seen as a significant limitation to its application. Linking up to some sort of prior isomeric separation strategy will be required to overcome this.

5.8 Nuclear Magnetic Resonance

Nuclear magnetic resonance (NMR) seems currently to be the method of choice for medical metabolomics (metabolomics). For plants, its application has so far been limited because of its currently much poorer sensitivity compared to MS-based methods. Nuclear magnetic resonance analysis is a favored choice for the major metabolites but many other important compounds will typically be missed as they occur in plant extracts at levels below the usual NMR detection threshold. It does, however, have the advantage that it is a more uniform detection system and can directly be used to identify and quantify metabolites, even *in vivo*.

6. Application of Metabolomics in Plants

Plant metabolomics has a potentially broad field of application in plant science in terms of monitoring crop quality characteristics or identifying potential biochemical markers to detect product contamination and adulteration. Metabolomics is expanding the general knowledge of plant materials and is illustrating how organisms function as integrated biological systems. Studies in plant biology involving the four different types of metabolite analysis (target analysis, metabolite profiling, metabolomics, metabolite fingerprinting) may be classified into following five broad research areas viz: (i) Bioengineering of metabolism, (ii) metabolic pathway analysis, (iii) deciphering the complex physiological processes, (iv) development of plant metabolomics methods, and (v) food science.

In bioengineering, metabolite analysis is typically restricted to only a few target compounds or select pathways are profiled to conclude the effectiveness of a certain treatment. If metabolite profiling or target analysis is used, hundreds of lines involving thousands of analyses can easily be scrutinized to guide the bioengineering efforts because both sample purification and instrumental methods can easily be optimized for this task. The result of metabolomic analyses is a series of measurements of metabolite levels, that is, snapshots of metabolism. However, this is just one way to study the control of metabolism. Metabolic snapshot data are usually not sufficient to directly derive enzyme activities and hierarchical structures of pathways and, even more importantly, the dynamics of carbon partitioning between the different organs or the different metabolic cycles. A range of possible scenarios may explain a finding of altered metabolite levels: anabolic reactions might be faster; the catabolic fate might be different or transport activities may have been changed. For vascular plants, a highly suitable way to analyze fluxes is to use NMR-based techniques. Physiological adaptation to environmental stress has been the focus of several studies. Investigating metabolic relationships has been the focus of a physiological study on sink–source transitions in plant. Novel insights into the control of metabolism go hand in hand with improved methods. For this reason, various research groups validated the usefulness of novel methods or improved protocols by using model plant experiments. Metabolomics has also found application in control of metabolism is food science. Food quality is easily and rapidly deteriorated by a number of different pests, and therefore it must be tightly monitored to prevent major losses. Metabolomics based methods enable a rapid survey during food storage by observing metabolic effects for disease diagnostics rather than trying to understand the biochemical or physiological control mechanisms.

7. Conclusion

Systems biology approach seems to be very valuable in answering the complex issues related to plant growth, development and their interaction with environment. A natural shift is taking place in the approaches being adopted by plant scientists in response to the accessibility of systems-based technology platforms. Metabolomics is one such field, which involves a comprehensive non-biased analysis of metabolites in a given cell at a specific time. Metabolomics now plays a significant role in fundamental plant biology. Plants collectively produce a huge array of chemicals, far more than are produced by most other organisms; hence, metabolomics is of great importance in plant biology. Metabolomics has gained importance in biotechnology applications, as exemplified by quantitative loci analysis, prediction of food quality, and evaluation of genetically modified crops. Systems biology driven by metabolome data will aid in deciphering complex traits of potato crop including tuberization, interaction with a range of environmental cues, nutritional improvement and storage behavior.

Chapter 17

Nutritional Value and Biochemical Analysis of Potato

Sushil S. Changan, Brajesh Singh, Somdutt, Arvind Jaiswal, D. Kumar and Milan K. Lal

ICAR-Central Potato Research Institute, Shimla, Himachal Pradesh

1. Potato: A Wholesome Food

Potato is considered as the most productive vegetable and provides a major source of nutrition and income to many population and communities. Due to its versatility in way of cooking *viz.*, boiling, baking, deep frying *etc.* potato became popular over the period of time and is being consumed by one and all. Potato is popularly known as the "Vegetable King". It may be consumed in the form of snacks (chips, fries and dehydrated products) by the rich, though most of the undernourished households consume potato as primary or secondary source of food and nutrition. In India, potatoes have been utilized largely for consumption as fresh potatoes and the major part of potato harvest (approx. 68.5 per cent) goes to domestic table consumption. The nutritional value of potato is well acclaimed and is known as a versatile, carbohydrate-rich and low-fat food. Freshly harvested potatoes contain about 80 per cent water and 20 per cent dry matter, out of this dry matter approximately 60-80 per cent is constituted in the form of starch. Its content of dry matter, edible energy and edible protein makes it a good choice for nutrients availability. On dry weight basis, potato protein content is similar to that of cereals and is very high in comparison to other roots and tubers. Potato is known to everyone as a supplier of energy but its ability to supply vital nutrients is vastly underestimated. Potato is an excellent source of complex carbohydrates, dietary fibres and vitamin C. It also contains a variety of health-promoting compounds, such as, phytonutrients that have antioxidative activity. Among these, important health-promoting compounds are carotenoids, flavonoids, and caffeic acid, as well as unique tuber storage proteins, such as patatin, which exhibit activity against free radicals. Potato is also a substantial source of ascorbic acid, thiamine, niacin, pantothenic acid and riboflavin. Due to the nutritional value of potato, it is highly desirable in human diet. The nutritive value of a potato containing food depends on the other components served with it and on the method of preparation. By itself, potato is not fattening and the feeling of satiety that comes from eating potato can actually help people to control their weight. However, preparing and serving potatoes with high-fat ingredients raises the caloric value of the dish. Since the starch in raw potato cannot be digested by humans, they are prepared for consumption by boiling (with or without the skin), baking or frying. Each preparation method affects potato composition in a different way, but all reduce fiber and protein content, due to their leaching into cooking water and oil, destruction by heat treatment or chemical changes such as oxidation. Range of various compounds reported in Indian potato varieties is depicted in Table 17.1.

Table 17.1: Range of Nutrients and Anti-nutrients in indigenous potatoes

Parameters (Units)	*Minimum*	*Maximum*
Dry Matter (%)	13 (K. Chipsona-4)	24 (K. Sherpa)
Specific gravity	1.032 (K. Lauvkar)	1.093 (K. Kuber)
Starch (%)	49 (K. Shailja)	86 (K. Surya)
Amylose (%)	11 (K. Gaurav)	37 (K. Shailja)
Total soluble sugars (mg/100g FW)	120 (K. Chipsona-1)	647 (K. Satlej)
Reducing sugars (mg/100g FW)	78 (K. Himsona)	388 (K. Pukhraj)
Sucrose (mg/100g FW)	31 (K. Badshah)	182 (K. Ashoka)
Glucose (mg/100g FW)	4 (K. Frysona)	244 (K. Dewa)
Fructose (per cent)	0.08 (K. Muthu, K. Lauvkar)	0.38 (K. Badshah, K. Dewa)
Inulin (%)	0.12 (K. Chipsona-1)	1.14 (K. Gaurav)
Cellulose (%)	1.1 (K. Chipsona-1)	5.1 (K. Naveen)
Protein content (%)	0.7 (K. Gridhari, K. Chandramukhi)	2.7 (K. Chipsona-3, K. Jyoti, K. Sindhuri)
In vitro protein digestibility (per cent)	54.90 (K. Safed)	99.93 (K. Naveen)
Amino acids (mg/100mg FW)	1.93 (K. Ashoka)	4.23 (K. Anand)
Lysine content (%)	1.70 (K. Chipsona-3)	15.09 (K. Girdhari).
Vitamin C (mg/100g FW)	20 (K. Lauvkar, K. Chipsona-4)	59 (K. Chipsona-1, K. Chamatkar)
Iron (mg/kg DW)	19 (K. Garima)	49 (K. Girdhari)
Zinc (mg/kg DW)	12 (K. Arun)	23 (K. Anand)
Total Phenols (mg/100g FW)	26 (K. Garima, K. Chipsona-3)	76 (K. Megha)

Parameters (Units)	Minimum	Maximum
Carotenoids (µg/100g FW)	23 (K. Himalini)	255 (K. Himsona)
Anthocyanin (µg/g FW)	0.09 µg/g FW (K. Lauvkar)	1.41 (K. Chipsona-2)
DPPH inhibition	3 (K. Garima)	89 (K. Lalima, K. Surya, K. Pukhraj)
Total glycoalkaloids (mg/100g FW)	24 (K. Kundan)	362 (K. Badshah)
Acrylamide (µg/kg in chips)	61 (K. Chipsona-2)	3252 (K. Girdhari)
Phytate (mg/100g FW)	2.89 (K. Giriraj)	60.84 (K. Red)
Fat	0.077 (K. Chipsona-2, K. Chipsona-3)	0.803 (K. Kundan)

2. Potato: A Low Energy Diet

Raw potato (on dry weight basis) provides about 80 kcal energy whereas, a boiled potato provides about 69 kcal energy per 100 g of weight. When eaten without added fat as in case of frying, potato is a good food for weight conscious people because of its low energy density. The energy value of potato is less than major food crops like rice, wheat, maize and sorghum. The energy value is lower than other tuber and root crops as well as food products from animal origin. Potatoes are an excellent source of complex carbohydrates. These carbohydrates take longer time for break down into glucose and result in energy that lasts longer. Complex carbohydrates are longer chains of sugars, such as starches and fiber. In potato, starch is the major carbohydrate and sucrose, fructose and glucose are the main sugars. Carbohydrates are the body's primary source of fuel for energy. The energy produced through potato gets stored as glycogen in muscle and liver and functions as a readily available energy during prolonged, strenuous exercise.

3. Potato: A Good Source of Carbohydrates

The major role of carbohydrates in nutrition is to provide energy. The complex carbohydrates present in potato are important to a healthy diet. Carbohydrates in potato are mostly found in the form of starch. On an average, potato contains 14-16 per cent of starch on fresh weight basis. The total carbohydrate content of potato (about 18.5 per cent) is lower than some important crops such as wheat, rice, corn, sorghum and beans. Sugars are the most basic carbohydrates, the building blocks of complex carbohydrates. Starch furnishes most of the energy supplied by the potato. Digestibility of starch influences the energy value of the potato and hence also the bulk of potato which must be eaten to supply a given amount of energy. The digestibility of potato starch is low in raw state but improves considerably after cooking or processing. Sucrose, fructose and glucose are the main sugars in potatoes. Potatoes are known to contain prebiotics such as resistant starch and inulin in low concentrations.

Table 17.2: List of protocols and References

Name of Protocol	*References*
Resistant Starch	Raigond P, Ezekiel R, Kaundal B (2014) Starch fractions of cooked potatoes at low temperature. Potato Journal. 41(1): 58-67.
Total Soluble Proteins	Lowry OH, Rosenbrough RJ, Fern AL, Randall RJ (1951) Protein measurements with Folin-phenol reagent. Journal of Biological Chemistry 193: 263-75.
Total Phenolics	Kumar A, Kumar D, Singh B (2007) Changes in processing quality of potato tubers occuring between haulm cutting and harvesting. Potato Journal. 34(3-4): 239-241.
Ascorbic Acid	Bandana, Sharma V, Raigond P,Singh B, Kaushik SK (2015) Ascorbic acid losses during storage of potato tubers. Potato Journal. 42(1): 76-79.M. Stan, M. L. Soran, and C. Marutoiu (2014) Extraction and HPLC Determination of the Ascorbic Acid Content of Three Indigenous Spice Plants. Journal of Analytical Chemistry, Vol. 69, No. 10, pp. 998–1002
Total Carotenoids	Dalamu, Singh B, Barwal S, Raigond P, Sharma R, Joshi A (2015) Assessment of phytochemical diversity in Indian potato cultivars. 72(3): 447-450.
Individual Phenols	Deuber H, Guignard C, Hoffmann L, Evers D (2012) Polyphenol and glycoalkaloid contents in potato cultivars grown in Luxembourg. Food Chemistry, 135: 2814-2824. Albishi T, John JA, Al-Khalifa AS, Shahidi F (2013) Phenolic content and antioxidant activities of selected potato varieties and their processing by-products.
Anthocyanin	Dalamu, Singh B, Barwal S, Raigond P, Sharma R, Joshi A (2015) Assessment of phytochemical diversity in Indian potato cultivars. 72(3): 447-450.
β- Carotene	Bhatnagar-Panwar M, Bhatnagar-Mathur P, Bhaaskarla VVA, Dumbala SR, Sharma KK (2013) Rapid, accurate and routine HPLC method for large-scale screening of pro-vitamin A carotenoids in oilseeds. Journal of Plant Biochemistry and Biotechnology. http: //dx.doi.rog/10.1007/ s13562-013-0239-1
Biguanide and Related Compounds	Raigond P, Kaundal B, Sood A, Devi S, Dutt S, Singh B (2018) Quantification of biguanide and related compounds (anti-diabetic) in vegetables and fruits. Journal of Food Composition and Analysis 74: 82-88.

4. Potato: Source of High Quality Protein

Potato is a very good source of high quality protein. Average protein content of potato is 2 per cent on fresh weight basis and about 10 per cent on dry weight basis. Potato protein content is lower than wheat, rice, corn, sorghum, and beans but is higher than other major root and tuber crops like sweet potato, yam and cassava. The total nitrogen of potato tubers can be divided into soluble protein, insoluble protein and soluble non-protein nitrogen. The insoluble protein fraction is mainly present in the peel. Soluble potato protein contains substantial levels of the essential amino acids. Free amino acids present in potatoes are totally available for absorption. Potato protein has an adequate ratio of total essential amino acids to total amino acids and a balance among individual essential amino acid concentrations to meet the needs of infants and small children. However, the digestibility of potato protein is relatively low in infants. Potato protein has a very high biological value since all the essential amino acids are present in it in a good proportion. The biological value in potato is higher than major cereals and higher than even proteins of animal origin like milk and beef. With its high lysine content, potato can supplement diets which are limiting in lysine. Potato has a clear advantage over cereals in India because of its ability to provide high quality protein. Diet which can fulfill only the energy requirement of body cannot support growth of children, if its protein content is below the recommended requirement. However, if a diet provides inadequate energy, its protein is metabolized as a source of energy rather than being used for growth. Therefore, diet should be well balanced in terms of energy and protein. Therefore, potato is a superb food which has correct balance between net protein calories and total calories adequate for all age groups. Nutritive value of potato protein is better than beef, wheat flour, rice and corn in terms of recommended nitrogen in diet of adult human beings (Ezekiel and Khurana 2003).

5. Potato: Say No to Fat

There is a common misconception that eating potato may cause obesity due to its high fat content which is not at all a true statement, since potatoes contain very little quantity of fat. The average fat content of potato is 0.1 per cent on fresh weight basis which is too low to have any negative nutritional significance. Fat content in potato is lower than major cereals like rice, wheat, maize and sorghum. The little fat present in potato contributes towards potato palatability. Major proportion (i.e nearly 60-80 per cent) of potato fat consists of unsaturated fatty acids and linoleic acid is the predominant one. The high content of unsaturated fatty acids increases the nutritive value of the fat present in potato. When eaten without added fat, potato is good for weight conscious people because of its low energy density. However, when fat is added to the fried or processed potato products, it becomes rich in energy and may certainly become a cause of concern. Especially excessive consumption of processed potato products such as chips and French fries containing up to 40 per cent fat may cause obesity (Ezekiel and Khurana 2003).

6. Potatoes: Good Source of Vitamins

Potato is one of the rich natural source of vitamin C or ascorbic acid as it contains 17-35 mg of ascorbic acid per 100 g tuber. Potatoes have high quantities of vitamin C than other vegetables like carrots, onion and beet root. When consumed in sufficient quantities, potatoes itself can meet all the vitamin C requirements of an individual. Potato is an important source of thiamine, niacin and pyridoxine and its derivatives (vitamin B6 group). It also contains pantothenic acid (vitamin B5), riboflavin and folic acid. B-vitamins are essential for general health and growth, since they are water soluble. However, it is recommended that potatoes should not be washed after peeling to prevent loss of vitamins. A small potato can deliver more than 20 per cent Daily Value of vitamin C and vitamin B 6.

6.1 Vitamin C

Potato contains 17-35 mg/100 g Vitamin C which is far high compared to corn, wheat, rice, sorghum and beans. This water-soluble vitamin acts as an antioxidant and stabilizes or eliminates free radicals, thus helping to prevent cellular damage. Potato as major source of Vitamin C provides protection against scurvy. Vitamin C enhances the absorption of iron in human body and may help support the body's immune system. The ascorbic acid content of potato declines when potatoes are stored, cooked or processed. Though potato loses some of its vitamin C during storage, substantial amounts are retained until it sprouts. In Indian scenario, the consumption of fresh fruits and vegetables is much less and for poor it is almost negligible. Potato being in reach of all income groups may serve as the main source of vitamin C, especially for poor people.

6.2 Vitamin B Complex

Potato is an important source of vitamin B complex. All B vitamins help the body to convert food (carbohydrates) into fuel (glucose), which is used to produce energy. These B vitamins, often referred to as B complex vitamins, also help the body use fats and protein. B complex vitamins are needed for healthy skin, hair, eyes and liver. They are also required for proper functioning of nervous system. Potatoes are important source of thiamin, niacin and pyridoxine. They provide 0.10 mg vitamin B1 (thiamine) per 100 g of freshly harvested potatoes. Thiamine is needed to release energy from carbohydrates. Insufficient intake of thiamine results in a disease called 'beriberi' affecting the peripheral nervous system (polyneuritis) and/or the cardiovascular system. Potato contains 0.01 mg riboflavin per 100 g freshly harvested potatoes. It is also used for treating riboflavin deficiency, acne, muscle cramps, burning feet syndrome, carpal tunnel syndrome and blood disorders. Being an important source of niacin (Vitamin B3) potato provide 1.5 mg niacin per 100 g freshly harvested potatoes. Niacin is involved in both DNA repair, and the production of steroid hormones in the adrenal gland. Pantothenic acid is an essential nutrient. Potatoes contain 0.3 mg pantothenic acid per 100 g of freshly harvested potato. Pantothenic acid is required to synthesize coenzyme-A (CoA), as well as to synthesize and metabolize proteins, carbohydrates and fats. Potatoes are good source of vitamin B6 (pyridoxine). Vitamin B6 plays important roles in carbohydrate and protein metabolism. It helps the body to synthesize nonessential amino acids needed to synthesize various body proteins. It is also a cofactor for several enzymes involved in energy metabolism, and it is required for the synthesis of haemoglobin – an essential component of red blood cells. Potatoes contain 14 mg folic acid per 100 g of freshly harvested potatoes. Folic acid is essential to numerous body functions. Human body needs folate to synthesize, repair and methylate DNA as well as to act as a cofactor in biological reactions involving folate. It is especially important in aiding rapid cell division and growth, such as in infancy and pregnancy. Children and adults both require folic acid to produce healthy red blood cells and prevent anemia. The role of potato as a source of some vitamins of B complex is vastly underestimated. 100 g of boiled potatoes (boiled with skin) can fulfill the daily requirement of thiamine, niacin, folic acid and pantothenic acid (Ezekiel and Khurana 2003).

7. Potato: A Cocktail of Minerals

Potato is a good source of important minerals and trace elements. 100g of potato contains approximately 40-65 mg of phosphorus. The phosphorous present in potato is more assimilable than the phosphorous present in other food crops because of the relatively small percentage of phytic acid in potato. The lower phytic acid content of potatoes enhances phosphorous bioavailability to human body and also helps in increased bioavailability of calcium, iron and zinc. The potassium content of potato is also relatively high *i.e.* 247-455 mg/100g fresh weight. Because of high potassium content, potatoes are not included in the diet of patients with renal failure. On the other hand, the sodium content of potato is very low and the content is 11 mg/100g fresh weight. Potatoes are a good source of iron and their iron content is comparable to most of the other vegetables. 100 g of cooked potato can supply between 6 and 12 per cent of daily iron requirement for children or adult. Moreover, due to high ascorbic acid content of potato, bioavailability of non-haem form of iron from potato is increased. Potatoes mixed with other food are also beneficial as it increases the bioavailability of iron from other foods also due to its high ascorbic acid content. Moreover iron availability from potato is higher compared to other foods such as kidney beans, wheat flour and bread, reason being the high proportion of iron from potato is soluble. Potatoes provide a good source of magnesium and magnesium content in potato is up to 22 mg/100 g fresh weight. Potato can be consumed with foods low in magnesium such as milk. Magnesium content of milk is one fifth to one tenth of potato. Hence potato is superior to milk in terms of magnesium, but consumed together they form best combination as milk is rich in calcium and potato provides magnesium. Zinc is an important trace element found in potato. Though zinc content of potato is not very high but its availability is high because of the low phytic acid content. Potato can supply at least part of daily requirements of trace elements like copper, manganese, molybdenum and chromium. Traces of boron, bromine, iodine, aluminium, cobalt and selenium are also present in potato. A small potato can deliver 10 per cent Daily Value of folate, manganese, magnesium and phosphorus. Therefore, potato being in reach of poorest of the poor can play a vital role in eradication of 'Hidden Hunger' which is also known as micronutrient malnutrition.

8. Potato: Its Dietary Fiber Helps in Digestion

Potato is a rich source of dietary fiber. Cellulose, pectin and pectin associated substances are higher in potatoes compared to cereal brans (Ezekiel and Khurana, 2003). Dietary fiber content in raw potato tuber ranges from 1-2 g/100 g fresh weight. Unpeeled potatoes contain more dietary fibers than peeled potatoes. The dietary fiber from potato tuber comes mainly from its cell walls that constitute about 1.2 per cent of the fresh weight of the tubers. To increase the dietary fiber intake, potatoes must be consumed along with peel. More than half of the dietary fiber in potato is in the form of pectic substances which improves the quality of potato dietary fiber and thus helps in lowering cholesterol levels. One medium potato with the skin contributes 2 g of fiber or 8 per cent of the daily value. Dietary fiber is a complex carbohydrate and is the part of the plant material that cannot be digested and absorbed in the bloodstream. Dietary fiber has been shown to have numerous health benefits, including improving blood lipid levels, regulating blood glucose, and increasing satiety, which may help in weight management.The main components of dietary fiber are non-starch polysaccharides (NSP), lignin, resistant starch and non-digestible oligosaccharides. Potatoes also contain resistant starch which is known as 'starch and starch degradation products that escape digestion in the small intestine of healthy individuals'. Resistant starch acts in similar fashion as fibers and is found naturally in foods such as legumes, bananas, potatoes and some unprocessed whole grains. In Indian potatoes resistant starch content is approximately 1.5 to 2 per cent in cooked and cooled tubers (Raigond *et al.*, 2014). Natural resistant starch is insoluble, fermented in the large intestine and a prebiotic fiber (i.e it may stimulate the growth of beneficial bacteria in the colon). Other types of resistant starch may be soluble or insoluble and may or may not have prebiotic properties. Resistant starch appears to exert beneficial effects within the colon. The amount of resistant starch found in potatoes is highly dependent upon processing and preparation methods. For example, cooking and then cooling potatoes leads to nearly a two-fold increase in resistant starch. Even processed potatoes (*e.g.* potato flakes) appear to retain a significant amount of resistant starch with the potential to confer health benefits. Resistant starch is considered as the third type of dietary fiber, as it can deliver some of the benefits of insoluble fiber and some of the benefits of soluble fiber.

9. Potatoes: A Rich Source of Antioxidants

Along with vitamins and minerals, potatoes contain a number of small molecules, many of which are beneficial phytonutrients such as phenols, flavonoids, kukoamines, anthocyanins and carotenoids.Coloured potatoes may serve as a potential source of natural anthocyanin pigments and also a powerful source of antioxidant micronutrients. Potato antioxidants have potential role in immune function and disease prevention. Yellow pigmented potatoes are known to have high carotenoids content such as lutein, zeaxanthin, violaxanthin and

Chapter 18

Viral Diseases and their Management in Potato Production

Ravinder Kumar

ICAR-Central Potato Research Institute, Shimla, Himachal Pradesh

1. Introduction

The potato (*Solanum tuberosum* L.) is an important crop worldwide. Potato production is not an easy task, as potatoes are affected by many kinds of viruses and their various strains belonging to different taxonomic groups. Viruses contribute to "running out", or degeneration, of seed stocks. Viral diseases are prevalent throughout the India but are most severe in North-eastern plains and plateaux where population of aphid vectors is high throughout the crop season. Tomato leaf curl New Delhi virus (ToLCNDV), potato leaf rolls virus (PLRV) and potato virus Y (PVY) are the most important viruses in India. Viruses once introduced may persist and spread in agricultural environments and under favorable conditions they may multiply rapidly and cause significant yield loss and economic damage. Therefore, sustainable potato production is possible only if the diseases are kept under check, especially in sub tropics where the weather is highly conducive for common viral diseases.

2. Major Viruses and Viroid of Potato

2.1 Apical Leaf Curl Disease

Apical leaf curl disease has recently emerged as one of the most devastating viral disease of potato in India. It was first observed during 1999 in northern India and the association of a begomovirus with this disease was confirmed through electron microscopy in the year 2001. Further the virus was confirmed as a strain variant of *Tomato leaf curl New Delhi virus* (ToLCNDV) on the basis of nucleotide sequence of genomic components (Usharani*et al.*, 2003). The incidence of apical leaf curl disease has been observed to be higher in early planted crop when temperature is high in October, than in November planted potato crop. Recently, 40-70 per cent infection was recorded in Indo-Gangetic plains of India. There are even reports of up to 100 per cent infection causing heavy yield losses. The disease incidence has positive correlation with whitefly population and whitefly infestation period.

2.1.1 Causal Virus

It is caused by a strain of *Tomato leafcurl New Delhi virus* (ToLCNDV) belonging to the genus, *Begomovirus* and family, *Geminiviridae*. It is a bipartite geminivirus consisting of two circular ssDNA molecules (DNA A and DNA B) of ~2.7 kb encapsidated in geminate particles. The DNA A and DNA B components of ToLCNDV-potato share >90 per cent and 86.6- 91.7 per cent identities, repectively with ToLCNDV infecting cucurbitaceous crops, tomato and okra (Jeevalatha *et al.*, 2012).

2.1.2 Symptoms

Infected potato plant show curling of apical leaves, crinkling and a clear mosaic symptom, later entire plant appears bushy and stunted due to reduced internodal distance. The primary infection in the field appears within about 40-45 days from the date of planting.

Figure 18.1: Symptoms of Apical Leaf Curl Disease on Potato Plants.

2.1.3 Transmission

Under field conditions the virus is transmitted by whitefly (*Bemisia tabaci*) in a persistent manner. Once an adult has acquired the virus by feeding, it retains the virus for long period. The females are more efficient in transmitting the virus than the males. It is also transmitted through seed tubers from one generation to the next generation.

2.1.4 Disease Cycle

Under field conditions, the whiteflies acquire the virus and transmit from infected plant to healthy potato plants. Once the plant is infected, the infection is systemic in a given plant and therefore the tubers are also being infected (serves as primary source of infection) with the virus. If this infected tuber is planted, then it spreads the virus into the upcoming seedling/plant and again the secondary spread is by whitefly as mentioned earlier. Therefore, whiteflies play a vital role in secondary spread of the virus. This entire cycle repeats till proper management practices are adopted.

2.1.5 Management Strategies

Diseases caused by viruses cannot be easily controlled by chemicals alone. Therefore, it has to be managed by adopting the following measure on co-operative basis and in an integrated approach:

2.1.5.1 Cultural Management

- ☆ Use virus-free seed tubers for planting.
- ☆ Place yellow trap (15X30cm^2) just above the canopy height @ 60 traps/ha at equidistance from each other.
- ☆ Prompt destruction of crop residues after crop maturity/harvest.
- ☆ Uproot and burn (roguing) the plants showing virus symptoms immediately.
- ☆ Manage the weeds at regular interval and avoid/remove the reservoir plants like okra, brinjal and datura.

2.1.5.2 Host Plant Resistance

- ☆ Kufri Bahar is tolerant to ToLCNDV-potato

2.1.5.3 Chemical Control

- ☆ Seed treatment with imidacloprid (200SL) @ 0.04 percent (4ml/10lit) for 10 minutes before planting.
- ☆ First spray with imidacloprid (200SL) @ 0.03 percent (3ml/10lit) at the time of emergence of crop.
- ☆ Second spray with thiamethoxam (25WG) @0.05 percent after 15 days of crop emergence.

2.2 Potato Leaf Roll

Potato leaf roll is one of the most prevalent viral diseases of potato in India. All Indian potato varieties are susceptible to this virus. Infected plants produce only a few, small to medium tubers. Yield loss normally ranges from 20-50 per cent in India but in extreme cases may be as high as 50-80 percent. However, it is lower in autumn season (7-16 percent) than in spring season (39-60 percent).

2.2.1 Causal Virus

Potato leafroll virus (PLRV) (Syn. = phloem necrosis virus) is the type species of the genus *Polerovirus*, in the family *Luteoviridae*. PLRV has small, isometric virions (23-25 nm dia) which are primarily confined to the phloem of the infected plants. At genome level, Indian isolates are closer to European and Canadian isolates (95.8 to 98.6 percent) than to an Australian isolate (92.9 percent to 93.4 percent similarity). In nature, potato is the principal host of PLRV yet several plant species, mostly from *Solanaceae*, are known as hosts *viz., Datura stramonium, Physalisfloridana, Solanum villosum*. It can also infect some non- solanaceous plants like *Amaranthus caudatus, Gomphrena globosa etc.* (Loebenstein *et al.*, 2001).

2.2.2 Symptoms

Depending upon the age of infection, the infected plants show two type of symptoms *viz.*, primary or secondary. The primary symptoms are confined to top young leaves, which usually stand upright, roll and turn slightly pale in certain cultivars. Most varieties, however, develop reddish/pink color in top leaves starting at the margins, sometimes accompanied with slight rolling of the leaflets. Secondary symptoms develop when the plants are grown from infected seed tubers. Such symptoms are rather prominent in older leaves, *i.e.* absent or less pronounced on younger top leaves. Infected plants have characteristic pale, dwarfed, and upright appearance with rolling of lower leaves that turn yellow, brittle and are leathery in texture. In some cultivars, a reddish or purple discoloration develops on the margins and underside of the leaves. Singh *et al.* (1982) reported occurrence of three groups of PLRV *viz.*, mild, moderate and severe strain in India based on their symptom severity on the test plant, *Physalis floridana*.

Figure 18.2: Primary Leaf Rolls Symptoms on Infected Potato Plants.

In storage, the tubers from the infected plants, in certain varieties develop phloem necrosis, but most Indian varieties do not develop this necrosis. Often non-viral false or pseudo "leafroll" symptoms are also seen in the field, induced either by purple top roll (phytoplasma), cutworm, canker, black scurf (fungi) or calcium deficiency, water logging, *etc.* Some varieties also have the characteristic of foliage showing rolling/cupping towards maturity.

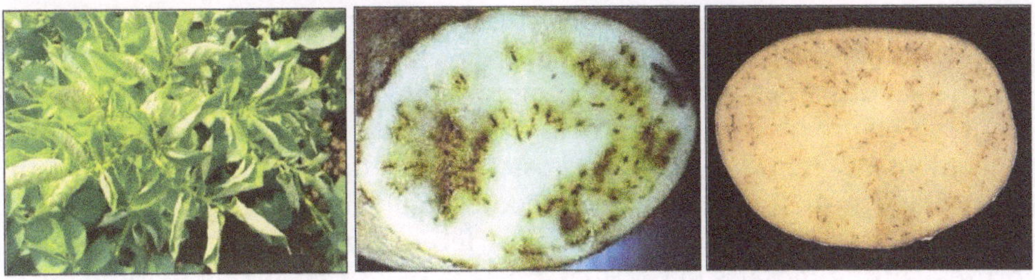

Figure 18.3: Leaf Roll Symptoms Due to Secondary Infection (Seed Borne) of PLRV.

2.2.3 Disease Cycle and Transmission

The virus is tuber borne and transmitted efficiently by aphid in a persistent manner (circulative type *i.e.* aphids normally need longer feeding periods on diseased and healthy plants). Primary infections are normally caused by the viruliferous aphids coming either from distant or nearby fields and/or diseased volunteer plants arising from the infected tubers. The green peach aphid (*Myzuspersicae*) is the most efficient and important vector of PLRV and also *Aphis gossypii.Macrosiphum euphorbiae* transmits potato strains less efficiently. *M. nicotianae* was found to be an efficient experimental vector. PLRV is not transmitted by mechanical inoculation seed or pollen but it is transmitted experimentally by grafting.

2.2.4 Management

- ☆ Use of virus free seed tubers.
- ☆ Multiply virus free seed only in aphid free areas/periods coupled with good sanitation practices, roguing of infected plants and dehaulming the crop as soon as the vector populations exceeds critical limit of 20 aphids/100 compound leaves.
- ☆ Application of suitable contact/systemic insecticides to control aphid vectors in field and also in stores.
- ☆ Tubers can be freed from PLRV infections by storing tubers in ordinary stores in summer months at 29 to 35 °C. Two hr daily heat treatment of tubers for 6 weeks at 40 °C inactivates PLRV. Hot water tuber treatment also (50 to 55 °C for 15-20 minutes) has also been found effective in eradicating the virus. In tissue culture, thermotherapy helps to eliminate PLRV from meristematic tissue of plant.

2.3 Potato Mosaics

Viral infection causes enormous economic losses particularly in the areas of tropics and semi tropics which provide ideal conditions for the perpetuation of viruses and their vectors. Any attempt to minimize the losses caused by these diseases must always be, preceded by a correct and precise detection and diagnosis of the causal agent. Nevertheless, it is very difficult to diagnose the plant virus infections as the symptoms may vary depending upon the plant variety involved, the environmental conditions and strain of the virus. Sometimes different viruses can cause similar symptoms in the same plant species and sometimes the disease could result from the synergistic effect of infection caused by two or more different viruses. The widest spread viruses causing mosaics in potatoes are usually PVY occurring worldwide. In combination with *Potato virus X,* it causes an even more destructive disease known as rugose mosaic. These viruses found wherever potatoes are grown.

2.3.1 Causal Agent

In general majority of the potato viruses like PVY, PVX, PVA, PVM, PVS, PAMV and PSTVd (viroid) causes mosaics, mottling/crinkle, necrosis, *etc.*, either individually or in different combinations. Whereas, *Potato leafroll virus* causes upward rolling and *Tomato leaf curl virus* causes curling of young leaves. These viruses generally infect potato and few other hosts from the family of *Solanaceae* like pepper *(Capsicum* spp.*)*, tobacco *(Nicotiana* spp.*)*, tomato *(Lycopersicon esculentum)* and other families like *Chenopodiaceae, Leguminosae, Compositae* and *Amaranthaceae.*

2.3.2 Symptoms

Mosaics are the most common symptoms expressed by plant viruses infecting potato plants. Overlapping symptoms are invoked by different viruses individually and/or in different combinations, such as typical greening or yellows, mild or severe/rugose mosaics, faint mottle, chlorosis of plants and discoloration or distortion of foliage. Rugosity of leaves is accompanied by interveinal chlorosis and/or interveinal puckering. Besides, there can be necrosis of the veins (lower side), petioles and even inward curling of lamina which may lead to premature defoliation. Mosaics and mottle are prominent in cool and dull weathers.

2.4 Latent or Faint Mosaics

2.4.1 Potato Virus X (PVX)

It is latent in many of the cultivars whereas the severe strains can cause mosaic symptoms and the appearance of mosaics is limited by the veins. Bright day light generally affects visual symptoms, which can be observed under low light or placing a white card below the leaf.

2.4.2 Potato Virus S (PVS)

Usually latent or very mild or barely perceptible mottle and faint vein banding. Some cultivars develop mild rugosity while few strains invoke bronzing of leaves and in few cultivars tiny necrotic spots can also be seen.

2.5 Mild Mosaics

2.5.1 Potato Virus A (PVA)

Slight and transient mosaics that are particularly visible in cloudy weather. Mosaic faint mottling, slight crinkling of leaves and inter-veinal and veinal chlorosis. These symptoms can be reinforced by placing a sheet of white paper under the leaf.

2.5.2 Potato Virus M (PVM)

Spoon-shaped leaves corresponding to a soft curling of the leaves (while the potato leaf roll virus produces "cracking" of the roll). Called leaf rolling mosaic or para crinkle. It occurs on developed plants, preferentially on the topmost leaves.

2.5.3 Potato Spindle Tuber Viroid (PSTVd)

The symptoms of PSTVd are very obscure and unusual. Generally, the plants are slightly stunted, erect, often with curling leaves darker green than healthy. Eyes are numerous in number with distinct eye brows. Tubers are cylindrically longer than normal with tapered ends with cracks upon infection with severe strain of PSTVd.

2.6 Severe Mosaics and Complex Diseases

2.6.1 Potato Virus Y (PVY)

Generally, the symptoms vary with the strains of the virus or variety of potato *i.e.*, plants will be stunted with mild or severe mosaic, veinal necrosis. Ordinary strain (PVY^{o}) and stipple-streak strain (PVY^{C}) causes necrosis, yellowing of leaflets and premature leaf drops. PVY^{NTN} strains cause mild mosaic and clearly visible raised necrotic ring spots on the tuber surface which later become sunken and skin cracking.

2.6.2 Crinkle (PVX+PVA)

Infection of PVX and PVA may lead to heavy blotching of distorted, erect leaves with wavy margins, severe stunting, mottling, necrotic spotting and streaking of leaflets.

2.6.3 Rugose (PVX+PVY)

The primary cause of rugose mosaic is due to the infection by PVY but in combination with PVX may result in severe rugose symptoms, stunted growth and leaf drop or curling with veinal necrosis in lower leaves and severe mosaic in upper ones.

2.6.4 Rosette (PVX+PVS+PVM or PVX+PVA+PVY+PVM)

Severe stunting of the infected plants with single or a few stems per hill, puckered, roughened leaflets with inward and yielding a few small tubers.

Figure 18.4: Potato Plants showing Symptoms of PVY.

2.7 Disease Cycle and Transmission

Tubers are the main source of infection and all mosaic evoking viruses are tuber perpetuated. Highly contagious viruses and viroid are readily transmitted mechanically by leaf, root or tuber contact of diseased and healthy plants, though the viroid also readily spreads through pollen and true seed. Contagious viruses are also transmitted by the cutting knife, as well as cultivating and spraying equipments. The other viruses are transmitted mainly by aphid species like *Myzus persicae* and *Aphis gossypii.*

2.7.1 Management Strategies

Diseases (mosaics) caused by viruses can be managed by adopting an integrated management as follows:

- ☆ To avoid the spread of contagious viruses/viroid, observe strict sanitation in the field and also in stores, right from harvest to planting.
- ☆ Disinfect all field equipments by dipping in or washing them either with 3 percent trisodium phosphate or calcium hypochloride (1 percent) solution.
- ☆ Essentially use disease-free seed stocks from approved or reliable sources.

- ☆ Rogue out the diseased plants carefully along with their tubers and dispose them away from the field at regular intervals during the growing season.
- ☆ Dehaulming the crop before the aphids cross critical level to enforce rigid control of the insect vector and
- ☆ Seed treatment with imidacloprid (200SL) @ 0.04 percent (4ml/10lit) for 10 minutes before planting.
- ☆ First spray with imidacloprid (200SL) @ 0.03 percent (3ml/10lit) at the time of emergence of crop.
- ☆ Second spray with thiamethoxam (25WG) @ 0.05 percent after 15 days of crop emergence.
- ☆ Combine thermo-and chemo-therapy followed by apical meristem culture for eliminating the viruses from seed tubers to produce pre-nucleus/mother seed stocks.

2.7.2 Persistent and Non-persistent Viruses

Potato viruses are transmitted by aphids in two basic ways. The virus is either non-persistent (stylet-borne) or persistent and circulative (they are ingested and persist in the aphid throughout its life). The most common persistent and circulative virus affecting potatoes in India is potato leaf roll (PLRV). Common non- persistent or stylet-borne viruses are PVY, PVA, PVS and PVM. Because the aphid has to feed for more than several minutes to transmit persistent-circulative viruses, control by insecticides is highly effective, whereas with a stylet-borne virus the virus is transmitted as soon as the aphid stylet penetrates the epidermal cells of the leaf- too soon for any insecticide to kill the aphid and prevent virus transmission.

3. Integrated Management of Potato Viruses

No single approach as outlined above will yield desirable result. A combination of them or most of them will be the only lasting solution. Schedule of integrated control of potato viruses include: inspection of seed production areas and rejection of fields with mosaic incidence higher than the prescribed level, killing of vines of seed crop at the recommended date or earlier and not to allow re-growth of the vines, destroy volunteer potato plants and weeds in and around the seed crop, monitor the population of vectors and application of insecticides to keep the aphid vectors below the critical level, use of properly disinfected tools, maintaining proper isolation of the seed crop from virus sources, use of the best quality certified seed tubers for planting, avoiding use of cut tubers as seed for seed crop, planting seed crop at a specified period to avoid exposure of the crop to the vectors, minimizing chances of virus spread through farm machinery and stop irrigation 10-15 days before harvest to allow skin curing.

Chapter 19

Fungal Diseases of Potato and their Management

Sanjeev Sharma

ICAR-Central Potato Research Institute, Shimla, Himachal Pradesh

1. Introduction

Worldwide, potato (*Solanum tuberosum* L.) is the major food crop for human consumption after rice, wheat, and maize. Globally 1.3 billion people eat potato as staple food which accounts for one third of freshly harvested potato. The capability of producing highest dry matter, energy, and edible protein per unit area per unit time makes potato a fastest growing food crop in developing countries which face constant risk of food deficit. China and India respectively are the leading potato producing countries in world and together they contribute to 38 per cent of global potato produce. Historically, potato is vulnerable to several threatening fungal (including oomycetes), viral, and bacterial pathogens which drastically reduce its production. Potatoes in India are grown under varied climatic conditions as a result the spectrum of insect-pests and diseases is very large. Therefore, management of diseases and pests is important to realize full potential of the crop. Scope of this chapter is limited to the important fungal diseases of potato, which causes considerable losses to the growers. A brief description of these diseases and their management is given in this chapter.

2. Foliar Diseases

2.1 Late Blight [*Phytophthora infestans* (Mont.) de Bary]

Most devastating disease of potato and can wipe-out the crop in a week's time under favourable conditions. Losses up to 85 percent have been reported if crop (susceptible cultivar) remains unprotected. In India, losses are more in hills where the crop is grown under rain-fed conditions as compared to the plains. Disease appears every year in epiphytotic form in hills whereas in the plains, although it usually appears every year but its intensity is low (traces to 25 percent). It is only in few years that it assumes epiphytotic form. Recently, late blight has become a serious problem in *kharif* grown potatoes and tomatoes in Karnataka state. The annual average losses to the tune of 15 percent have been estimated in the country.

2.1.1 Symptoms

Symptoms appear on leaves, stem and tubers. It appears on the leaves as pale green, irregular spots which enlarge into large water soaked lesions. Often, the spots have a purplish tinge. On the lower side, white mildew (cottony growth) ring forms around the dead areas. In dry weather the water soaked areas dry up and turn brown. On stems and petioles light brown elongate lesions develop often encircling the stem/petiole. Under favourable conditions, the whole vine may be killed and blackened and the disease spread rapidly killing the entire crop in a few days. Tubers are readily infected while in soil by rain borne spores from blighted foliage. Initially the tubers show a shallow, reddish brown dry rot that spreads irregularly from the surface through the flesh. At low storage temperatures, the lesions usually remain firm and frequently show a metallic tinge especially at the border of healthy tissues.

2.1.2 Epidemiology

Tubers carrying the pathogen are the real carriers and serve as the source of the disease in the subsequent season. In the plains, the pathogen over summers through infected seed tubers in cold stores. Appearance and build-up of late blight depend solely on weather conditions. There are specific requirements of temperature and humidity for initiation and further build-up of disease. Temperature requirements are different for fungus growth (16-20 °C), spore production (18-22 °C), spore germination (10-20 °C) and for infection and disease development (7.2-26.6 °C with optimum 18-22+1 °C). Spore germination and infection requires 100 per cent humidity and spores get killed under low humidity (<75 percent). Fungal spores are produced during the night and are sensitive to light. Cloudiness favours disease development.

2.2 Early Blight [*Alternaria solani* (Ell. and Mart.) Jones and Grout]

Early blight occurs in all the potato growing areas but is common in central India and plateau of Bihar, Jharkhand, Chhatisgarh and Maharashtra. The disease has been reported to cause significant losses (up to 20 percent) in *Kharif* crops in Ranchi and adjoining plateau region. In north-western and north-easten hills and plains, the disease appears regularly but in lesser significant form since late blight takes over.

2.2.1 Symptoms

Initially the symptoms occur on the lower and old leaves in the form of small (1-2 mm), circular to oval, brown spots. These lesions have the tendency to become large and angular at later stage. Characteristic 'target board' concentric rings of raised and depressed necrotic tissue can be observed, often with a chlorotic halo surrounding the lesion. The tuber symptoms comprise brown, circular to irregular and depressed lesions with underneath flesh turning dry, brown and corky. Lesions tend to enlarge during storage and affected tubers later become shriveled.

2.2.2 Epidemiology

The fungus can survive in soil and plant debris particularly in temperate climate. The infected tubers form the primary source of inoculum. The disease is favoured by moderate temperature (17-25 °C) and high humidity. Intermittent dry and wet weather is more conducive for early blight.

2.3 Phoma Leaf Spots (*Phoma exigua* Desm., *P. sorghina* Doerema, Doren and Kest.)

Leaf spots caused by *Phoma* spp. also occur widely both in hills and plains. Depending upon the severity, these leaf spots may cause significant yield losses.

2.3.1 Symptoms

Leaf spots due to *P. exigua* are larger (1-2.5 cm) with broad alternate light and dark concentric zones. Affected tubers have grey to greenish black depressed lesions (up to 3 cm) on the surface. Leaf spots due to *P. sorghina* are characterized by pin head size spots, which may be oval, circular or irregular (not exceeding 4 mm). Infected tubers show grey large lesions (up to 1.7 cm).

2.3.2 Epidemiology

These fungi can survive in soil and plant debris and on infected tubers during storage. The infected tubers form the primary source of inoculum. Infection usually appears on the lower leaves near ground level and results in the infection of young immature tubers if not covered by the soil. The disease is favoured by moderate temperature (17-25 °C) and high humidity.

3. Foliar Disease Management

The disease can be contained if farmers follow the integrated management schedule as follow:

3.1 Use of Healthy Seed

Only disease free seed should be used. Avoid seed from the field which has been infected by late/early blight in the previous year. The infected tubers should be thoroughly checked for the infection. The infected tubers should be removed and buried in the soil. This practice of sorting out infected tubers can also be done at the time of planting.

3.2 Use of Resistant Cultivars

Select varieties which have moderate to high degree of resistance to late blight.

3.3 Cultural Methods

Important cultural methods include:

- ✰ Selection of well drained soils for potato cultivation.
- ✰ High ridging to prevent exposure of infected seed tubers which serve as primary source of the disease. Besides, it helps in the reduction of new tuber infection.
- ✰ Scouting of the field for identifying primary infection foci and their destruction by removal of the infected plants after drenching them with recommended fungicides. Nearby plants should also be sprayed.
- ✰ As soon as the weather conditions become congenial for late blight, irrigation should be stopped wherever applicable. Only light irrigation may be given later, if required.
- ✰ Removal and burning of haulms of the affected potato crop help in reducing the inoculum in the field.
- ✰ Cultivation of solanaceous crops, being collateral hosts, nearby potato field must be avoided
- ✰ Destroy and remove the haulms from the field when late blight severity reaches >80 percent to reduce tuber infection.
- ✰ Harvest the crop 12-15 days after haulms cutting, sort out the infected tubers and store the seed after treating it with boric acid (3 percent).

3.4 Chemical Control

A spray schedule of minimum of four fungicide sprays is recommended for the management of late blight. However, the number of sprays may be increased or decreased depending on disease pressure.

- ✰ *I spray:* As a prophylactic measure, spray the crop with contact fungicides like mancozeb 75 percent WP (0.2 percent), propineb 70 percent WP (0.2 percent) or chlorothalonil (0.2 percent) as soon as the weather conditions become congenial for late blight, or about a week in advance of canopy closure whichever is earlier. Do not wait or allow late blight to appear and establish in the field. Always use a sticker @0.1 percent for proper sticking and uniform spread of fungicide on leaf surface.

- ☆ *II spray:* As soon as the disease is noticed in the field, apply any of the systemic fungicides *viz.*, cymoxanil-based (0.3 percent) or dimethomorph-based (0.3 percent) or fenamidone-based (0.3 percent) fungicides.
- ☆ *III spray:* Apply contact fungicides *viz.*, mancozeb (0.2 percent), propineb (0.2 percent) or chlorothalonil (0.2 percent) after 8-10 days of 2nd application of fungicides. However, if weather is highly congenial, repeat application of systemic fungicides may be resorted to.
- ☆ *IV spray:* Apply systemic fungicides or contact fungicides as mentioned above depending on disease severity and weather conditions.

Ensure thorough coverage of plants top to bottom with fungicides. Special attention should be given to lower leaves which need to be covered with fungicides.

4. Soil and Tuber Borne Diseases

Soil and tuber borne diseases are multifaceted in nature. Most of the pathogens have a very long soil phase and also carried through potato tubers. These diseases may cause disfiguring of tubers thereby impairing the quality, tuber rots in storage and transit, and wilts and stem rots in field.

4.1 Black Scurf (*Rhizoctonia solani* Kuhn)

4.1.1 Symptoms

Almost all plant parts are affected. The fungus attacks young sprouts through epidermis and produces dark brown lesions thereby killing the sprout before emergence, which result in gappy germination. Elongated reddish brown lesions develop on the stem at or below soil surface that may girdle the stem. When the girdling is complete the foliage curl and turn pinkish to purplish. Often aerial tubers are formed as a result of interference in starch translocation. Towards the end of the season, the fungus produces numerous hard, small, dark brown to black sclerotia on the surface of mature tubers. These sclerotia when get deposited continuously, form a black encrustations on the tuber surface. The fungus also causes foliage blight of potato.

4.1.2 Epidemiology

Seed tubers serve as the main source of the disease. In the hills, the fungus survives in the soil throughout the year and is a potential source of the disease. However, high summer temperatures are not conducive for the production of sclerotia and their survival. Therefore, *R. solani* has to over summer either as saprophytic mycelium or by infecting the crops grown during summer period. The soil temperature governs production of sclerotia on the tuber surface. The optimum temperature for growth of the fungus is 25-30 °C and for the germination of sclerotia is 23 °C.

4.2 Charcoal Rot [*Macrophomina phaseolina* (Tassi) Goid]

4.2.1 Symptoms

The pathogen produces three types of symptoms *i.e.* stem blight, charcoal tuber rot and dry tuber rot. The charcoal tuber rot phase is important under Indian conditions. The first visible symptoms are black spots (2 to 8 mm) surrounding the lenticels and eyes. As the disease advances, the tissue underneath the skin becomes uniformly black up to a depth of 2 to 5 mm. No sclerotia are formed.

4.2.2 Epidemiology

Both tubers and soil may serve as primary source of inoculum. However, soil is the main inoculum source. Soil temperature at or preceding harvest is the most crucial factor for disease development. Temperature below 28 °C almost completely checks the disease. Therefore, in sub-tropics, tuber rottage is less in crop lifted before middle of February. Disease buildup is faster in sandy-to-sandy loam soil as compared to clay soil.

4.3 Black Dot [*Colletotrichum coccodes* (Wallr.) Hughes (Syn.: *C. atramentarium* (Bek. and Br.) Traub.]

It is generally considered to be a surface blemishing disease of tubers, which downgrades potatoes, destined for table purposes and may affect seed tuber sales due to disease tolerance restrictions.

4.3.1 Symptoms

Most common symptoms appear on stem or tuber in the field. Infection of vascular tissue and girdling stem can induce yellowing and wilt like symptoms, which generally progress from plant apices to lower portions of the plant. Small, black, dot-like sclerotia (microsclerotia) are formed abundantly in stem lesions, particularly late in the growing season, and are visible to the naked eye. Sclerotia may form in internal tissues as well. On roots and stolons silvery brown lesions are formed on which characteristic microsclerotia are readily formed-aiding to diagnosis. Tubers infected with *Colletotrichum* develop dark, grayish lesions which appear similar to silver scurf. However, black dot lesions are more irregular, with undefined margins. They also usually contain microsclerotia which are smaller than those on stolons. Extensive tuber blemishes may increase tuber respiration, resulting in shriveling and tuber shrinkage.

4.3.2 Epidemiology

The pathogen overwinters as microsclerotia occurring free or on colonized plant debris in the soil. The fungus can persist in the soil for at least 8 years. The fungus may also overwinter as sclerotia on infected seed tubers and, therefore, infection of plants may be due to tuber –borne and/or soil borne inoculum. Conidia probably serve as the primary inoculum for infection. Conidia do not germinate at 7 °C, the optimum temperature for germination and infection is between 22 and 28 °C. Roots are the organs most susceptible to infection; stems generally become diseased only after the fungus is well established on the underground stem of the plant. Black dot is commonly associated with high temperature, poor soil drainage and sandy soils, and low nitrogen levels. In storage, infection and symptom development are favoured by warm, humid conditions.

4.4 Silver Scurf (*Helminthosporium solani* Dur. and Mont.)

It is a common storage disease and occurs wherever potatoes are grown. Silver scurf does not usually cause yield loss, but severe seed infection can affect vigour. The disease is also becoming important in potato processing, because crisps made from potatoes with severe silver scurf infection may result in blackened edges, making the product unmarketable. Fresh weight reduction of tubers may also occur due to excessive moisture loss from the tubers through lesions.

4.4.1 Symptoms

There are no above ground symptoms and on roots. However, lesions can be observed on stolons soon after tuber initiation. The most conspicuous symptoms are produced on tuber periderm. The lesions are roughly circular in size, expanding up to several centimeters. The edge of the lesion is regular. The disease gets its name because the lesions are mostly silvery in colour. In soil, established lesions expand rapidly within a few weeks of planting infected seed tubers. Lesions on progeny tubers spread slowly on the surface when in soil. The lesions are usually small at the harvest but enlarge during storage.

4.4.2 Epidemiology

Perpetuation of the disease takes place through soil as well as tuber borne inoculum. Therefore, transmission of silver scurf can occur through infected seed introduced into soil or through conidia present in soil. Conidia produced in storage conditions are released and carried to other tubers via circulating air. Under favourable conditions – moderate to warm temperatures (10-32 °C) and very high humidity or free water-conidia germinate on plant tissue by polar germ tubes and cause infection of tubers.

4.5 Fusarium Wilt and Dry Rot (*Fusarium* spp.)

4.5.1 Symptoms

Variety of symptoms is produced on potato including wilt, stem rot and damping off of seedlings. In wilting, lower leaves turn yellow and the affected plant dries off rapidly. Both stems and tubers at stolon end show vascular browning. In some cases wilting may be accompanied by rotting of stem base. It may cause damping off of seedlings if planted early in the season when temperature is high. Dry rot is a storage disease and does not become evident until 2-3 months of storage. Rot may occur in any part of the tuber but wounded site and stolon end are the most vulnerable. Initially the infected tissue develops slight depression, which increases, and the skin develops wrinkles in the form of irregular concentric circles. Underlying tissue assumes mealy and brown fungal mycelium.

4.5.2 Epidemiology

Infected tubers and soil are the primary source of inoculum. Dry rot development is affected by tuber age, tuber size, storage conditions, tuber damage and degree of curing. Dry rot infection gets aggravated 5-6 months after harvest. Store temperature ranging 20-28 °C is congenial for dry rot development. Wilt is mainly affected by soil temperature and relative humidity. High wilt incidence in early planted crop is mainly associated with high soil temperature.

4.6 Powdery Scab [*Spongospora subterranea* (Wallr.) Lagerh.]

4.6.1 Symptoms

The fungus attacks all underground parts of the plant without showing any adverse effect on plant growth. The damage to the tubers is however, more serious. The disease does not affect the potato yields but disfigures tubers, reducing its commercial value and renders them unsuitable for seed purpose. Pimple like spots appears on the surface of young tubers. These spots are circular, smooth and light brown which gradually increase in size and ultimately rupture, exposing a cavity containing a brown powdery mass of spore balls. Deep pustules of powdery scab resemble deep pitted common scab lesions. However, powdery scab pustules are filled with mass of fungal spore ball whereas common scab lesions are empty.

4.6.2 Epidemiology

The fungus over winters through spores in soil and on infected seed tubers. The spores germinate during crop season and produce zoospores in soil, which infect the tubers through lenticels or directly through epidermis. Soil temperature and moisture are the main factors affecting the disease. Low soil temperature (0-15 °C) coupled with high soil moisture is ideal for disease development. This disease is a high altitude disease and is seldom noticed below 2500 m amsl and its incidence increases with the increase in altitude.

4.7 Wart [*Synchytrium endobioticum* (Schilb) Perc.]

4.7.1 Symptoms

It is a disease of potato tubers and is usually not recognized in the field until the tubers are dug out. The disease is characterized by prominent warty protuberances resembling cauliflower or bunches of 'cocks comb' like proliferated outgrowth on tuber. Sometimes small greenish warty growths on the stalks may be observed near the ground level.

4.7.2 Epidemiology

Wart disease is both soil and tuber borne. Once the soil is contaminated with the resting sporangia, it becomes an important source for the spread of the disease, as winter sporangia are known to remain viable for many years. The chief means by which the disease spreads is through the transportation of material containing resting spores. The disease is worst in wet season. Both winter and summer sporangia can germinate over a wide range of temperature (12-18 °C) if the moisture is favourable.

4.8 Verticillium Wilt (*Verticillium alboatrum* Reinke and Berth.)

4.8.1 Symptoms

The infection starts from the roots and the fungus grows into the stem and colonizes the xylem vessels thereby disrupting the water and mineral supply to the aerial parts as a result plants remain stunted, lack vigour, lower leaves tend to droop and there is loss of turgidity. Vascular bundles of stem and tuber become brown. In tuber initial infection is seen as yellowish discolouration at the stolon end. In tuber, initial infection is seen as yellowish discolouration at the stolon end.

4.8.2 Epidemiology

Infected tubers and contaminated soil serve as the source of primary inoculum for *Verticillium* wilt. For the perpetuation of the disease the seed surface contamination has been reported to be more important than the internal seed borne inoculum. The pathogen requires comparatively low temperature and therefore, restricted to cooler parts of the country.

4.9 Sclerotium Wilt (*Sclerotium rolfsii* Sacc.)

4.9.1 Symptoms

Infection starts at the stem base in the form of 1-2 cm dark brown lesions, which gradually enlarge and encircle the stem base resulting in the collapse of plant. The pathogen produces white fungal mat and mustard sized sclerotia on the underground parts within the hyphal mat.

4.9.2 Epidemiology

All the wilt causing fungi survive in the soil and plant debris although infected seed tubers my also act as the primary source of inoculum. *Sclerotium* survives in the soil in mycelial as well as in sclerotial form. The fungus may also survive on collateral hosts. The fungus gets aggravated at high soil temperature (25-30 °C) and requires alternate periods of wet and dry soil. Flooding of soil kills *S. rolfsii* thereby reduces the wilt.

4.10 Sclerotinia Wilt [*Sclerotinia sclerotiorum* (Lib.) de Bary]

4.10.1 Symptoms

The disease occurs on the stem either at the soil line or at the junction with leaf petioles. Early symptoms on stems are the appearance of water soaked areas on which white fluffy mycelial growth subsequently develops, which gradually enlarge and encircle the stem base resulting in the collapse of plant. Rotting of the stem may extend up to 5 cm above the ground. In the later stages of symptom development, large, dark, compact resting sclerotia are formed in stem pith.

4.10.2 Epidemiology

S. sclerotiorum overwinters as mycelium in dead or living plants, but primarily in the form of sclerotia. In soil it can remain viable up to 5 years. It also requires comparatively low temperature (10-27 °C) and therefore, restricted to cooler regions of the country.

5. Management of Soil and Tuber Borne Diseases

Soil and tuber borne diseases primarily perpetuate through infected seed tubers and soil. Their management, therefore, requires elimination or lowering down of the inoculum load on the tubers as well as in soil. Management strategies therefore, have to be many fold for combating these diseases.

5.1 Crop Rotations and Green Manuring

When potato crop is planted year after year in the same field, the survivability of the pathogens and their build-up gradually increases over the years. Although, most of the diseases, which infect the potato crop, have wide host range, it is still possible to keep the pathogens population within manageable limits by practicing suitable crop rotations. It has been found that long-term rotation of maize or sun hemp with potato significantly reduced black scurf and charcoal rot incidence. Sesbania, sunhemp and pearl millet are also effective against black scurf. *Verticillium* wilt can be effectively managed if potato crop is grown after two years of Kulth cultivation. Intercropping of potato with maize, rotated with bean or radish was quite effective in the management of potato wart.

5.2 Amendment of Oil Cakes

Oil cakes have mostly been tried for the management of black scurf and *Fusarium* wilt. Build-up of the *Fusarium* population was least in groundnut cake amended soil followed by mustard cake and cotton seed cake.

5.3 Adjustment in Planting and Harvesting Time

Some of the soil and tuber borne diseases are temperature sensitive and can be effectively managed by altering the planting and harvesting dates. By advancing the harvest from February 16 to January 30, the incidence of black scurf was brought down by more than 50 percent. Similarly, harvesting of potato tubers before the soil temperature crosses 28 °C reduces charcoal rot incidence in endemic areas. By delaying the planting from October 1 to October 30 resulted in 36 percent reduction in *Fusarium* wilt.

5.4 Sanitation

Use of disease free seed, weed control and removal of diseased plants/ debris from field are some of the cultural practices that reduce soil and tuber borne inoculum.

5.5 Soil Solarization

Soil solarization by the use of transparent polyethylene sheet is an effective, simple and ecofriendly way of managing soil borne diseases. This method could be useful in tropical and sub-tropical plains where summer temperatures are very high and is practicised during the hottest period of the year. Solarization was superior to deep summer ploughing as it reduced black scurf incidence by 55.6 percent and russet scab by 58.4 percent.

5.6 Biological Control

Use of *Bacillus subtilis* (B-5) has been found effective against black scurf, common scab, Fusarium wilt and bacterial wilt. A combination of soil solarization and seed treatment with boric acid or *Trichoderma viride* improved black scurf.

5.7 Host Resistance

Disease resistant or immune varieties are the best methods to check soil and tuber borne diseases, however, such varieties are available only against few diseases. Varieties immune/resistant to wart disease are Kufri, Sherpa, K. Kanchn, K. Jyoti, Kufri Muthu, Kufri Bahar, Kufri Chmatkar, Kufri Khasigro, Kufri Kumar, Kufri Giriraj, Kufri Chipsona-2, Kufri Anand, Kufri Pukhraj, Kufri Jawahar and Kufri Sutlej. The early maturing varieties like Kufri Chndarmukhi and Up-to-Date are less prone to charcoal rot and may be cultivated in spring. Most of the varieties in India are susceptible to black scurf. However, Kufri Dewa, Kufri Bahar and Kufri Sherpa are comparatively less susceptible.

5.8 Chemical Control

Dipping of infected tubers in boric acid (3 percent) for 30 minutes or spraying on tubers especially before storage has been recommended for the management of tuber borne diseases (black scurf and common scab). Besides, many more fungicides like Pencycuron @ 0.25 percent, Penflufen 240SC @ 0.083 percent, Thifluzamide 24 percent SC @ 3.0 ml/10 kg seed and carboxin + thiram @ 2.5-3.75 g/kg seed have also been found effective against black scurf of potato.

6. Screening Techniques of Late Blight Resistance

Potato is a high value food crop contributing directly to hunger reduction as well as increasing income, especially in the developing countries. Potato cultivation is affected worldwide by late blight, the most destructive potato disease caused by an oomycete *Phytopthora infestans*, affecting the quality of tuber and yield as well. Resistance in plants is mainly of two types; vertical which is race specific, monogenic and expressed in the form of hypersensitive response while horizontal resistance is polygenic, race-non-specific, durable or quantitative. However, pathogen develops matching virulences and leads to breakdown of resistance in potato varieties. Therefore, instigating new resistance sources remains the ultimate objective to combat this notorious pathogen and for which robust screening methods are needed.

6.1 Production of *P. infestans* Inoculum

P. infestans inoculum is multiplied using tuber slice method. Tubers of any susceptible potato cultivar are surface sterilized with ethyl alcohol and cut into 1 cm thick slices using a sharp sterilized knife. Slices are placed in plastic petri dishes and inoculated with *P. infestans* by scratching the mycelium on slice surface with sterilized needle and incubated in air tight plastic boxes lines with moist foam sheet at 18±1 °C for a week in the dark. A thick white growth of sporangiophore with plenty of *P. infestans* sporangia would cover the tuber slices surface within 5-6 days.

6.2 Zoosporangial Germination

Tuber slices containing *P. infestans* zoosporangia are gently dipped in sterilized distilled water to dislodge the zoosporangia from tuber slice surface. This zoosporangial suspension is then kept at 12±2 °C for 60-90 min for releasing zoospores. The zoospore suspension is calibrated to a desired level ($6x10^4$ zoospores/ml) using haemocytometer for each isolates. This zoospore suspension is utilized for inoculating the leaves or tubers for screening the material for foliage/tuber resistance.

7. Screening for Foliage Resistance

7.1 Detach Leaf Method

Plants of genetic material to be screened are raised under glass house in the pots. On the 45th days after planting, fourth leaves from the top are plucked from each clone and placed in plastic trays on perforated plastic separators. Five leaves of each clone are collected and placed in the tray. Leaves of a susceptible variety are kept to serve as control. These leaves are inoculated with 20 ul zoospore suspension using an autopipette. *P. infestans* zoospores suspension is prepared and calibrated as described above. The trays are incubated at 18±1 °C for 5 days. High humidity in the trays is maintained with the help of moist foam sheet. Observations are recorded on the 6th day of inoculation measuring the leaf area infected. The accessions are categorized into highly resistant, resistant, moderately resistant and susceptible on the basis of following scale.

Lesion area (cm^2)	*Grade*
Up to 1.0	Highly resistant (HR)
1.1 to 2.5	Resistant (R)
2.5 to 6.0	Moderately resistant (MR)
>6.0	Susceptible (S)

7.2 Whole Plant Method

Whole plant screening is done in the screening chambers having controlled temperature, humidity and fluorescent light. Plants of genetic material to be tested along with a susceptible cultivar are raised in pots in the glass house. Forty days after planting, the pots are transferred to the screening chamber. Close the chamber and put on the humidifier for 1-2 hrs. so that leaf surface is completely wet. Prepare zoospore suspension of most complex races and spray the plants with inoculums. Keep the humidifier off for a night and then incubate the plants at 18±1 °C with 90 percent humidity. After 6 days, score the plants using the following scale:

Percent Area Infected	*Score*
Trace of infection	9
10	8
11-25	7
26-40	6
41-60	5
61-70	4
71-80	3
81-90	2
Collapsed	1

Malcolmson, 1976.

8. Tuber Resistance

8.1 Tuber Slice Method

In this method 3 tubers of each accessions to be tested, are surface sterilized with ethyl alcohol and cut into 1 cm thick slices aseptically by a sterilized knife. One slice from each tuber is kept in a sterilized petri dish. Three slices of a susceptible variety are also placed in a petri dish to serve as control. Tuber slices are inoculated with zoospore suspension using filter paper discs (0.3 cm^2) dipped in the zoospore suspension prepared and concentration adjusted as per methods given above. The petri dishes are kept in plastic trays lined with moist foam sheet and incubated at 18±1 °C for 5 days in dark. Length and breadth of the lesions are measured and lesion area calculated as per following formula:

Lesion area = $\pi/4$ x ab,

where a= length and b= breadth of the lesion. The accessions are categorised in 4 groups as per scale given under foliage resistance by detach leaf method.

8.2 Whole Tuber Test

After 7-16 days of harvest, give superficial injury to the tubers. After 24 hrs of injury place the tubers on wet filter paper in seed boxes. Incubate the tuber (i) By spraying the zoospore suspension of *P. infestans*; (ii) By placing the soaked filter paper discs on the lanticels or eyes. After inoculation, close the box to avoid drying and incubate at 18±1 °C. After two weeks score the tubers individually on the following scale:

Score	*0*	*1*	*2*	*3*	*4*	*5*
Percent of the surface with symptoms	0	5	5-10	10-25	25-30	>50

9. Field Screening

Field screening is generally done on a large population of plants. For this purpose disease should be recorded in each cultivar right from its appearance till the maximum build up of the disease at regular (weekly) intervals. This data may be used to calculate the Vander Plank 'r' (apparent infection rate), AUDPC (Area under disease progress curve, or mean values. However, out of them AUDPC has been considered more reliable for categorizing the cultivars according to their resistance grades.

Chapter 20

Soil and Tuber Borne Diseases of Potato and their Management

Vinay Sagar

ICAR-Central Potato Research Institute, Shimla, Himachal Pradesh

1. Introduction

Several soil and tuber borne diseases are known to affect potato crop and result in heavy losses. These diseases not only reduce quality and market value of the produce but also spread to new un-infested fields through infected seed potato tubers. Diseases such as black scurf, common scab and powdery scab disfigure potato tubers and reduce their market value where as dry rots, charcoal rots and soft rots spoil the harvested produce during storage causing significant losses. Potato wart is a serious disease which once established is difficult to eradicate and thus require quarantine measures. Bacterial wilt or brown rot of potato is another serious disease which has threatened potato cultivation in many potato growing areas of the world including India. The important soil and tuber borne diseases and their management are discussed here in brief.

2. Fusarium Dry Rot

2.1 Causal Organism and Symptoms

Fusarium dry rot is an important post-harvest disease of potato tubers. It causes significant losses in storage and during transit of both seed and table potatoes.The initial symptoms on potato tubers appear as small brown lesions on surface. These lesions subsequently enlarge, appear dark, sunken and wrinkle producing white, pink, or blue pustules. In later stages, a cavity often develop in the centre of the concentric ring and whitish, pinkish or dark brown growth of fungal mycelium may become visible.Rotten tubers may shrivel and get mummified (Figure 20.1).

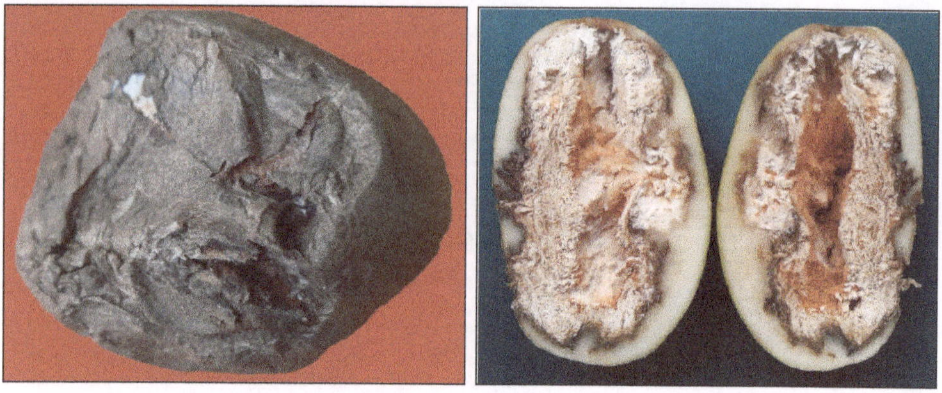

Figure 20.1: Symptoms of *Fusarium* Dry Rot on Tubers.

As many as thirteen species of *Fusarium viz., F. sulphureum*(syn. *F. sambucinum*), *F. coeruleum, F. oxysporum, F. avenaceum, F. culmorum, F. acuminatum, F. crookwellense, F. equiseti, F. graminearum, F. scirpi, F. semitectum, F. sporotrichioides* and *F. tricintum*(Corda) Sacc. cause dry rot of potato. In India, at least ten species of *Fusarium* have been reported to cause dry rot of potato in stores. Infected or rotting tubers are main source of spread of the inoculum of *Fusarium* spp. and results in soil infestation. The pathogen may also get introduced to new locations through contaminated soil, farm implements, through wind and irrigation water *etc.* The pathogen cannot infect intact periderm and lenticels of tubers however, cuts and wounds created during harvest, grading, transport and storage predispose them to infection. Development of disease is also affected by moisture and temperature. The fungus grows well between 15 to 28 °C, however, disease development continues at low temperature in cold stores especially dry rot due to *F. sambucinum*.

2.2 Management

Some level of *Fusarium* dry rot is almost always present in commercially available seed and it is not possible at present to be 100 percent sure that a seed lot is completely free of dry rot. Practicing the following procedures will help prevent dry rot.

- ☆ Use only healthy, disease free/certified seed tubers.
- ☆ Clean and disinfect seed storage facilities thoroughly before receiving seed tubers.
- ☆ Treat tubers with 3 percent boric acid (either dip tubers for 30 minutes or spray ensuring full coverage of tuber surface) and dry under shade before cold storage (for seed purpose only).

- ☆ Seed tubers should be stored at 4-5 °C, 85 to 90 percent RH and be kept ventilated.
- ☆ Monitor stored tubers often for dry rot. Grade out rotten tubers when tubers are removed from storage for marketing/planting.
- ☆ Disinfect seed cutting and handling equipment often and make sure that cutters are sharp enough to give smooth cut that heals easily.
- ☆ Preferably plant whole tubers only. If cut seed tuber is used, treat the seed with fungicide (tuber dip in mancozeb 72WP @ 0.25 percent for 10 minutes) and dry under shade before planting to control seed piece decay and sprout rot.
- ☆ Adopt sanitation measures to avoid soil contamination through farm implements, irrigation water *etc.*
- ☆ Reduce soil inoculum through crop rotation and eliminating volunteer potatoes which can reduce the risk of dry rot.
- ☆ Delaying harvesting for about two weeks after haulm destruction when skin of the tubers have matured and avoiding injury to tubers during harvest, handling and transportation minimize the dry rots.
- ☆ Harvested potatoes should be stored at around 13 to 18 °C under shade and moderate humidity for two to three weeks for bruises to heal before putting the potato in cold stores.

3. Black Scurf

3.1 Causal Organism and Symptoms

The disease affects tuber quality by the development of sclerotia on progeny tubers, thus their market value is reduced. Yield losses up to 35 percent primarily due to reduced crop stand have been reported. The disease can kill eyes during storage; affect stems before emergence (stem canker) and form black scurf (black irregular lumpy encrustations of fungal sclerotia) on progeny tubers after senescence. Infected sprouts become blackened and may not emerge in severe cases. Reddish brown lesions develop on stems and often girdle them. Partial or complete girdling of the stems results in rosetting of plant tops, purple pigmentation, upward chlorosis or rolling of leaves. Formation of aerial tubers in axis of leaves due to interference with starch translocation, are also observed in the infected plants (Figure 20.2).

The disease is prevalent in most potato growing regions of the world including India. In India, it has spread widely in most potato growing regions.

The disease is caused by a fungal pathogen *Rhizoctoniasolani* Kuhn (perfect basidial stage: (Frank) Donk). The pathogen is both soil and seed borne but the disease spreads to new growing areas primarily through sclerotia-covered seed tubers. Soil-borne infection emerges later in the season since the fungus needs

Figure 20.2: Stem Canker (a), Rosetting of Plant Tops (b) and Black Scurf (c) Caused by *Rhizoctonia solani*.

some time to grow into proximity with its potato host. Optimum temperature for germination of sclerotia is 23 °C and for development of stem lesions is 18 °C. Sclerotial development on tubers is initiated depending on environmental conditions. Maximum development of sclerotia takes place in the period between dehaulming and harvest of the crop. Late harvested crop show more black scurf incidence. The host range of *R. solani* is very wide and the pathogen causes diseases of many economically important plants belonging to Solanaceae, Fabaceae, Asteraceae, Poaceae and Brassicaceae as well as other ornamental plants and forest trees.

3.2 Management

The practices like use of healthy seed tubers free from sclerotia, seed tuber treatment with 3 percent boric acid (dip for 30 minutes/spray) before cold storageor pencycuron (@ 0.25 percent) as spray on tubers at planting, planting in relatively dry and warm soil with shallow covering of seed tubers to achieve rapid crop emergence with less opportunity for the fungus to attack, two to four year crop rotation with cereals and legumes, soil solarization with transparent polyethylene mulching during hot summer months especially in Indian subtropical plains have been found very effective for control of the disease. Bio-control agents such as *Trichoderma viride*, *T. harzianum* and fungicides such as benomyl, thiabendazole, carboxin, and azoxystrobin have been reported effective for control of the disease. A continuous use of treated seed for 2-3 crop seasons has been found to completely check the disease. Wherever possible, crop rotation of 1-2 years may be followed.

4. Charcoal Rot

4.1 Causal Organism and Symptoms

The disease is favoured by high soil moisture coupled with temperature exceeding 28 °C. The affected tubers rot both in field and during storage and can cause severe losses under unusually warm wet weather. The fungus attacks the growing plants and tubers at harvest and storage. Affected plants in field exhibit stem blight or shallow rot similar to black leg and cause the foliage to wilt and turn yellow. Early symptoms on tubers develop around eyes, lenticels and stolon end where a dark light grey, soft, water soaked lesion develop on the surface of the tuber. Subsequently, the lesions become filled with black mycelium and sclerotia of the pathogen (Figure 20.3). Secondary organisms may develop in such lesions especially under wet conditions causing significant losses.

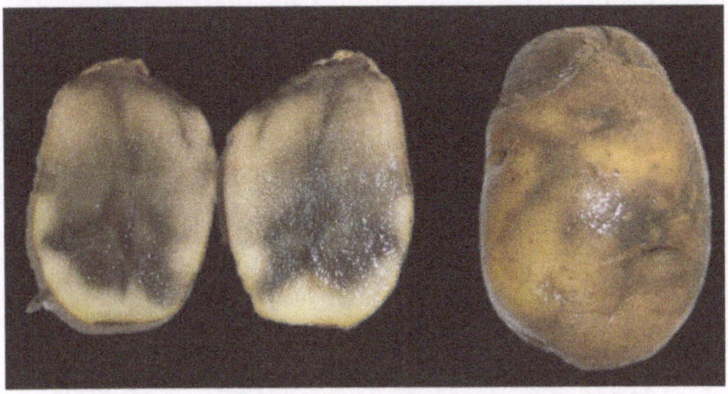

Figure 20.3: Charcoal Rot of Potato Tubers.

The disease is incited by a fungus *Macrophomina phaseolina* (Tassi) Goidanich. Black, smooth, hard 0.1 to 1.0 mm sized sclerotia of the fungus develop within roots, stems, tubers and leaves. *M. phaseolina* is a weak parasitic soil fungus and over-winters in soil as micro-sclerotia in plant debris, weeds and alternate host crops. The pathogen spreads through the infected seed tubers and through the infested soil carried along with the implements. Poor plant nutrition and wounds predispose the plants to charcoal rot. Temperature around 30 °C or above are very favourable for infection, the rot is slow at 20 to 25 °C and stops at 10 °C or below. Fungal growth stops in tubers placed in cold stores but it resumes the growth after cold storage. The fungus is found on underground parts of an extremely wide range of plants, both cultivated and wild.

4.2 Management

Charcoal rot used to be a number one problem in Bihar and eastern Uttar Pradesh, was eliminated by adjusting the dates of planting and harvesting. The disease is active only when soil temperature goes beyond 28 °C. Therefore, planting early maturing cultivars, frequent irrigations to keep down the soil temperature and harvesting before the soil temperature exceeds 28 °C can reduce the disease incidence in eastern plains of India. Rotation with non-host crops and use of seed from disease free area, avoiding cuts and bruises at harvest can also be followed to reduce disease incidence.Grade out the rotten/diseased tubers at harvesting. Harvested potatoes should be stored at around 13 to 18

°C under shade and moderate humidity for two to three weeks for bruises to heal before putting the potato in cold stores.

5. Wart

5.1 Causal Organism and Symptoms

Potato wart is an important and serious disease worldwide. The disease once established is difficult to eradicate since the resting sporangia can survive inter-host periods for up to 20 years. Rough warty mostly spherical outgrowths or protuberances appear on buds and eyes of tubers, stolons, or underground stems or at stem base (Figure 20.4). Wart may appear occasionally on above ground stem, leaf or flowers. Underground galls are white to light pink when young and become brown or light black with age. Above ground galls are green to brown or black. The wart tissues are soft and spongy. Tubers may get completely replaced by warts which desiccate or decay at harvest.Potato wart has been reported in Asia, Africa, Europe, Oceania, North America, and South America. In India, the disease occurs in Darjeeling hills and has been managed by enforcing strict quarantine legislation.

Figure 20.4: Warty Outgrowths on Tubers of Potato.

The disease is caused by (schilberszky) Percival, a member of *Chytridales*. Numerous pathotypes of the fungus exist and at least 43 pathotypes have been described from Europe. Wart is favoured by periodic flooding followed by drainage and aeration since free water is required for germination of sporangia and dispersal of zoospores. Mean temperature below 18 °C and annual precipitation of about 70 cm favour disease development.

The pathogen spreads from one locality to another through infected seed tubers, infested soil adhering tubers, machinery and other carriers of contaminated soil. Resting sporangia survive passage through the digestive track of animals fed with the infected potatoes, and the contaminated manure, therefore, can disperse the inoculum. Earthworms have been found to serve as means of inoculum dispersal. The resting sporangia can also be dispersed by wind-blown soil or by flowing surface water. Resting sporangia of *S. endobioticum* are endogenously dormant and can remain viable for 40 to 50 years at depths of up to 50 cm in soil. Potato is the principal host of the pathogen, although, experimentally a number of species of *Solanaceae* takes infection upon artificial inoculation.

5.2 Management

The disease has been successfully managed by sanitation, long crop rotation (*e.g.* maize/cabbage/pea followed by bean or radish), growing resistant and immune varieties (*e.g.* Kufri Jyoti, Kufri Sherpa, Kufri Bahar, Kufri Kanchan, Kufri Khasigaro *etc.*) and by enforcing strict quarantine legislation in India. Periodic surveys are however, required to monitor viability of the pathogen in soil and efficiency of the quarantine measures.

6. Bacterial Soft Rot

6.1 Causal Organism and Symptoms

Bacterial soft rot can cause significant loss of potato tubers at harvest, transit and storage. Losses due to poor handling of the produce, poorly ventilated storage or transit may go up to 100 per cent. Soft rot bacteria usually infect potato tubers which have been damaged by mechanical injury or in the presence of other tuber borne pathogens. On tubers, the disease appears as small soft water soaked spots around lenticels which enlarge under high humidity.A brown to black pigment may develop around the lesion (Figure 20.5). The soft rot affected tubers become slimy and foul smelling and brown liquid oozes out from the affected tubers. The tuber skin remains intact and sometimes the rotten tubers are swollen due to gas formation. At harvest, many small rotten tubers with

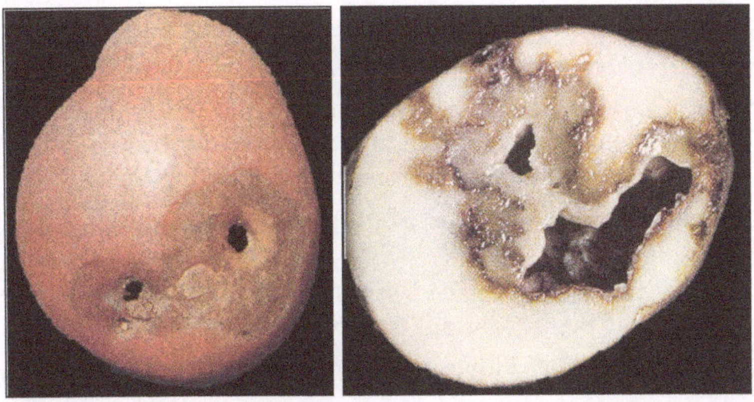

Figure 20.5: Soft Rot Symptoms (Left and Right) and Potato Tuber Cut to Show Tissues Affected by Soft Rot (Right).

intact skin can be seen. The disease affect the crop at all stages of growth but it is more serious on potato tubers under poor storage conditions especially in warm and wet climate.

Pectobacterium atrosepticum (syn. *Erwinia carotovora* sub sp. *atroseptica*), *Pectobacterium carotovorum* sub sp. *carotovorum* (syn. *Erwinia carotovora* subsp. *carotovora*), *Dickeya* spp. (including *D. dianthicola, D. dadantii, D. zeae*) (syn. *Erwinia chrysanthemi*) have been found to cause soft rot. Soft rot bacteria may be carried latently in lenticels, wounds and on surface of tubers without any visible symptoms and spread to healthy tubers in stores, during seed cutting, handling and planting. Water film on surface of tuber which cause proliferation of lenticels and creates anaerobic conditions and injury on surface of tuber predispose potatoes to soft rot. Decaying tubers in soil could serve as source of contamination for healthy tubers. Tubers harvested in wet soil, poor ventilation in transit and storage promotes the rot. In warm climates, where one potato crop follows another or where only short rotation cycles are applied, the bacteria can pass easily from one crop to the next, especially in poorly drained soil. The bacteria can survive at places where rotten potatoes and vegetables are dumped.

6.2 Management

Soft rot bacteria are carried deep inside the tuber, in lenticels and surface wounds making it difficult to eradicate. These quiescent bacteria proliferate in high moisture condition and require water film that cause anaerobic conditions leading to disease development. Surface injury predisposes the tubers to soft rot infection. Based on ecology and epidemiology of the disease following management practices have been worked out:

- ✰ Avoid excess irrigation, provide proper drainage and restrict nitrogen dose to minimum (150 kg/ha).
- ✰ Adjust planting time to avoid hot weather during plant emergence. Harvest the crop before soil temperature rises above 28 °C.
- ✰ Harvest the crop only when tuber skin is fully cured.
- ✰ Avoid injury to tubers and sort out bruised/injured tubers.
- ✰ Store harvested potatoes at around 13 to 18 °C under shade and moderate humidity for two to three weeks for bruises to heal before putting the potato in cold stores or transporting them.
- ✰ Treat tubers (meant for seed purpose) before storage with 3 percent boric acid for 30 minutes and dry under shade.
- ✰ Clean and disinfect seed storage facilities/transporting means thoroughly before storing/transporting tubers.
- ✰ Store the produce either in well-ventilated cool stores or cold stores.

7. Common Scab

7.1 Causal Organism and Symptoms

Common scab causes superficial lesions on surface of potato tubers. The disease seldom affects yield but economic losses may be high as the affected tubers fetch low price in market due to their poor appearance. Seed lots exceeding 5 percent incidence are rejected by seed certification agencies (in India) causing huge loss to seed industry. Scab begins as small reddish or brownish spot on the surface of the potato tubers and its initial infection takes place during juvenile period of tubers. Infection takes place mainly through lenticels and surrounding periderm turns brown and rough. Lesion becomes corky due to elongation and division of invaded cells. Under Indian conditions multiple kinds of symptoms have been recorded and they are grouped as (1) a mere brownish roughening or abrasion of tuber skin (2) proliferated lenticels with hard corky deposition, might lead to star shaped lesion (3) raised rough and corky pustules (4) 3-4 mm deep pits surrounded by hard corky tissue (5) concentric series of wrinkled layers of cork around central black core (Figure 20.6).

Figure 20.6: Various Types of Scab Symptoms Caused by *Streptomyces* Species on Potato Tubers.

At least 13 different *Streptomyces* spp. *viz., Streptomyces scabies, S. acidiscabies,S. turgidscabis, S. collinus, S. griseus, S. longisporoflavus, S. cinereus, S.violanceoruber, S.alborgriseolus, S. griseoflavus, S. catenulae*have been found to cause common scab on potato worldwide. *Streptomyces* spp. may be pathogenic or non-pathogenic. The pathogenic species produce thaxtomins which are phytotoxins and cause hypertrophy and cell death. Ability to produce thaxtomin toxin is strongly correlated with the pathogen's pathogenicity.

Potato is physiologically most susceptible to *Streptomyces* spp. in the period following tuber initiation. *Streptomyces* species infects the newly formed tubers through stomata and immature lenticels. Once the periderm has differentiated, tubers are no longer susceptible to the pathogen. The pathogen is both seed and soil borne. It can survive in soil for several years in plant debris and infested soil. Soil conditions greatly influence the pathogen. Favourable conditions include pH between 5.2 to 8.0 or more, temperature in the range of 20 to 30 °C and low soil moisture. The pathogen is aerobic in nature and maintaining high soil moisture for 10 to 20 days after tuber initiation can help in reducing the common scab. The organism is a tuber-borne and is well-adapted saprophyte that persists in soil on decaying organic matter and manure for several years. Infected tubers serve as source of inoculum in the field, giving rise to infected progeny tubers.

7.2 Management

The pathogen is difficult to eradicate because of long survival both on seed tubers and in soils. Therefore, following practices to minimize the inoculum and creating adverse condition for pathogen spread/disease development are recommended:

- ☆ Use only disease free seed tubers.
- ☆ Give tuber treatment with boric acid (3 percent for 30 min.) and dry under shade before cold storage (for seed purpose only).
- ☆ Irrigate the crop repeatedly to keep the moisture near to field capacity right from tuber initiation until the tubers measure 1 cm in dia.
- ☆ Maintain high moisture in ridges at least for a few weeks during the initial tuberization phase.
- ☆ Follow crop rotation with wheat, pea, oats, barley, lupin, soybean, sorghum, bajra, and adopt green manuring to keep the disease in check.
- ☆ Plough the potato fields in April and leave the soil exposed to high temperatures during summer (May to June) in the North Indian plains.

8. Brown Rot/Bacterial Wilt

8.1 Causal Organism and Symptoms

Brown rot or bacterial wilt is one of the most damaging diseases of potato. The disease causes wilting of plants in standing crop and also causes rot of infected tubers in field, storage and transit. The earliest symptom of the disease is slight wilting in leaves of top branches during hot sunny days. The leaves show drooping due to loss of turgidity followed by total unrecoverable wilt

(Figure 20.7.1a). Bacterial wilt in field can be distinguished from other fungal wilts by placing the stem cut sections in clear water (Figure 20.7.1c). Within a few minutes, a whitish thread like streaming can be observed coming out from cut end into water. This streaming represents the bacterial ooze exuding from the cut ends of colonized vascular bundles. The same test can also be carried out to see infection in tuber. In tubers, the vascular tissues of a transversely cut tuber show water soaked brown circles where dirty white sticky drops appear in about 2-3 minutes. In advanced stages of wilt, bacterial mass may ooze out from eyes (Figure 20.7.2).

Figure 20.7.1: Symptoms of Bacterial Wilt (a); Brown Discoloration of Stem Tissues (b); Bacterial Streaming in Clear Water from Stem Cut Section of Potato (c) Infected with *R. solanacearum*.

Figure 20.7.2: Vascular Browning of Tubers (Left) and Emergence of Bacterial Mass through Eye Due to *R. solanacearum*.

The disease is caused by a Gram-negativebacterium *Ralstonia solanacearum*. The pathogen infects roots of healthy plants through wounds. Infected tubers and plant debris in infested soil are two major sources of inoculum. Mean soil temperature below 15 °C and above 35 °C do not favour the disease development. Soil moisture influences the disease in at least four ways; (i) increasing survival of the bacterium in the soil, (ii) increasing infection (iii) increasing disease development after infection, and (iv) increasing exit of the bacterium from host and spread through the soil. *Ralstonia solanacearum* is capable of causing brown rot in a wide range of soil types and levels of acidity. In majority of the cases, the disease has been reported in acidic soils (pH 4.3 to 6.8) and only in a few cases in alkaline soils. Under favorable conditions, potato plants infected with *R. solanacearum* may not show any disease symptoms. In this case, latently infected tubers used for potato seed production may play a major role in spread of the bacterium from infected potato seed production sites to healthy potato-growing sites.

8.2 Disease Management

Absolute control of the disease is difficult to achieve due to seed and soil borne nature of the pathogen, however economic losses can certainly be brought down considerably using the following eco-friendly means:

8.2.1 Healthy Seed

Use of healthy planting material can take care of almost 80 percent of bacterial wilt problem. We fortunately have bacterial wilt free areas in western and central Indo-Gangetic plains. Do not cut seed tubers as cutting knife spreads pathogen from diseases to healthy tubers. Also cut tubers contact disease from soil easily.

8.2.2 Field Sanitation

Where the field is already infested, the best way to minimize the disease is to adopt the following agronomic practices:

Crop rotation: Follow 2-3 years' crop rotation using crops like maize, cereals, garlic, onion, cabbage and sanai. Do not rotate vegetables like brinjal, ginger, chillies and other solanaceous crops. Paddy and sugarcane although are not host, still they carry pathogen and contribute to the disease perpetuation.

Avoid tillage operations:Pathogen enters in plants through root or stolen injuries. Such injuries cannot be avoided during intercultural operations. Therefore, restrict tillage to the minimum and it is advisable to follow full earthing-up at planting.

Off-season management of field: The pathogen perpetuates in the root system of many weeds and crops. Clean the field from weeds and root/foliage remnants and burn them. The pathogen in remnants can be exposed to high temperature above 40 °C in summer in plains and plateau and low temperature below 5 °C in hills by giving deep ploughing. This may cause extinction of pathogen from the field.

8.2.3 Early Planting

Early planting of crop in February, reduces the wilt in north-eastern and north western hills. In eastern and Nilgiri hills, planting the autumn crop in the month of September prevents wilt by 80 percent.

8.2.4 Chemical Control

Application of stable bleaching powder @ 12 kg/ha at the time of potato planting in furrow along with fertiliser reduces pathogen population from field and gives effective control.

Based on intensive ecological and epidemiological studies at Central Potato Research Institute, Shimla, the following integrated practices are recommended for checking the bacterial wilt in different agro-climatic zones of the country.

Zone I: Non-endemic areas (Gujarat, Maharashtra, north-western and north-central plains). This zone mainly is characterised by hot and dry summer (April-June) with scanty vegetation; temperature can go up to 40-43 °C. Bacterial wilt is no more a major problem. Therefore deep ploughing in summer and use of disease free seed is adequate for the disease control.

Zone II: North-western mid hills (up to 2200 masl), north-eastern hills and the Nilgiris. This zone comprises typical complex, diverse and risks prone areas. It is characterised by mild summer, profuse vegetation and maximum temperature range of 26-30 °C. Winter temperature may go as low as 3-6 °C. Many weed hosts can provide perpetual niche for colonisation and survival. The use of disease free seed plus application of stable bleaching powder @ 12kg/ha mixed with fertiliser at planting, ploughing the field in September- October and exposing the soil to winter temperature are adequate for disease control. The application of bleaching powder can be substituted by 2 year crop rotation with crops like wheat, barley, finger millet, cabbage, cauliflower, knol-khol, carrot, onion, garlic *etc*. Early planting and early harvesting are also recommended.

Zone III: Eastern plains and Deccan plateau. Here crop is cultivated as short day crop during winter months (October-March). Day temperature sometimes reaches to 38 °C. Heavy precipitation occurs due to western disturbances. The area is relatively rich in vegetation. Eastern plains and Deccan plateau have many symptomless carriers of the pathogen. Therefore, management of the disease is most difficult. However, the disease can be kept under check with practices like use of disease free seed, application of bleaching powder, blind earthing-up and ploughing in March and leaving the soil exposed to summer temperatures during April- May and crop rotations along with clean cultivation.

Zone IV: North-western high hills above 2200 masl (excluding Kumaon hills). This zone has a temperate climate with severe winters; daily temperature range from –10 to 5 °C during December–January. Snow is common during these months. Bacterial wilt is not endemic and the use of disease free seed alone is adequate.

9. Diagnostics of Bacterial Wilt and Common Scab Diseases

9.1 Collection of Bacterial Wilt and Scab Infected Plants/Tubers

Potato plants/tubers infected with typical symptoms of bacterial wilt/ common scab disease (Figures 20.8, 20.9a) are obtained from various potato growing regions. At each collection site, tubers (brown rot/scab infected) or stem pieces (6 to 10 cm) of wilt infected plants showing typical brown rot/ scab/wilt symptoms are collected, labelled properly and brought to laboratory immediately for isolations.

Figure 20.8: Bacterial Wilt Symptoms (a); Brown Discoloration of Stem Tissues (b); and Bacterial Streaming in Clear Water from Stem of Potato Plant Infected with R.solanacearum**; Symptoms of Common Scab on Potato Tubers (d,e).**

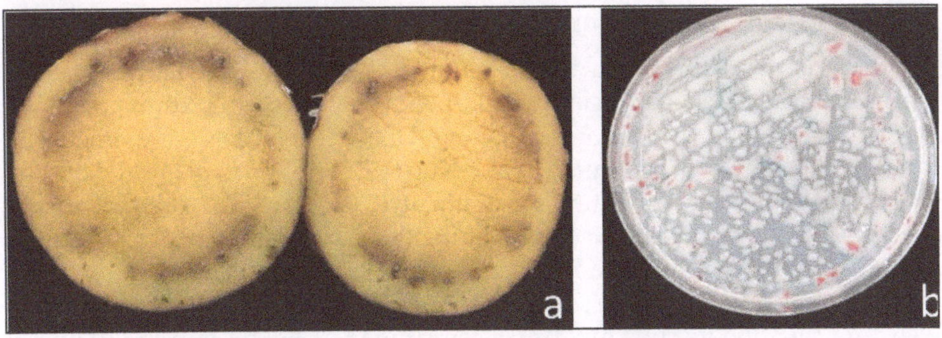

Figure 20.9: Grey-Brown Discoloration of Vascular Tissues and Bacterial Ooze in Potato Tuber Infected by *R. solanacearum* (a); Typical *R. solnacearum* Colonies on TZC Agar Medium (b).

9.2 Isolation and Maintenance of *Ralstonia solanacearum* Isolates

Potato tubers/stem pieces are washed thoroughly in running tap water for 5 to 7 minutes to remove soil adhered to their surfaces and air dried. The samples (stem pieces and tubers) are then surface disinfected with 70 percent ethanol, peeled, subsampled and macerated in sterile distilled water. Macerates are streaked on Kelman'striphenyltetrazolium chloride (TZC) agar medium (Kelman, 1954) with slight modification (Peptone, 10 g litre^{-1}; glucose/glycerol, 2.5 g/5 ml litter $^{-1}$; Casamino Acids, 1 g litre^{-1}; yeast extract, 1 g litre^{-1}; agar, 15 g litre^{-1}; TZC, 50 mg litre^{-1}; pH 7.0-7.1). Plates are incubated at 28±2 °C for 48 to 72 hrs. This medium is useful for distinguishing *R. solanacearum* among other bacteria during isolation, and for distinguishing virulent (wild type) colonies from avirulent mutant ones during purification of cultures. Bacterial colonies developing the typical irregular mucoid colonies are again transferred to fresh TZC medium for further purification. Well separated typical wild type *R. solanacerarum* colonies (Figure 20.9b) are further transferred to medium modified by exclusion of TZC for multiplication of inoculum free of formazan pigment (which is slightly bacteriostatic).Two loop full of bacterial culture are then transferred in 2 ml of double distilled sterile water and the cultures stored at 20±2 °C.

9.3 TZC Stock Solution

One gram of 2,3,5 triphenyltetrazolium chloride is dissolved in 100 ml of distilled water, placed in a light proof capped bottle, autoclaved for only 8 minutes or sterilized by filtration and stored refrigerated. To each 200 ml of melted, somewhat cooled Kelman's basal medium *i.e.* Casamino Acid Peptone Glucose/Glycerol Agar (CPG) medium, one ml of TZC stock solution is added to give a final concentration of 50 mg litre^{-1}.

10. Isolation of Genomic DNA

Bacterial growth from a well separated colony on TZC agar is used to inoculate 1.5 ml of CPG broth in 2.0 ml Eppendorf tubes. The cultures are grown at 28 °C for 48 hours with vigorous shaking. Total genomic DNA is extracted by rapid method as described by Chen and Kuo (1993). 1.5 ml of saturated culture is harvested with centrifugation for 3 min at 12,000 rpm. The cell pellet is re-suspended and lysed in 300 µl of lysis buffer (40 mMTris-acetate pH 7.8, 20 mM sodium-acetate, 1 mM EDTA, 1 percent SDS) by vigorous pipetting. To remove most proteins and cell debris, 100 µl of 5M NaCl solution is added andmixed well, and then the viscous mixture is centrifuged at 12,000 rpm for 10 min at 4 °C. After transferring the clear supernatant into a new vial, an equal volume of chloroform is added, and the tube is gently inverted at least 50 times when a milky solution is completely formed. Following centrifugation at 12,000 rpm for 3 min, the extracted supernatant is transferred to another vial and the DNA is precipitated with 100 percent ethanol, washed twice with 70 percent ethanol, dried in speed-vacuum, and re-dissolved in 50 µl of TE buffer. If required, RNA could be removed by adding RNAase in the lysis step for 30 min at 37 °C.

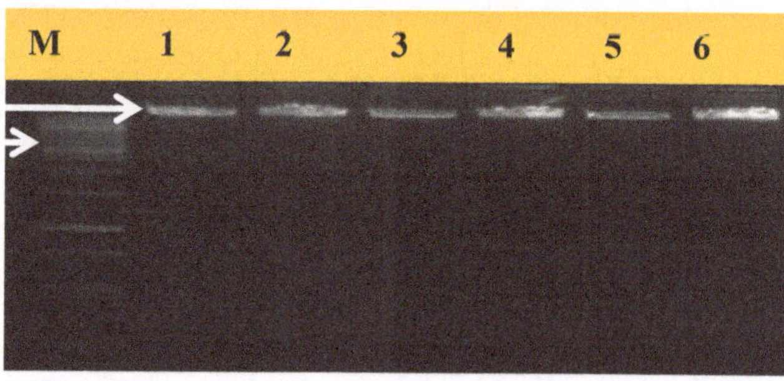

Figure 20.10a: Bands of Total Genomic DNA.

Table 20.1: Stock Solutions for Preparation of Lysis Buffer for DNA Extraction

Sl.No.	*Name of Stock Solution*	*Required Concentration*	*Volume (µl) Required for 10 Samples @ 300µl Lysis Buffer per Sample*
1.	Tris-acetate (1.0M)	40mM	120
2.	Sodium acetate (1.0M)	20mM	60
3.	EDTA (0.1M)	1mM	30
4.	SDS (10 percent)	1 percent	300
5.	Distilled water	-	2490
	Total		3000
6.	NaCl (5.0M)	100µl (added after lysis to remove most proteins and cell debris)	

10.1 Purity Check and Quantification of DNA

For purity check and quantification, 5 µl template DNA is loaded with 2 µl of 6X loading dye on 1.0 percent agarose gel in Tris–Acetate EDTA buffer with 0.04 µl ml^{-1} of ethidium bromide at 100V for 1hr and visualized under ultraviolet (UV) light. DNA samples showing intact bands are selected for PCR amplification.

10.2 Characterization of *R. solanacearum* Isolates into Phylotypes

Phylotype identification of each isolate is done as described (Fegan and Prior 2005; Prior and Fegan 2005). Multiplex PCR (Pmx-PCR) is carried out in 25 µl final volume of reaction mixture, containing 1× Taq Master Mix (PCR buffer, 1.5 mM MgCl2, 200 µM of each dNTP, 50 mM KCl, 10 mM Tris-HCl and 1.25 U of Taq DNA polymerase.), 6 pmoles of the primers Nmult:21:1F, Nmult:21:2F, Nmult:22:InF, 18 pmoles of the primer Nmult:23:AF and 4 pmoles of the primers 759 and 760 (Opina *et al.*, 1997). The following cycling programme is used in a thermal cycler: 96 °C for 5 min and then cycled through 30 cycles of 94 °C for 15 s, 59 °Cfor 30 s and 72 °C for 30 s, followed by a final extension period of 10 min at 72 °C.

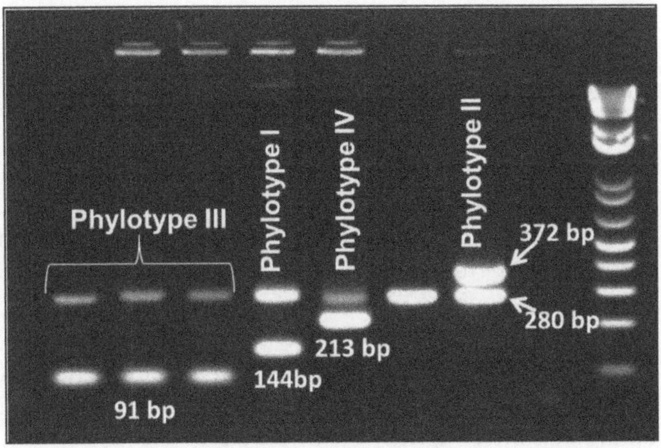

Figure 20.10b: Pmx-PCR using Phylotype-Specific Primers showing PCR Products of 280 bp (*i.e. R. solanacearum*) Amplicons for all Isolates, 91bp (Phylotype III, lane 1-3), 144 bp (phylotype I, lane 4), 213 bp (Phylotype IV, lane 5) and 372 bp (phylotype II, lane 7) Amplicons.

A 5 µl aliquot of each amplified PCR product is then subject to electrophoresis on 1 percent agarose gel, stained with ethidium bromide and visualized on a UV-trans-illuminator. This Pmx-PCR produces the following phylotype-specific PCR products: a 144-bp amplicon from phylotype I strains; a 372- bp amplicon fromphylotype II strains; a 91-bp amplicon from phylotype III strains; and a

213-bp amplicon from phylotype IV strains (Figure 20.10b) and *R. solanacearum* specific PCR product of 280-bp.

Table 20.2: List of Primers Used for Multiplex PCR

Sl. No.	*Primer*	*Primer Sequence*	*Expected Band Size*	*Remarks*
Primers for multiplex PCR				
1.	Nmult:21:1F	52 -CGTTGATGAGGCGCGCAATTT-32	144 bp	Phylotype I (Asiaticum)
2.	Nmult:21:2F	52 -AAGTTA TGGACGGTGGAAGTC-32	372 bp	Phylotype II (Americanum)
3.	Nmult:22:InF	52 ATTGCCAAGACGAGAGAAGTA-3	213 bp	Phylotype IV (Tropical)
4.	Nmult:23:AF	52 -ATTACGAGAGCAATC GAAAGATT-32	91 bp	Phylotype II (African)
5.	Nmult:22:RR	52 TCGCTTGACCCTATAACGAGTA-3		Amorce reverse unique
6.	759R	5'-GTCGCCGTCAACTCACTTTCC-3'	280 bp	Universal *R. solanacearum* specific primers
7.	760F	5'GTCGCCGTCAGCAATGCGGAATCG-3'		

10.3 Inoculation

Ralstonia solanacearum isolates are inoculated to susceptible cultivars of potato (cv. K. Jyoti), tomato (cv. Pusa Rubi), brinjal (Pusa Purple Long) *etc.* for pathogenicity test or screening of varieties. The pathogenicity test or screening of varieties is conducted on potted plants (healthy plants grown in sterilized soil) in glass/green house (temp. 20-30 °C) using 3-5 replicates for each isolate/cultivar. A bacterial suspension of an isolate is prepared in distilled sterilized water using 48 h old growth on casamino acid peptone glucose (CPG) agar medium. The plants are inoculated by stem stab method when they are 15-20 cm tall. Usually 3rd of 4th axil bud from the top is inoculated by injecting/placing a droplet of suspension (15-20 µl) on injury. Plants stabbed with sterilized distilled water serve as control. The inoculated plants are incubated at 20-30 °C temperature and observed for wilt appearance till 40 days of inoculation.

11. Isolation and Maintenance of *Streptomyces* Isolates

Common scab infected potato tubers are washed thoroughly in running tap water for 5 to 7 minutes to remove soil adhered to their surface and air dried. The tubers are surface disinfected with 70 percent ethanol and small tuber pieces containing scab lesions are cut out and surface disinfected for one minute in 1 percent sodium hypochlorite solution. After several rinses in sterile distilled water, the tissues (about 1g) are ground in a pestle mortar in 5 ml of sterile water.

Serial dilutions are plated onto starch casein agar (SCA) medium (Soluble starch 10g, casein 0.3g, potassium nitrate (KNO_3) 2g, di-potassium hydrogen orthophosphate (K_2HPO_4) 2g, magnesium sulphate ($MgSO_4 \cdot 7H_2O$) 0.05g, calcium carbonate ($CaCO_3$) 0.02g, ferrous sulphate ($FeSO_4 \cdot 7H_2O$) 0.01g, agar 20g, distilled water to make the final volume 1 litre, pH 7.2). Plates are incubated at 28±1 °C, and individual characteristic powdery (mycelial) bacterial colonies are picked and serially transferred until a pure culture is obtained. Potato dextrose agar medium is used for routine sub culturing of *Streptomyces* strains. Strains are maintained as spore suspension in 20 percent glycerol at –20 °C.

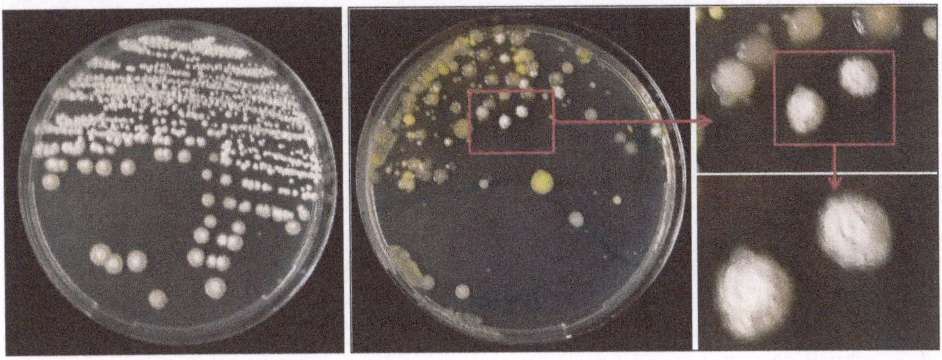

Figure 20.11: Typical Leathery Colonies of *Streptomyces* on Starch Casein Agar giving Powdery Appearance and Pure Culture of *Streptomyces* (Right).

11.1 Extraction of Genomic DNA

Bacterial growth from a well purified culture on SCA medium is used to inoculate 1.5 ml of starch casein broth in 2.0 ml Eppendorf tubes. The cultures are grown at 28±1 °C for 48-72 h with vigorous shaking (200 rpm). Genomic DNA is isolated by a versatile quick-prep method for genomic DNA of Gram-positive bacteria with some modifications. Mycelia (1–2 ml) grown in a starch casein broth shake culture are centrifuged, rinsed with TE and re-suspended in 0.4 ml SET buffer (75 mMNaCl, 25 mM EDTA, 20 mMTris, pH 7.5) (Table 20.3). Lysozyme is added to a concentration of 1 mg ml^{-1} and incubated at 37 °C for 0.5–1 hour. Then 0.1 volume 10 percent SDS and 0.5 mg Proteinase K ml^{-1} are added and incubated at 55 °C with occasional inversion for 2 h. One-third volume 5 M NaCl and 1 volume chloroform are added and incubated at room temperature for 0.5 hour with frequent inversion. The mixture is centrifuged at 4500 g for 15 minutes and the aqueous phase is transferred to a new tube using a blunt-ended pipette tip. Chromosomal DNA was precipitated by the addition of one volume 2-propanol with gentle inversion, washed twice with 70 percent ethanol, dried in speed-vacuum, and re-dissolved in 100 µl of distilled water. The dissolved DNA is treated with 20 mg RNase A ml^{-1} at 37 °C for 1 hour. Samples are extracted in the same volume of phenol/chloroform/isoamyl alcohol (25:24:1) and precipitated with 2.5 volume cold ethanol and 0.1 volumes

3 M sodium acetate. The pellets are washed with 70 percent ethanol, dried and dissolved in 50 µl TE or distilled water.

Table 20.3: Stock Solutions Used for Preparation of Lysis Buffer for DNA Extraction

Sl.No.	*Name of Stock Solution*	*Required Concentration*	*Volume Required for 10 ml of SET Buffer*
1.	NaCl (5.0M)	75 mM	150 µl
2.	EDTA (0.1M)	25 mM	2500 µl
3.	Tris (2.0 M)	20 mM	100 µl
4.	Distilled water	-	7250 µl
	Total		10,000 µl

11.2 Purity Check and Quantification of DNA

For purity check and quantification, 5 µl template DNA was loaded with 2µl of 6X loading dye on 1.0 percent agarose gel in Tris–Acetate EDTA buffer with 0.04 µl ml^{-1} of ethidium bromide at 100V for 1 h and visualized under ultraviolet (UV) light. DNA samples showing intact bands were selected for PCR amplification.

11.3 Sequencing of the 16S rRNA Gene

The 16S rRNA gene was amplified from DNA of *Streptomyces* isolates by PCR using the primers 16S-1F (5′-CATTCACGGAGAGTTTGATCC-3′) and 16S-1R (5′AGAAAGGAGGT GATCCAGCC-3′) (Takeuchi *et al.*, 1996). Amplification is carried out in 50 µl of reaction volume containing 10 mM TrisHCl (pH 9.0), 50 mM KCl, 0.1 percent Triton X100, 1.0 to 2.0 mM $MgCl$, 250 µM each dNTP, 10 pmol each primer, *Taq*DNA polymerase at 5 U/µl, 20 to 40 ng DNA template, and MilliQ water to bring the volume to 50 µl. The thermal cycler is set to the following conditions: initial denaturation at 95 °C for 3 min; followed by 40 cycles of denaturing at 95 °C for 20 s, annealing at 60 °C for 30 s, and extension at 72 °C for 2 min; and ending with a 4 °C hold. The PCR products of 1,500 bp (expected size) are electrophoreses in 1 percent agarose gel in Tris–Acetate EDTA buffer with 0.5 µl ml^{-1} of ethidium bromide at 105 V for 1 hour and visualised and photographed under ultraviolet (UV) light.

For sequencing of PCR products, amplified products are further purified using Qiagen Mini elute PCR purification kit following manufacturer's guidelines. The concentration of each product is estimated by gel electrophoresis with a low DNA mass ladder(Fermentas 1 Kb) and diluted with ultra-pure water to give a final concentration of 10-20 ng $µl^{-1}$ for sequencing. PCR products are then cycle sequenced using forward primer in thermal cycler (Gen-AmpR PCR System 9700 of M/S Applied Biosystem). The cycle sequenced products are further purified. DNA samples then sequenced by an Applied Biosystem

(Foster City, CA, USA) Model, AB310 Genetic Analyzer). The 16S rRNA gene sequences analysis are performed using off-line software CLUSTAL X, Org. CLUSTAL W or X is a multiple sequence alignment program which calculates the best match for the selected sequences, and lines them up so that similarities and differences can be seen. Sequence data of all the 48 strains are manually edited and compared with the sequences from NCBI through BLAST programme. Sequences are aligned with the reference sequences from NCBI and phylogenetic analysis is done using CLUSTAL X software (Thompson *et al.*, 1994).

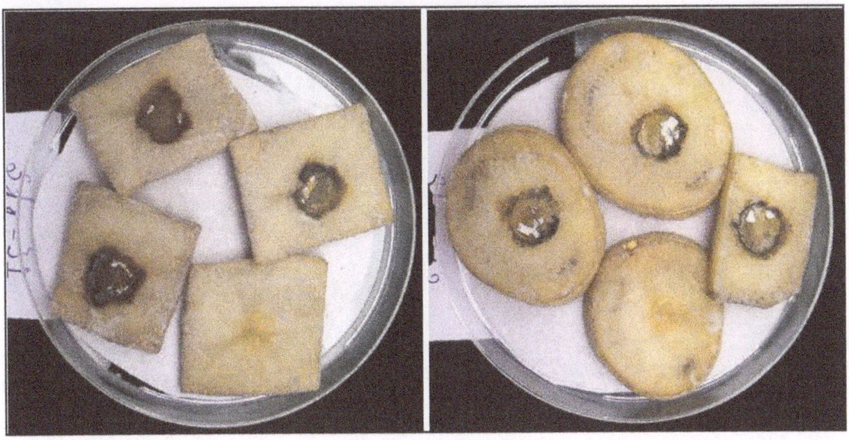

Figure 20.12: Pathogenicity Assay on Tuber Slices of cv. Kufri Chandermukhi: Necrotic Halos Produced by the Plugs from 5 Days Old Oat Meal Agar Cultures of *Streptomyces* strains, Slices without Halo are Control.

11.4 Pathogenicity Test

Pathogenicity of *Streptomyces* strains is examined through Tuber Slice Assay as described by Hao *et al.* (2009).Potato tubers (cv. Kufri Chandermukhi) free from surface diseases are washed and sterilized for 1 min with 10 percent bleach (0.0625 percent NaClO). A tissue disk (20 mm in diameter by 7 mm in height) is bored from the tuber and placed on moist filter paper in a petri plate (100mmx 15mm). Pathogenicity of *Streptomyces* strains is examined as described by Loria*et al.* (1995). *Streptomyces* sp. inoculum is prepared from the cultures grown on oatmeal agar medium for 5 to 7 days. Agar plugs from *Streptomyces* cultures (3 mm-diameters) are placed at the centre of the potato tuber disks replicating thrice. Agar plugs from oat meal agar placed on potato tuber slices served as control. The disks were incubated in a moist closed container at 28±1 °C in the dark and observed for necrosis of tuber slices.

Chapter 21

Seed Potato Production Scenario in India

Rajesh K. Singh[1], *Tanuja Buckseth*[2], *J.K. Tiwari*[2], *A.K. Sharma*[2], *Vinod Kumar*[2], *Manoj Kumar*[2] *and N.K. Pandey*[2]

[1]*ICAR-Indian Institute of Vegetable Research, Varanasi, Uttar Pradesh*
[2]*ICAR-Central Potato Research Institute, Shimla, Himachal Pradesh*

1. Introduction

India is the second largest producer of potato in the world contributing about 12.52 percent of production from 11.29 percent area under potato with an average yield of 22.31 t/ha. Potato is grown in almost all the states in India under diverse agro-climatic conditions. About 90 percent of the total potato area is located in the sub-tropical plains, 6 percent in the hills and 4 percent in the plateau region of the peninsular India. Indo-Gangetic plains account for about 76 percent of the potato area and about 87 percent of the potato production in the country. Availability of assured quality seed is the most important production constraint in all the potato growing regions of the country.

Seed is the basic and vital input in agriculture and its timely availability in adequate quantity as per demand decides the strength and growth of agricultural economy in the country. Seed has played a very important role in India's green revolution and shall continue to be the important component in the days to come. There is a saying "as you sow, so shall you reap", which traditionally relates to the quality of seed determining the production, has been the wisdom of our forefathers. The traditional farmers' practices are to save their own seeds generally selected from healthy plants from fields which are free from diseases and pests, harvest them separately, clean, cure and then store in containers and inspect regularly for any kind of damage; or they get quality seed from other seed producing farmers on barter basis, indicating thereby the importance of seed in their production system.

There is limited accessibility of quality material in case of vegetatively propagated plants in India. Generally, tubers are used as seed in potato overwhelming various problems like low multiplication rates, increased production cost various seed borne diseases befitting the risk of degeneration over the years. Such tubers when planted yield low and diseased plants in future generations. This was coined as 'degeneration' or selenity by the Dutch workers in 1972. Such selenity acts as a major constraint in producing clean potatoes. Potato production is constrained by various factors that include availability of suitable varieties, agro- techniques, quality seed and storage infrastructure, especially in tropics and sub-tropics. But most important being the availability of quality seed as it has a direct bearing on crop yield. Potato seed is bulky and the rate of multiplication is slow (5-6 times), this poses a unique challenge to seed production and seed supplying agencies. Since the crop is multiplied vegetatively using tuber as seed, it gets degenerated very fast necessitating replacement of the seed every year (ideally) or after every two years. Further the seed is either unavailable or out of the reach of farmers owing to high price. The seed related issues are further aggravated because of restricted accessibility of reasonable seed producing regions quality seeds can alone add to the increased yield by 15-20 percent.

2. Seed Potato in India

For seed production in India, sincere efforts were made to maintain healthy standards of potato seed stocks with in 1933 in UP (Uttar Pradesh) by Agriculture Department. They attempted to increase the pathogen free seed supplies of Dunbar Cavalier and Majestic cultivars imported from England. This aided in anchoring around 100 quintals of Majestic and 250 quintals of the other assortment *i.e.* Dunbar Cavalier at Almora and Nainital hills of Garhwal district. To conquer the issue of dormancy, a second round of multiplication was done in the mid hills of Bhawali. But this resulted in degeneration of stocks due to occurrence of bacterial brown rot disease in seed tubers and thus the scheme failed (Pushkarnath 1976).

In 1935, a thought was envisioned to establish an undeniable research foundation in India. This provoked the opening of a Potato Breeding Institute in Shimla and two seed creation farms at Bhowali (Kumar slopes, UP) and Kufri (Shimla Hills) as a piece of Indian Agricultural Research Institute, Delhi. Taking a look at the issue of seed degeneration, a research plan under the Indian (Imperial) Council of Agricultural Research was developed in 1941. It surveyed the various diseases affecting potatoes (Vasudeva and Azad 1952). Potato cyst nematode, popularly known as golden nematode is a 'major constraint to potato production throughout the world and this quarantine pest is restricted to Nilgiri hills of South India. It was first detected in1961 at Ootacamund and brisk execution of isolate estimates confined its further spread to other potato developing areas.

3. Hills, Ideal for Seed Potato Production

A project with the name 'Production of partially disease free potatoes' was set up. The fundamental approach was to clean up all the seed crops present in fields which could be obviously distinguished for viral infection. It brought about threefold increase in the yield of Darjeeling Red Round variety. But it was soon realized that potato seed generation on a huge scale requires production and development of India's own varieties. As the climatic conditions in India were entirely different from North America and Europe from where the potatoes were imported in. The experience additionally brought the acceptance that an association which is specialized in science and excels in the craft of seed production was required for commercial production of potato.

An overview was portrait by Davies in 1934 on the different reasons for seed potato degeneracy, and it was the aphids particularly *Myzus persicae* which was observed to be the fundamental driver. Additionally, with the advancements winning around then, hills were observed to be the perfect location for seed potato production which was pathogen free. According to the guidelines provided by him for the location and maintenance of seed potatoes in hilly areas, the principal seed production station for potatoes was established in the high hills of Kufri (Shimla hills, Himachal Pradesh). It satisfied the states of lower temperature in summers (underneath 20 °C) with the inability of aphid migrants to alight the potato crops and their colonization. It was additionally risky to grow potato crops for seed purpose when aphid population transcends above 20 on 100 leaves. Till 1935, the seed potato was being imported from various European countries on yearly basis, but during Second World War, European countries put a blanket ban on the export of potato seed to India.

4. Establishment of ICAR-CPRI

European varieties were imported to India and endeavors were made to acclimatize them to the agro-climatic conditions prevailing in India. But soon deterioration of seed stock took place due to absence of any seed potato production program. With this, it was realized that India can't rely on exotic varieties alone and therefore, needs to have its own program for research and development of potato. In 1945, a scheme for developing Central Potato Research Institute was advanced by Sir Herbert Steward, then Agricultural Advisor of Government of India. The plan was implemented by Dr. S. Ramanujam in 1946 and it was the first occasion when that scientific orientation was given to the seed potato production. In 1946-1947, 1072 mounds of seed potatoes of three varieties- *Arran consue, up-to-date and Kerrs pink* were distributed to local farmers. However, it took three years for the entire scheme to attain a concentrated shape in the form of CPRI, Patna with its first Director Dr. S Ramanujam. The three units functioning independently were (i) Potato Breeding Station, Shimla (ii) Seed Certification Station, Kufri (iii) Potato Multiplication Station, Bhowali were now converged into CPRI. The administration of Himachal Pradesh developed a research scheme for producing disease free seed tubers. This venture effectively delivered clones of 'Up-to-Date" which was imported from Northern Ireland. As hills were the only source of pathogen free seed, in 1956 the headquarters were shifted to Shimla. With the course of time, trade was successfully established between the hills and plains. In between, 1956 to 1983, a few local research stations were built up in various potato developing zones of the country to address local issues of potato development. The organization has seven territorial research stations situated at Jalandhar (Punjab), Patna (Bihar), Modipuram (Uttar Pradesh), Kufri (Himachal Pradesh), Gwalior (Madhya Pradesh), Shillong (Meghalaya) and Ootacamund (Tamil Nadu). Potato seed creation was confined to high slopes till mid 1960's.

5. Overdependence on Hill Seeds

Viruses are transmitted by vectors known as aphids, consequently potato seed started to be produced in the high hills which almost free from aphid during the potato cropping period. This increased the strain on the indigenous seed multiplication systems being practised in the hills. Subsequently, after 22 years in 1966 another research station was setup in Fagu (Kufri, H.P.), as all hilly areas have not been found suitable for production and maintenance of healthy seed potatoes which can meet the standards. The issue of rapid degeneration had made the plains fully reliant on the hills in order to satisfy their prerequisite for seed. Thus, the entire pressure was laid onto the hills for quality seed potato production which offered ascend to an extensive no. of issues, for example:

(i) Limited areas having elevation of 750 m or above which are technologically suitable for seed production.

(ii) Seeds produced in hills cannot be utilized in plains for sowing in October owing to dormancy.

(iii) Limited means of transporting the freshly harvested tuber but dormant seed tubers over long distances resulting in delay in planting.

(iv) Crop conditions due to irrigation practices: In the Indian hills *viz.*, Kashmir valley, Kinnaur and Lahaul Spiti potato planting takes place during the summer seasons followed by harvesting towards the end of monsoons. However, there was no uniform distribution of rainfall which is critical for the growth of plant. Therefore, in case of scanty rainfall, it's very difficult to provide required irrigation due to highly uneven topography, exceptionally porous nature and non-availability of ground water. Another factor which adds is limited area for cultivation and the seed not being of the right physiological age to be immediately used in plains.

Also, a high rate of degeneration makes the seed break down after a couple of multiplications. Since the eastern, north-eastern, Deccan and South western parts of the nation, having hotter climatic conditions, and were not suitable for customary quality seed production. To overcome these issues, it was important to fabricate another framework for seed production as well as certification in plains which could limit the reliance on slopes. In 1971, an undertaking under the name All India Coordinated Potato Improvement Project (AICPIP) with its headquarters in CPRI, Shimla was set up.

6. Development of Seed Plot Technique for Plains

Till mid-1960's the seed potato cultivation was confined to high hills only as it had low aphid population during cropping season. Potato was cultivated as a spring crop in the Indo-Gangetic plains which was totally exposed to aphids bringing about degenerated seeds. Different analyses were led by CPRI for seed potato creation in various agro-climatic areas. A noteworthy achievement underway of value potato seed was culminated through the introduction of '' Seed Plot Technique' in 1965. This made India, the only country in South Asia with its own breeding program for potato. This aided in beating seed decline because of aphids as the seed crops could be grown during very little or no aphid population in the plains of Northern India. Due to the advancement of Seed Plot Technique, the significant production of disease free seed moved from the slopes to the fields in plains. Despite the fact that the potato was presented as a late spring crop in temperate hills, and it at last turned into a noteworthy winter crop for the plains. The vast stretch of fertile belt reaches out from Punjab to Assam, the seed produced through seed plot method gave 30-40 percent higher yields free from many soil and tuber borne diseases. It was adequate for producing enough breeder seed for the nation every year.

Seed Plot Technique benefitted Indian agriculturists as well as aided in sparing expansive measure of cash which was generally spent on the import of foreign seeds. The primary guideline of this technique relied on developing healthy seed potato utilizing low intensity of aphid infestation period in between October and starting of January incorporated with coordinated pest and disease management. Though, roguing and dehaulming of the seed crop needed to be completed by the last week of December or mid of January before the aphid's infestation reach to the economic threshold level. This strategy opened a window in sub-tropical plains for production of quality seed. The seed production could be taken up in bigger zones under sub-tropical plains including Uttar Pradesh, Bihar, Punjab, Haryana and parts of Madhya Pradesh. Farmers could undoubtedly receive seed of right physiological age at the exact time of planting, with no issue of dormancy.

The selection of seed plot system prompted enormous development in cold stores for storing the produce which in the following season could be used for autumn planting. As a result of the advancement of seed plot method, potato tuber moth (PTM) got eliminated and seed rottage diminished. Thus, seed plot technique met the seed prerequisite as well as prompted higher productivity with minimum losses.

7. Virus Indexing

Exploiting the seed plot framework, a proficient consistent strategy for outfitting the quality of healthy seed stocks was envisioned after 1970 through clonal choice, tuber indexing and stage-wise field multiplication of healthy tubers. Indexed tubers concept for production of disease free potato seed took shape once diagnostic methods were developed for major potato viruses. In this method, tubers are selected from storage/growing crop that are uniform in type, free from detectable diseases, and within an acceptable size range. Each tuber is coded for identification, and an eye, preferably from the stolon end, is removed and planted in the greenhouse. Resultant plants are evaluated for vigour and viral diseases. The tubers which are free from all the viruses are used for production of disease free seed by multiplying them in the field for four (Stage 1, 2, 3, 4) generations. This has been made feasible utilizing ELISA (Enzyme Linked Immune Sorbent Assay) with its first use in 1984, by CPRI. In this method, tubers are first screened against different infections like PVX, PVA, PVY PVS, PLRV, PVM, PVY^N and PVY^C using ELISA. Tuber indexing is a system of testing the picked tubers to be free from any disease by growing up their eye plugs in net houses. Another essential technique grasped for virus detection and elimination is Immunospecific Electron Microscopy which is in use since 1987. Scotland pioneered this concept which has since been used world over for potato seed production, including India. The new diagnostics developed over the years have improved the efficiency of the system.

8. Seed Potato Production Systems

8.1 Conventional System

The conventional method of seed potato production is to repeatedly propagate a sample which has been proved to be free of pathogens in a system called clonal multiplication. The clonal multiplication system has been practiced effectively in the Netherlands, South Africa, Kenya and India. This potato seed production system is laborious, expensive and time consuming. Because of a low multiplication rate (usually 1:6), it takes many years to produce significant quantities of seed to meet the demand of potato industry. Another constraint is the degeneration of seed stocks from one generation to another due to accumulation of bacteria, fungi, viruses and viroid's. Modern seed programmes tend to minimize the use of clonal systems or use a combination of clonal and rapid multiplication systems. The production of potato seed under conventional system has not been effective in avoiding or reducing the buildup of pathogens and has consequently led to reduced quality potato seed and low crop yields. In this system, seed potato growers select better quality tubers for seed and discard those of poor quality. The diseased and healthy plants are identified and separated and the healthy tubers are used for the next season's production. If the mother potato plant becomes infected with a disease during the growing season, each of the new offspring is likely to be infected as well. During the growing season, growers peep seed fields visually for signs of disease and remove infected plants through the process of rouging. However, visual inspection, particularly for primary infection, is unreliable, time consuming and requires a well experienced eye. Inevitably, the quality of seed potato produced in subsequent generation's declines substantially. The conventional method of propagation is one of the slowest methods of seed multiplication. This method has also shown to be done at certain time of the year specific particularly in tropical and sub-tropical regions where potato is a winter crop. In addition, the method requires a seed producer to have enough land if he has to enter into commercial seed production. This, however, is associated with high labour cost in managing big fields.

Thus, the only way-out to overcome the above said limitations is augmentation of seed potato production through hi-tech seed production system to improve the quality and also to reduce the field exposure.

8.2 Hi-tech System

Hi-tech seed potato production including micropropagation, aeroponics, ARC is a potential alternative to overcome the everlasting shortage of quality seed potatoes in the country, as the. Due to the advantages that seeds under hi-tech system are produced from typically pathogen tested basic material and can be produced in any region under controlled conditions. Seed potato production involving micro-propagation is finding favours among the seed potato entrepreneurs. The new/Hi-tech system of seed potato production involves raising of in vitro plantlets followed by production of mini-tubers under protected conditions and their subsequent multiplications in the field.

9. Way Forward

The conventional seed potato production technology based on "seed plot technique" is successfully going on since last four decades. As there is limited scope to increase quantity of breeder seed at ICAR-CPRI farms, due to limitation of additional availability of land for seed production therefore, possibilities are being explored with the help of SAUs/KVKs/Pvt. farmers to identify the new areas of seed production, multiplication of breeder seed into FS-I, FS-II and Certified Seed under MoU and to produce seed through hi-tech system with the help of entrepreneurs/private companies. In areas having high disease pressure, the new system of seed potato production based on micro-propagation and aeroponics has the advantage of better health status of seed stocks due to the reduced number of field multiplications over the conventional (clonal multiplication) system. However, there are also options for simplifying the seed potato production systems for adaptation to low-technology situations, which has greater scope and relevance towards the increasing trends of potato production in developing countries.

Chapter 22

Production of Disease-Free Planting Material through Hi-Tech System including Aeroponics

Tanuja Buckseth, R.K. Singh, J.K. Tiwari, Vinod Kumar, A.K. Sharma and N.K. Pandey

ICAR-Central Potato Research Institute, Shimla, Himachal Pradesh

1. Introduction

Use of high yielding varieties and quality planting material are two important ingredients for harnessing benefits of modern production technologies in vegetatively propagated crops like potato (*Solanum tuberosum* L.). Low rate of multiplication, high seed rate, progressive accumulation of degenerative viral diseases, perishability and bulkiness are inherent problems associated with potato seed which result in non-availability of adequate quantities of quality planting material at affordable price. Inadequacy of quality planting material at affordable cost and high seed rate are often reflected in 40- 50 percent cost of cultivation on potato seed alone. To circumvent some of these problems, several modifications such as tuber indexing for virus freedom, seed multiplication stages and seed certification standards have been developed and integrated with conventional potato seed production programmes. The advent of tissue culture, in which virus-free plants can be produced through meristem culture, maintained indefinitely under controlled conditions and multiplied in artificial media under sterile conditions in the laboratory throughout the year irrespective of growing season, has revolutionized seed production in potato world over. In addition to quality assurance through meristem culture, the micropropagation, microtuber and minituber production techniques ensure enhanced multiplication rate in initial stages of seed production.

Hi-tech seed potato production is a potential alternative to overcome the everlasting shortage of quality seed potatoes in the country. Due to the advantages that seeds under hi-tech system are produced from typically pathogen tested basic material and can be produced in any region under controlled conditions. Seed potato production involving micro-propagation is finding favours among the seed potato entrepreneurs. The new/Hi-tech system of seed potato production involves raising of *in vitro* plantlets followed by production of mini-tubers under protected conditions and their subsequent multiplications in the field.

2. Micropropagation

Integration of meristem culture along with micropropagation and traditional multiplication can circumvent inherent problems associated with potato seed production. Micropropagation is an asexual reproductive method that allows production of selected genotypes *en masse* for commercial planting (Sarkar *et al.*, 2011). The most significant advantages offered by micropropagation are: (1) a large number of disease-free propagules can be obtained from a single plant in short period within a limited space; (2) propagation can be carried out under disease-free conditions throughout the year without intervening seasonal constraints; and (3) a constant flow of disease-free plants is possible from the original in vitro stock developed through meristem culture. Micropropagation technology has been successfully used to speed up initial stages of potato seed production programs by supplementing the traditional seed programs with disease-free plants obtained from pathogen-free micro propagated shoots or micro-cuttings or Microtuber or minitubers

2.1 Micropropagation of Disease Free Mother Plant

Soon after varietal release, 10-20 healthy uniform tubers are selected and planted under controlled conditions in the pots in poly/net house for indexing against the viruses. The ideal temperature for plant growth as well as virus multiplication should be 20-25 °C. The plants are tested by ELISA for virus freedom after 6 to 7 weeks of planting or 6 to 8 leaf stage. The infected plants with viruses during ELISA testing should be destroyed and only the healthy plants should be retained for further testing by polymerase chain reaction (PCR) for virus freedom. The infected plants obtained during PCR testing are removed. Finally, healthy plants obtained during series of testing will be used as mother plant for micropropagation.

2.2 Development of Healthy Mother Plants from Virus Infected Plant

Sometimes we may not be getting even a single plant completely free from viruses after releasing of the variety. In such situation, meristem tip culture coupled with thermotherapy has become a powerful and successful tool for virus elimination from infected plants and has been successfully applied in potato for development of virus-free plants. Supply of healthy planting material is the backbone of seed potato industry. The only option to avoid losses caused by viruses is the selection of virus-free mother plants employing meristem tip culture followed by reliable and sensitive virus diagnosis techniques besides breeding virus resistant varieties.

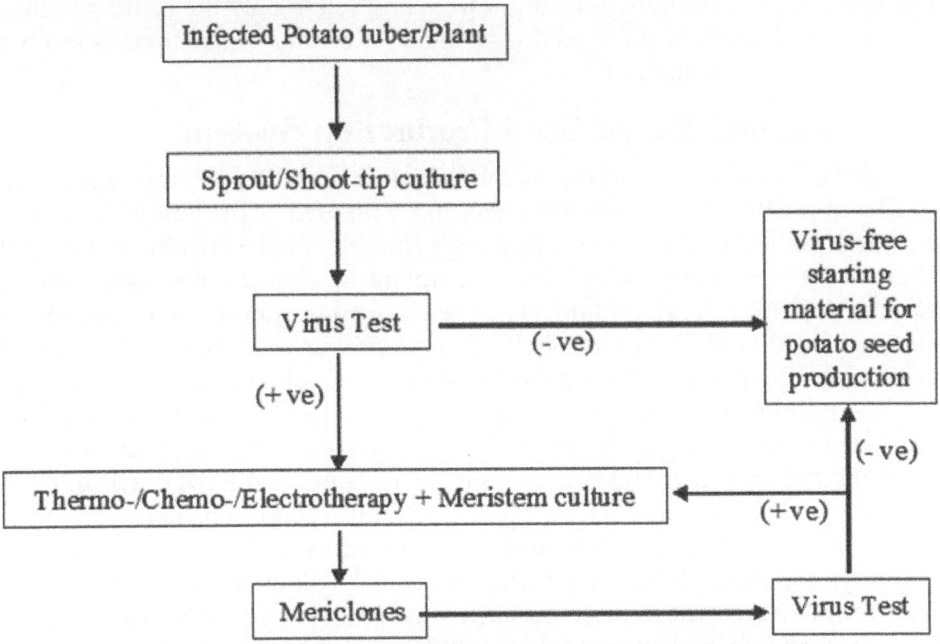

Figure 22.1: Schematic Presentation of Virus Elimination in Potato.

The plants are tested against potato viruses and viroids like PVX, PVS, PVA, PVY, PVM, PLRV, PALCV and PSTVd through ELISA, EM and PCR. In case, no plant is found free from the virus infection then the plants that are infected with minimum number of viruses are selected for meristem tip culture. Using nodal/sprout cuttings, the *in vitro* stocks of selected plants are developed and further sub-cultured in Ribavarin (20 ppm) modified MS media for chemotherapy. This culture is then given thermotherapy at 37 °C and 16 h photo period (120-200 $\mu mol/m^2/s^{-1}$) in the culture room for nearly 20 days. Using stereomicroscope, the apical/axillary meristem (0.2 to 0.3 mm) is excised from *in vitro* plants aseptically with the help of sterile scalpel, needle and blade. The excised meristem is grown in the test tubes containing MS medium with growth regulators and incubated in the culture tubes at 25 °C and 16 h photo period (120-200 $\mu mol/m^2/s^{-1}$) in the culture room until the meristem germinates. The meri-clones are then sub-cultured through nodal cutting after it attains a height of 4-5 cm and the pedigree is maintained. The fully grown mericlones should be tested against potato viruses like PVX, PVS, PVA, PVY, PVM, PLRV, PALCV and PSTVd through ELISA, EM and PCR. The virus-free cultures should be sub-cultured once in every 3-4 weeks so as to get more number of virus-free microplants. The microplants should be hardened for 2 to 3 weeks in the poly/net house before planting in the pots filled with peat moss under mist in poly house. The plants are further tested against all above said potato viruses through ELISA, EM and PCR after 40-45 days of planting. Remove the infected plants obtained during testing and retain only the healthy plants. Finally, healthy plants obtained during series of testing will be used as mother plant for micropropagation.

3. Microplant Based Seed Production System

Three to four weeks old microplants are transferred to protrays filled with sterilized peat moss. The microplants can be planted in portray with root or without root (cuttings). For planting with root, the media sticking to the root should be properly washed off. After transplanting, drenching is done with the mancozeb (0.25 percent) solution. The portrays are then transferred to the growth chambers and kept in dark for 48 h subsequently in 16 h photoperiod for 2-3 days. Once the plantlets are established in portrays (4-5 days), these portrays are transferred to hardening chamber and kept at 27 °C for 10-15 days. The hardened plantlets should be removed from portrays along with peat moss and transplanted on nursery beds in mixture of soil, sand and FYM (2:1:1) in rows at 30 x 10 cm spacing under insect proof net house condition. 5 percent of the plants are tested by ELISA. Rogue out all virus infected plants, off-type plants, abnormal and stunted observed during inspection. Allow the microplant crop to mature and harvest the minitubers. Each microplant shall yield 8-10 minitubers. Seed crop should be harvested 15 to 20 days after haulms cutting when the tuber skin is hardened. The seed tubers thus produced are minitubers. Curing is done by keeping the seed tubers in heap for about 15 to 20 days in a cool shady place. After curing, the seed tuber should be graded into >3 g and treat with 3 percent boric acid solution for 10-15 minutes to prevent surface borne pathogen inoculum. Minitubers harvested from microplants (Generation-0) are called as nucleus seed. Store the minitubers in country store in hills while cold store at 3-4 °C in the plains. Minitubers weighing >3 g will be planted in Generation-l in the field during next season. Whereas, <3 g minitubers may be recycled once again in Generation-0 under controlled poly/net house conditions if the crop meets the G-0 criteria, the produce can be used for raising G-1 crop in the field.

4. Microtuber Based Seed Production System

The microplants are tested for virus freedom before initiating microtuber production. The virus-free stock plants are mass multiplied through nodal cuttings on semisolid MS medium in culture tubes (25 x 150 mm) following the standard procedure upto 10 cycles. 3-4 weeks old explants are transferred into 250 ml conical flasks or culture bottles containing 25-35 ml liquid MS medium without agar. The culture tubes are incubated at 25 °C and 16 h photo period (120-200 $\mu mol/m^2/s^{-1}$) in the culture room. After 3-4 weeks of incubation, the

Figure 22.2: Integration of Tissue Culture in Potato Seed Production.

(a) Rapid micropropagation of virus-free mericlones on semi-solid medium. (b) Multiplication of micropropagated plants in liquid medium for production of microtubers. (c) Potato microtubers developed under dark in MS medium supplemented with 10 mg l^{-1} BA plus 80 g l^{-1} sucrose. (d) Harvested microtubers. (e) Pathogen-free high-density potato crop raised either from in vitro plants or microtubers in net house. (f) Harvested minitubers from high-density net house crop. (g) Healthy field crop raised from minitubers harvested in previous season.

unutilized liquid propagation medium is decanted from the conical flask/culture bottle under aseptic conditions and 30 ml of microtuber induction medium is poured into it. The microtuber induction medium is based on MS basal media supplemented with 10 mgl^{-1} N^6-benzyladenine (BAP) and 80 gl^{-1} sucrose/ commercial sugar. After adding induction medium, the cultures are incubated under complete dark condition at 15 °C for 60 to 90 days depending on the genotype. Microtubers develop epigeally at the apical as well as axillary buds of the shoots. In general, 15 to 20 microtubers weighing 50-300 mg are produced in each flask/culture bottles. Before harvesting, greening of the microtubers is done in the culture room by incubating microtuber induced cultures under 16 h photoperiod (approximately 30 µmol m^{-2} s^{-1}light intensity) at 22-24 °C for 10 to 15 days. Then carefully remove the cultures along with microtubers from conical flasks or culture bottles and manually harvest the green microtubers. Avoid damaging the microtubers, especially the thin periderm during harvest. The harvested microtubers are then washed and treated with 0.25 percent mancozeb for 10 minutes, and allowed to dry in the dark at 20 °C for 2 days. Grading of microtubers in <4m, 4-6mm and >6 mm should be done while packing. Pack the treated microtubers in perforated polythene covers and store in a refrigerator for 4-5 months until planting. Take out the microtubers from the refrigerator after about one month before planting for breaking the dormancy.

5. Aeroponic Seed Production System

The aeroponic system offers the potential to increase production in terms of number of minitubers per plant from three to four times. Aeroponics is the process of growing plants in an air mist environment without the use of soil or an aggregate medium. Aeroponic system mainly consist of an electrical unit, lightproof growth chambers, nutrient solution chamber, high pressure pump, filters and spray nozzles. Interiors of growth chambers are covered with black lining to avoid any admittance of light to the root zone of plants. For aeroponics, *in vitro* grown 15-21 days old microplants are required to be hardened before shifting to this system. For hardening, microplants are allowed gradually to acclimatize the outside conditions and are kept at 25-27 °C for 7-10 days. These hardened plantlets of about 15 cm height are planted in the holes made in the roof of the grow boxes of the aeroponic unit. All essential nutrient elements required for the plant growth are dissolved in water in nutrient tank. With the help of an automatically operated pump, the nutrient solution is regularly sprayed inside the chamber for desired duration of on/off time. By misting the solution round the clock 100 percent relative humidity was maintained inside the root zone. Rooting starts in a week and developed inside the growth chamber. Stolon and tuber formation initiated at different intervals depending upon cultivar. Sequential picking was done at regular interval. Tubers were harvested when they attained the desired size of 3-10 g. Roots, stolon's and tubers develop inside the chamber and leaves are exposed to light. Nutrient solution is replenished from time to time and desired pH of 6.0 of the solution is maintained. Tubers are harvested sequentially as they attain the desired size. These minitubers are sequentially cured at decreasing humidity and temperature and stored at 2-4 °C and used for planting in the next generation.

The aeroponic minitubers are thus harvested are called as generation-0 (G-0) and are strictly planted under net house in Generation-1 (Gen. 1) at a spacing of 30 x 15 cm depending upon the size of mini-tuber. The produce of Gen. 1 is called as pre-basic seed. The produce of Gen. 1 (pre-basic seed) is further multiplied in generation-II (Gen. 2) at a spacing of 45-60 x 20 cm and the produce of Gen. 2 is called as breeder seed or basic seed.

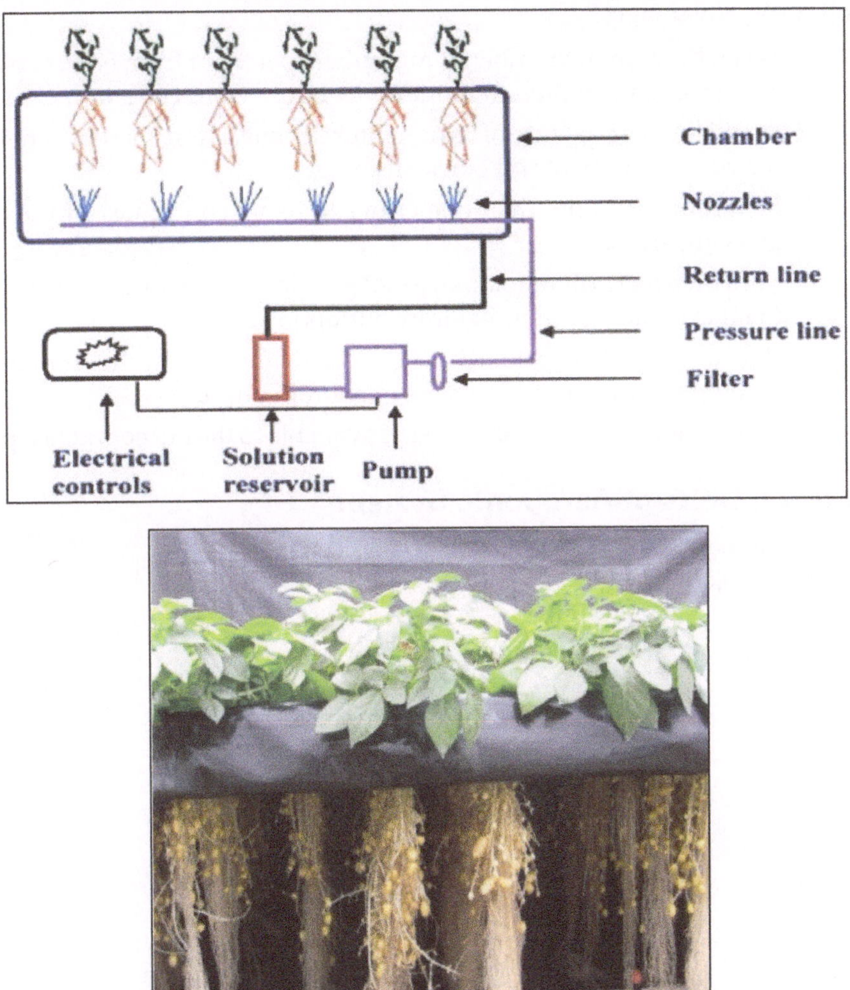

Figure 22.3: Aeroponics in Potato. Diagrammatic presentation of aeroponic system (upper) and mini-tubers as developed in aeroponic chamber (lower).

5.1 Aeroponics Unit

Aeroponics is a process of growing plants in an air mist environment without the use of soil or any other aggregate media. The basic principle of aeroponics is to grow plants in a closed environment by periodically spraying roots with a nutrient solution containing all the required elements for plant growth and tuberization. Aeroponic system is one important component of producing disease free seed of potato. Before this technology came into practice, potato mini tubers from *in-vitro* plants were produced in soil or other solid media (like coco peat, peat moss, perlite or mix of different media) under net house conditions. This technology is replacing this traditional method because of following advantages:

- ✰ Seven times more number of minitubers per plant from *in-vitro* plants as compared to cultivation under net house (soil) conditions.
- ✰ Avoidance of exposure of tissue culture plants to soil, thereby reducing the susceptibility of soil borne diseases.
- ✰ In aeroponics more than one crop can be taken in most of the seed growing areas.
- ✰ Mini tubers are of uniform size (2-3 gm) and can easily be transported to far off places at minimal transportation cost.
- ✰ Root and shoot zone conditions can be maintained at different levels of temperature and RH to have maximum output.
- ✰ Seed of new varieties can be made available to the farmers at fast pace.

5.2 Basic Units of Aeroponic System

- ✰ Tissue culture lab
- ✰ Hardening unit

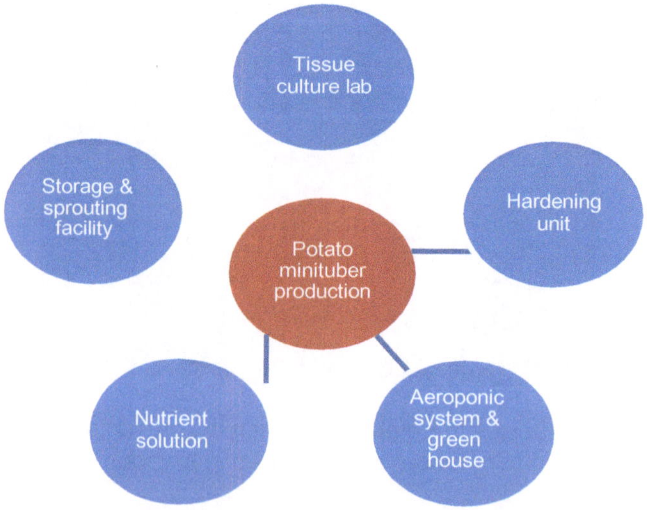

Hardening is a process of preparation of the plants for a natural growth environment. In tissue culture lab ideal conditions are created to encourage rapid growth. Due to the controlled nature of their maturation, the plantlets often do not have fully functional dermal coverings. This causes them to be highly susceptible to disease and inefficient in their use of water and energy.

5.3 Components of Seed Production System through Aeroponics

5.3.1 Aeroponic Unit

Aeroponics is a system of growing plants by keeping root zone humidity hundred per cent throughout the crop cycle by spraying nutrient solution directly over the roots. This units generally consists of:

- ✰ Green house with following subsystems
- ✰ Fan and pad coolingsystem
- ✰ Air circulationsystem
- ✰ Misters and foggers - To increase the relative humidity or lower the temperature in greenhouse
- ✰ Light control systems
- ✰ Grow boxes
- ✰ Solution tank and pH and EC monitoringsystem
- ✰ Nozzles/misters forroots
- ✰ Plant holdingsystem
- ✰ Pumps and solution delivery system
- ✰ Electrical control panel and monitoring instruments

5.3.2 Storage and Sprouting System

Potato mini tubers produced through aeroponics are highly watery in nature, have open lenticels and have low dry matter. If stored under inappropriate conditions tubers lose weight quickly and may have some rotting issues. To maintain good health of mini tubers storage system consists of two units.

- ✰ UNIT I - Temp 15-16 °C, RH 90-95 percent, lights for greening of tubers
- ✰ UNIT II – Temperature 2-3 °C, RH 95-97 percent
- ✰ Chitting (sprouting) room should have temperature (15-20 °C), RH (80-85 percent) and diffused light.

5.3.3 Automation of Aeroponic System

For round the clock running if system following components are used for automation-

- ✰ Sensors for temperature, pH, light and RH

- Timers for running of pumps, nozzles, fans, chillers and other components There can be many more components for automation as per requirements.

6. Maintenance of Greenhouse

6.1 Maintenance of Cooling Pads

- Dry the pads once every day in theevening.
- Drain sump every 24 hours ofuse.
- Replenish about 10 percent of circulating water each time to avoid scaling ofminerals.
- Use fungicide (1 gm/liter) in pad water every week to avoid fungusgrowth.
- In case of heavy salt deposition, open the pads and clean with soft brush.
- Replace pads when required.

6.2 Maintenance of Fans

- Lubricate bearingsregularly
- Tighten V belts as perneed
- Lubricatelouvers
- Clean fan blades regularly. Fan efficiency can be reduced by 30-50 percent with dusty blades, faulty louvers and dusty net/jali.
- Clean the fan covering net/*jali* every alternate day with brush while fan isrunning

6.3 Maintenance of Aeroponic System after Completion of Crop

- Do complete overhauling of pumps.
- Flush all the lines with water and then with hydrogen peroxide sol (2 ml/l)
- Open all the filters and nozzles and clean as explained.
- Repair leakage points if any.
- Change styrofoam sheets if required
- Clean all the components of pad cooling system
- Lubricate exhaust and circulation fans and replace faulty valves
- Repair/replace faulty wormgears
- Clean control panel from inside with air blower or soft brush
- After completing all maintenance, wash all the units, empty tanks let these dry completely

☆ Pack grow boxes, pumps, and all the instruments with black polythenesheet.

7. Nutrient Composition for Tuber Production under Optimum Nitrogen Conditions

Sol. No.	*Salt*	*Conc. of Final Solution [mM]*	*Amount for 250 Litre*	*Nutrient*	*Concentration (mM)*
1	NH_4NO_3	0.5000	10.01 g	N	2.00
2	KH_2PO_4	0.5000	17.01 g	P	0.5
3	K_2SO_4	1.0000	43.57 g	K	2.50
4	$CaSO_4.2H_2O$	0.5000	21.52 g	Ca	1.00
5	$MgSO_4.7H_2O$	1.0000	61.62 g	Mg	1
6	NaCl	0.1250	1.83 g	S	2.50172
7	Fe-EDTA	0.0062	0.57 g	Na	0.1251
8	H_3BO_3	0.004	8 ml	Fe	0.0062
9	$MnSO_4.H_2O$	0.0016	3 ml	Mn	0.0016
10	$ZnSO_4.7H_2O$	0.00008	2 ml	Zn	0.00008
11	$CuSO_4.5H_2O$	0.00004	2 ml	B	0.004
12	$Na2MoO_4.2H_2O$	0.00004	2 ml	Cu	0.00004
13	$CoCl_2.6H_2O$	0.00004	2 ml	Mo	0.00004
15	$Ca(NO_3)2.4H_2O$	0.5000	29.52 g	Cl	0.1251
				Co	0.00004

8. Apical Rooted Cuttings (ARC)

A novel low cost and farmer's friendly technology called 'Apical Rooted Cutting' (ARC) has recently been standardized by ICAR-CPRI, Shimla for hi-tech seed production in potato. ARC involves raising of *in vitro* plantlets in nursery and then transplanting of rooted cuttings in glass/net-house (Buckseth *et al.*, 2019). Rather than allowing tissue culture plantlets to mature and produce minitubers, multiple cuttings are produced from an individual plantlet. Studies has shown remarkable results which suggest that the technology can be used for production of quality planting material at low cost in seed deficit areas. The everlasting shortage of seed potatoes can be overcome through ARC on account of faster rate of multiplication.

9. Way Forward

Seed potato production involving micro-propagation (tissue culture) techniques can overcome many of the problems associated with the conventional multiplication system. The everlasting shortage of seed potatoes in most of the potato growing nations can be overcome through micro-propagation techniques

Figure 22.4: Step-wise Seed Multiplication through Apical Rooted Cuttings.

on account of faster rate of multiplication. Besides, rapid multiplication, disease freedom on account of multiplication of disease free mother stocks under controlled conditions followed by reduced number of field exposures as compared to conventional multiplication system is an added advantage of seed potato production through tissue culture techniques. Due to these numerous advantages, the new hi-tech system of seed potato production involving micro-propagation is finding favour among the seed potato entrepreneurs.

Chapter 23

Farm Mechanization and Drone Technology in Potato

Brajesh Nare and Sukhwinder Singh

ICAR-CPRI, Regional Station, Jalandhar, Punjab

1. Introduction

Potato is a short duration crop, having highest yield per day per unit area as compared to other major crops. Being a relatively short duration crop in the plains of India, advance potato production technologies assumes a special importance in order to accomplish various farm operations in limited available time. Manual production of potato is time consuming and tedious job, it requires about 1600-2000 man-hrs per hectare for all the operations. With the emergence of new crop rotations and enhanced cropping intensity, many farmers experience scarcity of time and labour. Efforts have been made at various levels to mechanize both production and processing to reduce the manpower requirement, drudgery and cost of production. Various mechanical tools, implements and machinery have been developed for land development, fertilizer application, seed treatment, planting, weeding, harvesting, grading, sorting, storage and on-farm processing of potato. Besides these mechanical technologies, drone technology is also being developed for crop health monitoring and precise application of farm inputs *viz.*, insecticides, pesticides, herbicides, fertilizers, *etc.* in potato crop.

Mainly, there are three types of power sources available for farm operations *viz.*, *Mechanical Power:* is used in India ranged from 0.5 to 70 hp in tractor, mini-tractor, self-propelled machines, *etc.*; *Animal power:* mostly bullocks are used as animal power for farm operations. And *Human power:* except tillage operation, human power is being used in all the other farm operations in hilly regions.

2. Field Preparation

Potato requires good tilth soils to ensure covering of stolon's/tubers and also to provide adequate aeration for respiration of roots and tubers. To get the good yield, soil should be loose and friable with good drainage and aeration. Field preparation in the plane areas for potato cultivation is generally practiced with the use of tractor operated ploughs, disc harrow, cultivator-planker, rotavator, *etc.*

Figure 23.1: Green Manuring with Disc Harrow, Deep Tillage with Subsoiler Improves Infiltration and Final Field Preparation with Rotavator.

Subsoiler is an important primary tillage implement. It breaks the hard pan below the soil surface and provides good drainage environment. After 5-7 days of pre-sowing irrigation, 2-3 cross harrowing or 3-4 tiller operations followed by two planking make the required seedbed for a good crop of potato. Among the primary and secondary tillage implements, the rotavator is most popular for preparation of field which provides fine level of soil tilth. All these implements are easily available in the market and are being used for other crops also.

In addition of these implements, self-propelled machines *viz.*, power tiller, power weeder, mini tractors with various attachments are also available for field preparation especially for hilly regions of NEH. Power tiller with combination of various implements *viz.*, three tyne cultivator, plough, rotory tiller, *etc.* can be used for soil bed preparation (Figure 23.2). The other machines like power weeder and mini-tractor are also getting popularity in hilly regions

Figure 23.2: Seed Bed Preparation with Power Tiller (*Source*: Kumar and Kumar 2018, CSISA, 2014).

3. Marking and Fertilizer Application

Fertilizer type, quantity and placement are the important factors for good production of potato crop. Precise and judicious use of fertilizer is of prime importance to ensure its efficient utilization and to reduce the cost of farm input and environmental risk. After seed bed preparation, a fertilizer drill-cum-marker (Figure 23.3) or fertilizer broadcaster can be used for fertilizer application.

Figure 23.3: Tractor Operated Fertilizer Applicator-cum-Marker.

Fertilizer drill-cum-line marker is a machine which ensures uniform placement of fertilizer in furrows and marks the line impressions at 600 or 650 mm fixed row spacing for tuber placement. With this machine fertilizer rate can be adjusted between 50-300 kg/ha with a field capacity of 4.0 ha/day. However, marking in small fields in hilly regions is also done with the use of hand marking tools (Figure 23.4). These tools are manually pool type can be used for marking of rows before placement of tuber on the prepared field.

Figure 23.4: Tractor Operated Fertilizer Applicator-cum-Marker.

In case, fertilizer unit is attached with the planter than the use of fertilizer drill can be avoided. In potato crop, fertilizer should be placed 5-7 cm below the seed tubers and should not come in direct contact with the seed. Band placement helps in improving efficient utilization of fertilizer which results in improved production. Generally, fertilizers are hygroscopic in nature, hence it is recommended to properly clean the applicator after use.

4. Seed Tuber Planting

About 80-85 thousand seed potato has to be planted per hectare for good yield of the crop. Uniform placement of tubers in prepared soil bed at specified spacing and depth is also important for better emergence as well as production of the crop. Uniform plant emergence can be achieved by planting of seed material with tractor operated potato planters (Figure 23.5). Function of the planter is to open a furrow, meter the seed tubers with a suitable metering mechanism, place the tubers in to the furrow and cover them by a soil layer of about 6-8 cm. These machines maintain uniformity of planting in terms of row to row (60 cm), plant to plant spacing and depth (6-8 cm) of planting. Various types of machinery are available in India for planting of potato *viz.*, Rotating drum type; Revolving magazine type; Belt and cup type and Picker wheel type

First two are semi-automatic type planters in which one person per row puts seed tubers in to the seed metering cells of the machines. Function of these machines is to open a furrows drop the seeds in to the soil through seed delivery tubes. Seed tubers are dropped in lines and a ridge of soil mass is formed over the tubers. Automatic planters are becoming popular nowadays due to their distinct advantages over the other types. In this type of machine labour is required only for filling of hopper. All other activities like lifting individual tubers, dropping them into rows and then covering with soil mass are carried out by the machine itself.

Figure 23.5: Potato Planting with Semi-automatic and Automatic Planters.

However, the seed tuber can also be planted manually by marking of rows on the marked field. After placement of the tubers, these can be covered with soil using and hand ridger (Figure 23.6). These hand tools are especially designed for hilly regions and small land holding farmers. A potato specific CPRI-spade has been designed by ICAR-CPRI for weeding, earthing up and harvesting operation in potato crop. This tool is light in weight, handle less amount of soil mass which results in less drudgery to the farm workers.

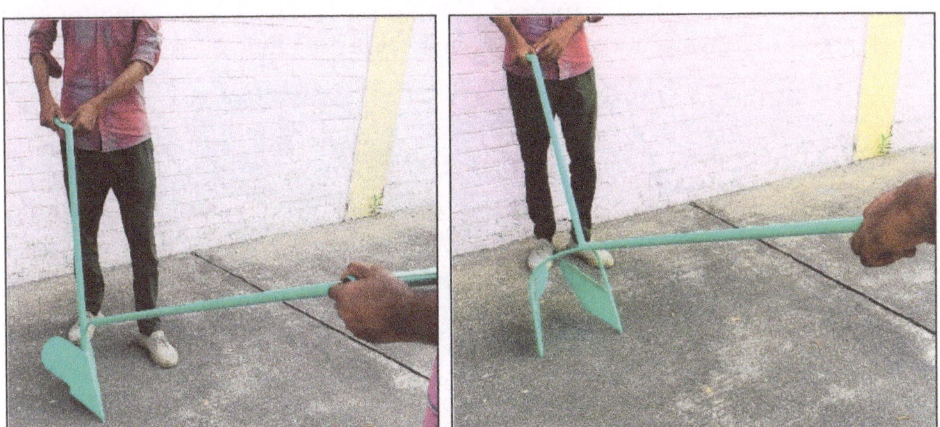

Figure 23.6: Ridger for Potato.

5. Weeding, Top Dressing and Earthing

Mechanical and chemical weed control methods are used in most of the crops. Potato being a tuber crop, is highly responsive to inter-row-cultivation. Three row tractor operated inter-row-cultivator is operated in 22-25 day crop when plant height is 8-10 cm. After operating this weeding machine, three or

Figure 23.7: CPRI-spade (*Patented Design-305966*).

five row ridger (Figure 23.8) is used for earthing up and remaking the ridges. Ridger accumulates sufficient soil mass around the plants. For light soils and low weed intensity areas another multipurpose machine is available which perform weeding, furrow opening, fertilizer application and re-earthing operation simultaneously. Additionally, some innovative hand tools are also developed by ICAR-CPRI for weeding operation in potato and other wide-row crops. This package of hand tools is especially designed for women farm workers which can be used for intercultural operations in potato and other crops. Package of tools includes L-hoe, T-hoe, trapezoidal blade weeder and TS- hoe, (Figure 23.9) which are light in weight, small in size and wide area in less time.

Figure 23.8: Interculturing and Fertilizer Top Dressing before Earthing Operations in Potato Crop.

Figure 23.9: CPRI-Hand Tools for Weeding Operations (*Patented designs*).

6. Plant Protection

Potato crop needs to be protected from different disease carriers, insects and pests by applying respective chemical formulations. Commonly the liquid chemicals are sprayed on crop canopy to protect the crop. Many times delay in chemical spray can cause huge crop losses. Particularly, in case of late blight disease. Therefore, timely and uniform application of plant protection chemicals is important, which is possible with high capacity tractor operated sprayers. Multinozzle boom sprayer, having 12-16 nozzles, is also useful for potato crop. Tractor operated gun sprayers is very popular in plane areas due to their high

Figure 23.10: A View of Spraying Operation with Power Sprayers.

capacity ease of operation and maintenance (Figure 23.10). However, for small holdings, manually or battery operated knapsack sprayer can be used which has field capacity of 0.5 ha/hr.

Immediately following this introduction is a discussion of the various types of UAVs which is done in Section 2. This is followed by the applications of UAVs in crop farming and in livestock in sections 3 and 4 respectively. Advantages of UAVs and corresponding challenges are discussed in Section 5, the chapter ends with the 6th and concluding section.

7. Potato Digging

Before starting harvesting operation haulms are removed manually sickles or with chemical methods. Potato digging is a cumbersome process and involves a lot of manpower. About 600-700 man-hours/ha are required for manual digging of 300-400 quintals of potato from huge soil mass of around 10,000 quintals (one kg potato form 30 kg of soil). Traditionally, in manual harvesting spade or *khurpa* is used which results up to 10 percent tuber cut or bruise which is a huge loss to the farmers and nation. Some efficient hand tools (Figure 23.11)

Figure 23.11: CPRI Hand Tools for Potato Digging.

are developed by CPRI for harvesting of potato which can be used in hilly region or by small land holding farmer. These tools handle less amount of soil as compared to spade hence, less draft force is experienced by the farm workers. Animal drawn plow is another tool for potato digging. This method is faster as compared to manual digging but in case of large scale farming with this method harvesting is delayed and fields become dry. Tractor operated diggers are fast, economical and cause least damage to the produce (Figure 23.12).

Figure 23.12: Animal Drawn Potato Digger.

Digger elevator is an efficient machine which exposes 90-95 percent tubers in optimum field conditions. This machine has to be maintained properly as there are many parts which move in soil and cause more wear and tear. Multipurpose passive blade potato digger (Figure 23.13) is another less expensive equipment. It exposes about 80-85 percent potatoes and is easy to maintain. It can be used in early crop (60 days without cutting haulms) and also can successfully work in dry field conditions.

Figure 23.13: Tractor Operated Potato Digger Elevator and Low Cost Blade Type Digger.

8. Potato Combine Harvester

A two row potato combine harvesters has been developed at CRPI to perform potato digging and picking operation combinely (Figure 23.14). Some other firms have also developed single and two row potato combine harvesters in recent years. Functions of the combine harvester machine are to dig the potato tubers from 15-20 cm deep soil, separate the tubers from soil clods, convey the tubers to the sorting platform and collect them in to a hopper. Further it can be unloaded in to a transporting trolley or on a heap. Four workers have to be employed with these machines to sort the rotten tubers or soil clods passing over the sorting table. These machines are usually powered by a 50 to 60 hp tractor.

Figure 23.14: Two Row Potato Combine Harvester.

9. Grading

Graded, sorted and properly packed potatoes always fetch good price in the market. Manual grading is not uniform and it requires a lot of time and energy which bring down the overall returns. Various manual as well as power operated graders have been developed for size grading of the potato tubers. Rubber belts having different size of round or square shaped perforation are generally used in the grading system (Figure 23.15). Lot of labour have to be still required in these high capacity graders for feeding sorting and packaging of the potatoes. Hand cranking type manual graders are also available for small land holding farmers. This machine is capable to grade the tubers in three sizes.

10. Drone Technology in Potato

Drone is also known as Unmanned Aerial Vehicle (UAV). It is a type of aircraft that operates without a human pilot on-board. The drone used for

Figure 23.15: Rubber Belt Screen Type Potato Sorting and Grading Machine.

agricultural activities is known as agriculture drone. Drones are designed to carry the sensors that can provide real-time information about the crop status, so that the crop management can be done efficiently and precisely. There is a huge potential for the application of drones in Agriculture. Use of drones is very advantageous in the case of pesticide spraying, substituting time consuming, labor-intensive and hazardous conventional methods.

Figure 23.16: Spraying of Pesticides in Potato Crop using Drone.

Crop spraying is much safer and cost-effective by its autonomous and pre-programmed run on specific schedules and routes. Drones are also programmed to self-adjust its path of run, altitude and speed GPS and other sensors. Trails have been carried out at CPRI stations for application of chemicals with the use of drone. From these experiments, it was observed that a drone can cover one hectare of area in 15-20 minutes in spraying operation. No crop damage found during spraying of chemicals on potato crop.

10.1 Advantages of Drones

- ☆ High speed of operation result in high field capacity
- ☆ No crop damage
- ☆ Greater penetration in the crop canopy due to the down thrust of drone rotors
- ☆ Spraying can be possible in tall crops like maize, sugarcane, *etc.*
- ☆ No human exposure to harmful chemicals
- ☆ Auto navigation system facilitates accurate movement and precise application.
- ☆ Spraying operation can be performed in irrigated or water logged field where the movement of tractor, machinery or human is not possible.

Further information relating to water access, changing climate, wind, soil quality, the presence of weeds and insects, variable growing seasons, and more can all be monitored with UAVs. Artificial intelligence and machine vision can also be combined with high resolution multispectral imaging technologies to identify soil conditions, plant health and crop yield prediction. Farmers can use the information by taking preventive action and management practices to diseases management and application of inputs to the crops.

11. Summary

Potato is a crop with huge output per unit area. For production, handling and processing of potato it takes lot of time and energy if all the operations are completed manually or with the help of animal drawn equipment. A number of efficient farm machinery and equipment are available in India to perform timely field operations like seed bed preparation, fertilizer application, planting, weeding, earthing, digging, grading and seed treatment. Therefore, in this era of competitive agriculture, for efficient use of expensive inputs like seed and fertilizer, farmers should opt for precise application of the farm inputs through drone and mechanical solutions like potato planters, fertilizer applicators, welders, diggers, graders and other machines. Of course some of these machines are expensive and are not in the reach of small farmers. In that case group of farmers can purchase the machine/drones or an individual farmer can have it and can go for custom hiring for more returns. Similarly, cooperatives can be a useful idea for having drone's/farm machinery/processing machineries at village or cluster levels. Promotion of small scale potato machines, tools, implements, drones and post-harvest machines for farm mechanization holds a great potential to enhance the productivity of the crops with reduced human drudgery.

Points to be taken care before purchasing of machines/implements/tools

- ☆ Size of machine should be in accordance of the farm size, availability of power source and topography of the fields.

- Weight and draft force required by machine should be matched with the available power source.
- Ease of operation and less maintenance.
- Machine should be available in local market.
- Spare parts of the machine should be available in the nearby market.
- Availability of local mechanic for repair.

Chapter 24

Agro-techniques for Quality Seed Potato Production

Ashwani Kumar Sharma

ICAR-CPRI, Kufri-Fagu Unit, Shimla, Himachal Pradesh

1. Introduction

Potato being a vegetatively propagated crop, the perpetuation of viral, mycoplasmal and other soil borne pathogens brings down the quality of seed stock and results in poor yields in subsequent multiplication of such seeds. Keeping in view the problems of hills grown seed and the limited area available for healthy seed production in hills, efforts were made to produce healthy seed in plains. Surveys were conducted in important potato growing areas of the country by CPRI from 1952 onwards. During the survey a remarkable consistency in the aphid build up was found. In north-western and central Indo-Gangetic plains, the aphid population remained very low during October to December months. Taking full advantage of very low aphid population coupled with the use of improved cultural practices; it was made possible to produce healthy seed in plains also and the technique is known as "Seed Plot Technique". During the last about six decades, seed production has shown spectacular progress in technology development and production strategies. The improvement in quality and quantity is commendable. At present 94 percent of the total quantity of seed is produced in eight states of sub-tropical plains and remaining 6 percent in hilly states. India is the only leading country in Asia which has developed scientific seed production technology for sub-tropics by taking advantage of low aphid period and absence of soil borne diseases and insect-pest.

2. Basic Requirement for Seed Production

Basic requirements like suitable areas, soils, fertilizers, manures and suitable varieties *etc.* play a vital role in a successful seed production programme. A quality seed can be produced in areas and fields which are free from serious soil borne pathogens and pests. High hills above 7000′ msl are suitable for nucleus and breeders seed potato production. Indo-Gangetic plains of Punjab, Haryana, North-Western parts of Uttar Pradesh, Madhya Pradesh and Bihar are suitable for seed production under low aphid period from October to March and are the primary source of high quality seed. Parts of North-western Uttar Pradesh, Madhya Pradesh, Bihar and entire West Bengal, Odisha, Rajasthan and Gujarat are only secondary source of seed production due to brown rot and aphids. There are two seasons for seed production *i.e.* summer and autumn. In summer, seed is produced under long days and rain fed condition in high hills from January to October whereas, in autumn, the seed is produced in sub-tropical plains under short days and irrigated conditions from October to March. The optimum temperature for foliar growth is 18- 22 °C and 10-16 °C is good for tuberization.

For the production of Quality Seed Potatoes, it is essential that all the components or practices as defined in Seed Plot Technique are strictly adhered to beside others. The Seed Plot Technique and its various components are thus described as under:

3. Seed Plot Technique (SPT)

The technique, of growing seed potato crop during low aphid period with healthy seed from October to January/February coupled with the use of insecticides, roguing and dehaulming from the last week of December up to second week of January, was developed by the CPRI, Shimla in 1959 and is called as "Seed Plot Technique". Quality seed production was possible under this technique in sub- tropical plains by advancing the date of planting from December end to first week of October.

3.1 Characteristic Features of SPT

- ☆ There should be a low aphid or aphid free period of 75 days after the planting of crop.
- ☆ Adopt 2-3 years crop rotation to take care of soil borne pathogens.
- ☆ Seed crop should be grown in isolation of 25 meters from ware crop.
- ☆ Seed should be procured from reliable sources and must be free from viruses and soil borne pathogens. Cold stored seed of right physiological age should be used.
- ☆ Crop should be planted 10th October in Punjab, 25th October in Haryana, Rajasthan, Western Uttar Pradesh and 5th November in Eastern Uttar Pradesh, Bihar, West Bengal, Odisha.

- ☆ Systemic insecticide such as Thimet 10G/Thiamethoxam to be applied in split doses of 10 kg/ha at the time of planting and earthing up against sucking insects.
- ☆ Pre-sprouted seed with multiple sprouts may be used, which ensures quick, uniform and early germination.
- ☆ Inspect the seed crop thrice at 45, 60 and 75 days during growing season to remove off type of diseased plants.
- ☆ Spray the crop with mancozeb @ 2.5 kg/ha at 10 days' intervals from 3rd week of November and spray Curzate M-8 @ 3.0 kg/ha or Ridomil Gold @0.25 percent as and when late blight is observed.
- ☆ Spray the crop with Imidacloprid 17.8 percent SL (0.4L/ha) or Thiamethoxam alternatively at 15 days' interval from 1st week of December to control the insect vectors.
- ☆ Irrigation should be with hold in 3rd week of December *i.e.* 7-10 days before haulms killing in north-western plains and 1st week of January in north-eastern plains.
- ☆ Haulms killing to be done in the end of December in Punjab, Haryana, Western Uttar Pradesh; by 10th January in Central Uttar Pradesh, Madhya Pradesh and by 15th January in eastern Uttar Pradesh, Bihar, West Bengal and Odisha.
- ☆ Harvest the crop, 15-20 days after haulms killing when the fields are in workable condition and tuber skin is hardened. Cure the crop in heaps in a cool shady place for about 15-20 days.
- ☆ Treat the produce with commercial grade Boric Acid 3 percent solution for 20 minutes to prevent surface borne pathogens. Dry in shade and fill in bags, sealed and labeled properly and cold store.

3.2 Major Components of SPT

3.2.1 Selection of Field

For raising a potato seed crop, soil should be free from perennial weeds and soil-borne pathogens such as scabs, brown rot, wart and black scurf and pests like cut worm (*Agrotis ipsilon*). For minimizing the perpetuation of soil-borne diseases, it is desirable to adopt 2-3 years crop rotations preferably with cereals. As far as possible, the seed crop should be grown in a field where potato has not been grown for last two years. A well-drained, light textured sandy loam soil with neutral to slightly alkaline soil pH is preferred.

3.2.2 Hot Weather Cultivation

In the Indo- Gangetic plains, opening the soil by deep tillage and keeping it exposed to extreme high temperatures during hot summers reduces the incidence of soil-borne diseases and controls weeds and cutworms. Deep ploughing the field in April end and keeping it open in May and June with one or two more ploughings serve the purpose of hot weather cultivation.

3.2.3 Green Manuring

Raising and ploughing under the 45-55 days old green manuring crops of sunhemp (*Crotalaria juncea*) or dhaincha (*Sesbania aculeate*) during rainy season at least one month before potato planting in the plains and growing of French bean or other leguminous crops in the hills as green manure is beneficial in reducing pest and disease incidence. Green manuring improves the soil fertility and water holding capacity to benefit growth and yield of potato.

3.2.4 Tillage

Potato seed production demands minimal mechanical interference in the standing crop to check spread of viruses through physical contact. Therefore, field should be cleaned of stubble and perennial weeds by adequate tillage operation before planting. If the field is relatively free from weeds, minimum tillage can be practiced to save on the fuel, time and money to reduce cost of cultivation. Minimum tillage combined with chemical weed control is best suited for seed production as it ensures minimum interference in standing crop. 2-3 ploughings followed by planking's make the soil loose and friable suited for potato planting.

3.2.5 Isolation of Seed Plots

The seed crop should be separated from crops meant for ware purpose by a distance of at least 5 meters to avoid mixture and spread of viral diseases. Isolation is also required between different varieties of the seed crop.

3.2.6 Varieties

Table 24.1 **Zone-specific varieties** and crop period for seed production are as under:

Zones	*Crop Period*	*Varieties*
North-Western hills	Summer (15th April–Oct.)	Kufri Jyoti, K. Himalini, K. Girdhari, K. Kanchan, K. Chandramukhi,
North-Western plains	Autumn (1st week Oct. – Feb./March)	Kufri Jyoti, K. Chandramukhi, K. Badshah, K. Pukhraj, K. Surya, K. Khyati, K. Satluj
North Central plains (MP)	Autumn (3rd week Oct.– Feb./March)	Kufri Jyoti, K. Chandramukhi, K. Sindhuri, K. Pukhraj, K. Lauvkar, K. Chipsona I, III, IV, K. Surya, K. Lalima
North Central plains (UP)	Autumn (2nd week Oct – Feb./March)	K. Bahar, K. Sadabahar, K. Anand, K. Surya, K. Chipsona I, III, IV, K. Frysona, K. Gaurav, K. Badshah,
North-Eastern plains	Autumn (4th week Oct.– Feb./March)	K. Jyoti, K. Chandramukhi, K. Kanchan, K. Sindhuri, K. Lalima, K. Arun, K. Ashoka, K. Pukhraj, K. Surya, K. Chipsona I, III, IV

3.2.7 Seed Source

For on farm multiplication of quality seed by the farmers, it is essential to obtain healthy, disease free, true-to-the-type and treated seed from reliable source, preferably from a government agency. The foundation or certified seed tubers should be used to start with and stocks should be replaced every 3-4 years.

4. Pre-sprouting

The pre-sprouting treatment of the seed tubers before planting ensures multiple, stout and healthy sprouts, which helps in quick emergence, uniform stand and early maturity of the crop. It also improves yield, number of tubers and proportion of seed-sized tubers in the produce. For pre-sprouting, withdraw the seed tubers from cold store 10 days before planting. Keep it in pre-cooling chamber of cold store for 24 hours. Whereas, in the hills, take out the tubers from country stores during first fortnight of March. Spread the seed tuber in thin layer preferably on cemented or pucca floor in a ventilated room under diffused light. If tubers are already over-sprouted due to malfunctioning of cold store or any other reason remove such sprouts. It helps in checking apical dominance. Sort out the blind, hairy sprouted, rotten and diseased tubers. Transfer the well-sprouted tubers in the trays to the field for planting. While transferring the sprouted tubers to the trays and at planting, care should be taken to avoid damage to the sprouts.

5. Planting Method

Use of cut tubers is prohibited and only whole tubers are used for seed crop. In the plains inter-row spacing for planting seed crop manually, bullock and tractor drawn implements may be kept at 40-45, 50-55 and 60 cm, respectively. Planting of seed-sized tubers of 35-50 g at 60 cm inter and 20cm intra-row spacing is best for seed potato production. Plant the large size tubers by increasing the intra-row spacing from 20 to 30 cm depending upon the size of seed tubers. Small-sized tubers are planted at reduced intra-row spacing. The approximate intra-row spacing for <25, 25-50, 50-100 and >100 g seed size is 15, 20, 25 and 30 cm with inter-row spacing of 60cm, respectively. Seed tubers are placed at a depth of 5-7 cm from top of the ridges made manually or by tractor drawn implements.

6. Manure and Fertilizers

Application of well rotten farmyard manure (FYM) @ 20-25 t/ha in absence of green manuring is beneficial for seed crop. It improves soil physical condition, soil fertility and water holding capacity of the soil. The FYM should be incorporated into the soil 20-25 days before planting. Nitrogen needs of seed crop are 25-30 percent lower than the ware crop. Excess N increases the yield of undesirable extra-large size tubers and produces dark green foliage masking the symptoms of viral and mycoplasmas diseases and detection of infected plants becomes difficult during roguing. In the plains, the crop requires a basal dose of 75kg N, 60-80kg P2O5 and 100-120 kg K2O/ha at planting, whereas, 100 Kg each of N, P2O5 and K2O is the recommended basal dose for the hills. Top dressing of 75 kg N/ha at earthing up in plains and 20 kg N/ha in the hills given through urea is adequate. The basal dose of N can also be applied through urea by incorporating it into the soil at least 48 hours before planting during the field preparation. It comes over the constraints in using costly nitrogenous fertilizers of ammonium sulphate (AS) and calcium ammonium nitrate (CAN).

7. Intercultural Operations and Weed Control

The objectives of intercultural operations in potato are weed control, earthing up for firming up the ridges to prevent exposure of growing tubers and for application of split dose of Nitrogen and Thimet insecticide. However, operations involving human, animal and implement movement in standing seed crop should be minimal to prevent transmission of plant viruses through physical contact. It would be better if these operations are completed by 20-25 days after planting when plants attain the height of about 10- 15 cm and foliage cover is still small. The split dose of Nitrogen and Thimet should be applied at hoeing and earthing up about 5cm away from the plants. The hoeing must not be delayed beyond 30 days in plains and 45 days in the hills after planting to avoid damage to the plant roots, foliage and stolon's, which may adversely affect the number of tubers resulting in reduction in yield. Pre-emergence weedicides like Metribuzin @ 0.75 kg/ha, Oxyfluorfen @ 0.15 kg/ha, Linuron @ 0.5 kg/ha, Alachlor @ 1.5kg/ha and Isoproturon @ 0.75 kg/ha applied 2-3 days after planting are effective. Pre-emergence herbicides are most effective when applied in moist soil. Therefore, if soil is dry apply herbicides after first irrigation as soon as it is possible to enter the field. In case pre-emergence herbicides are not used, spray paraquat @ 0.5 kg/ha at about 5-10 percent plant emergence of potato provided sufficient weeds have appeared, as it kills only emerged weeds. Chemical weed control eliminates manual hoeing. Thus, full earthing up at planting combined with chemical weed control effectively minimizes undesirable physical intervention in standing seed crop.

8. Water Management

Pre-sowing irrigation/rains before land preparation is beneficial for early and uniform emergence. If pre-sowing irrigation is omitted at the time of field preparation, irrigate the crop immediately after planting. First irrigation following planting should be light to minimize damage to the newly formed ridges. Heavy irrigation before emergence leads to anaerobic conditions resulting in rottage of seed tubers, gappy emergence and reduced tuber yield. Second irrigation is given a week after first irrigation. Subsequently, irrigate the crop at 7-10 days' interval depending upon the soil and weather conditions. Avoid flooding over the ridges and irrigate only up to 2/3 height of the ridges.

As far as possible, irrigate during morning and evening hours. In a normal seed crop, 6-8 irrigations are required. Light and frequent irrigations are much better than heavy irrigations given less frequently. Excess moisture makes lenticels prominent due to rupturing and seed tuber quality is impaired. It also promotes certain diseases. Stop irrigation at about 10 days before dehaulming in light soil and 15 days in heavy soils. Moisture stress restricts re-growths after dehaulming and hastens curing of peel of seed tubers.

9. Plant Protection

Additional plant protection measures against aphid and other vector transmitting viral diseases are required in the seed crop. Application of granular systemic insecticide, Thimet 10G @ 20 kg/ha at the time of earthing up takes care of jassids, leafhoppers and white flies at early stages of growth up to 30-35 days. After appearance of aphids, two sprays of imidacloprid 17.8 percent S.L. @ 0.003 percent may be repeated at an interval of 12-15 days depending on duration of the crop and level of infestation. Drenching of the ridges with chlorpyriphos 20EC @ 2.5 litre/ha effectively controls cutworms attack during the early stages of the crop. In white grub prone areas, Chlorpyriphos 20EC @ 2.5 litre/ha should be applied either after mixing with sand or can be sprayed on the ridges before the final earthing up.

For control of early and late blight, one prophylactic spray of Mancozeb @ 0.2 percent is given. It may be repeated at an interval of 7-14 days depending upon the weather condition. It will also take care of other foliar diseases like phoma blight, *etc.* In case of persistent and severe attack of late blight, spray of systemic fungicides like curzate M-8 @ 0.3 percent or Ridomil Gold @ 0.25 percent may be given and repeat the spray after 7-10 days if required.

10. Inspection and Rouging

Inspection of the seed crop to remove or rogue out the off type and diseased plants showing mosaic, mottling, veinal necrosis, crinkling and rolling of leaves, marginal flavescence and purple top roll symptoms is essential. The first roguing is done 35-40 days after planting and the second is done 20 days after the first roguing while the third roguing is done 5-7 days before dehaulming. At each roguing, make it sure to remove the tuber and tuberlets of rogued plants.

11. Dehaulming

Removal of haulms of the seed crop is essential by 5-25 January in the plains before aphids (*M. persicae* and others) population reaches the critical level of 20 aphids/100 compound leaves. Cut the haulms by 15th August in the hills. Dehaulming is done by manually cutting with the sickle close to the ground or by spraying non-selective herbicide Paraquat @ 0.5kg/ha. In the plains, if haulms are removed manually, it is preferred to keep the vines/haulms on the ridges to protect exposed tubers from high temperature and direct sunlight.

Re-growths of leaves if any are also cut after a week of dehaulming, because the tender and succulent leaves are more attractive to aphid vector.

12. Harvesting and Curing

Start digging 10-15 days after dehaulming when peel is firm to withstand handling operations. In the Indo-Gangetic plains potato digging beyond February promotes rottage due to soft rot and charcoal rot. Digging may be done either manually by spades or by mechanical potato digger. Exercise care to avoid bruising of tubers during harvesting, handling and transportation. After harvesting, keep potato tubers in heaps on raised beds for about 15 days for hardening of peel and shedding of adhered soil from tubers surface. Heaps of about 1.5 m high and 3.5 m broad at the base and variable length as needed are convenient, effective and economical. Cover the heaps with paddy straw or tarpaulins.

13. Grading and Seed Treatment

Proper size grading of tubers in the produce of seed crop is beneficial. It also helps in controlling the seed rate effectively by adjustment of spacing according to tuber size. Before grading, surface dry the produce and sorts out all cut, cracked and rotten tubers. Seed tubers are usually graded into four grades, *viz.*, extra small (<25 g), small (25-50 g), medium (50-100 g) and large (>100 g). Seed tubers should be dip-treated with boric acid (3 per cent) solution against tuber borne diseases of common scab and black scurf for 25-30 minutes. The fresh boric acid solution can be used for 20 times for treatment, provided tubers were washed clean with water. Dip treatment with organo-mercurial compound, Emisan 6 @ 0.25 per cent for 20 min is also effective, but is considered hazardous and may be avoided. The treated tubers should be thoroughly dried in shade before bagging and storage.

14. Packing and Storage

Keep the seed tubers in 50 kg gunny bags and store in cold store (4 °C) in the plains latest by the end of February. The produce in the high hills can be stored in country stores by the end of February. The treated seed bags should be properly sealed and labeled "Poisonous" to avoid human consumption mistakenly.

Chapter 25

Functional Genomics for Improving Nitrogen Use Efficiency in Potato

Jagesh K. Tiwari, Tanuja Buckseth, Rasna Zinta, Kanika Thakur, R.K. Singh, Devendra Kumar, V.K. Dua and V. Kumar

ICAR-Central Potato Research Institute, Shimla, Himachal Pradesh

1. Introduction

Nitrogen is the most important nutrient for plant growth and development including potato. Potato is the third most important food crop after rice and wheat in terms of human consumption. Potato is a resource intensive crop and requires high N fertilizer (150-240 Kg N/ha) to produce tuber yield (30-50 t/ ha) in India. Despite the high cost of N fertilizer, potato crop uptakes nearly 50 percent of the total N applied and excess lost in the environment (Trehan 2009). Since Indian soils are generally deficient in organic matter thus unable to meet crop N demand, application of inorganic N fertilizers is inevitable to meet the N requirements of potato. Potato being is a shallow rooted crop and mostly irrigated cultivations are practiced on sandy-loam soils with excessive application of N fertilizers increases the chance of nitrate leaching, contamination to the groundwater and high cost of cultivation. Though, field based approaches have been applied in potato for improving NUE for successful tuber production (Vos 2009). However, improving N use efficiency of plant is one of the key options to minimize N losses, save the cost of production and improve the environmental quality to achieve sustainable crop yield with less (or without) N fertilizer.

Substantial genetic variation has been observed for NUE-associated traits in cultivated and wild species of potato. Most studies focused on agronomic managements for improving NUE in potato, except a few gene expression of nitrate transporter, ammonium transporter, nitrate reductase, nitrite reductase and asparagine synthetase. Recently, we have developed an aeroponics system for seed potato production till full crop season (Buckseth *et al.*, 2016) and highlighted its importance to study N biology applying integrated genomics, physiology and breeding approaches (Tiwari *et al.*, 2018). Moreover, knowledge about precision phenotyping in aeroponics, traits associated with NUE, root traits, genes involved in N metabolism and other parameters are very important to dissect the underlying N mechanism to improve NUE in potato. Hence, we highlight genomics and agro-physiological approaches to be used to enhance NUE in potato in near future.

2. Nitrogen Use Efficiency (NUE)

Nitrate is the predominant form of nitrogen in soil. Though, potato plant uptakes mostly nitrate from soil, ammonium is also important for potato growth and development. Stolon initiation and its growth towards tuber formation are critical stages in potato, which are influenced by N, day length, temperature and other growth factors. NUE may be defined as tuber yield per unit of N supply (fertilizer and residual soil N). A few sample NUE parameters are given as below:

Table 25.1: A Few NUE Parameters in Potato

NUE Parameters	*Description*
Agronomic use efficiency (AUE)	Kg tubers produced/kg N supply
Nitrogen use efficiency (NUE)	Plant dry matter accumulation/crop N supply = (NUpE) x (NUtE)
N uptake efficiency (NUpE)	Plant N accumulation/crop N supply
N utilization efficiency (NUtE)	Plant dry matter accumulation/plant N accumulation
Harvest index (HI)	Tuber dry matter accumulation/plant dry matter accumulation
N harvest index (NHI)	Tuber N accumulation/plant N accumulation

3. Approaches of Improving NUE

Nitrogen use efficiency of plant can be achieved through application of a few approaches like increasing nitrogen uptake use efficiency, increasing nitrogen utilization efficiency, and increasing remobilization efficiency.

4. Agro-physiological and NUE Associated Traits

NUE parameters need to be determined like plant biomass, root system architecture, chlorophyll content, photosynthesis rate, dry matter, plant N content, and NUE-associated traits (NUE, NUpE, NUtE, NHI, HI *etc.*) *etc.* In N metabolism, various enzymatic activities, sugar, amino acids, protein and other metabolites are also important traits in potato to study NUE. Measurement of a large number of traits is required for improving NUE in potato as given below.

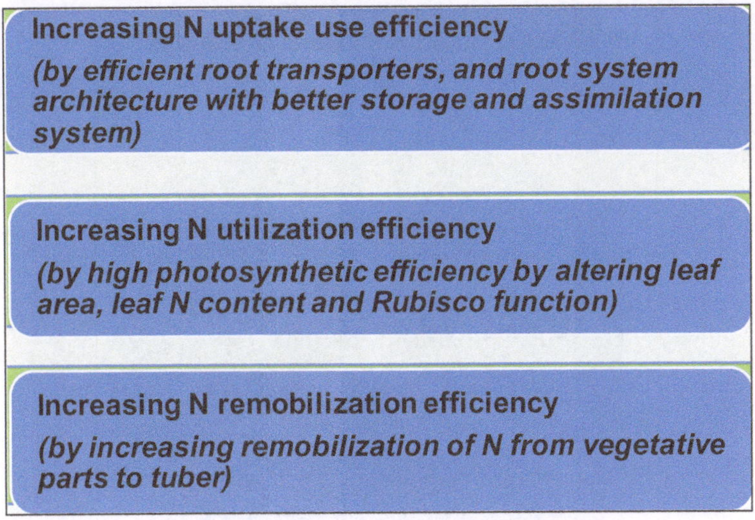

Figure 25.1: Approaches for Improving Plant NUE in Potato.

5. Root System Architecture

Measure of root system architecture is important to determine plant performance under different N regimes, which help in N uptake and later utilization by plant. The software like WinRhizo is widely used for root system architecture study in plants.

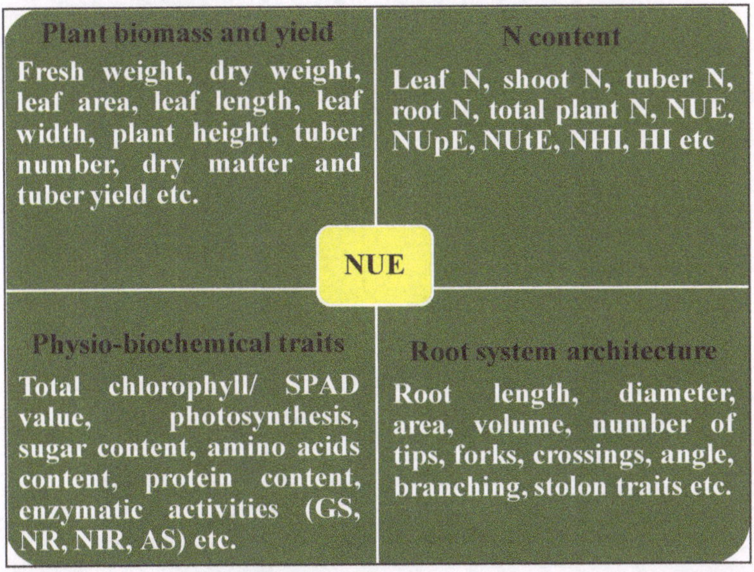

Figure 25.2. Agro-Physio-Biochemical Traits to Study NUE in Potato.

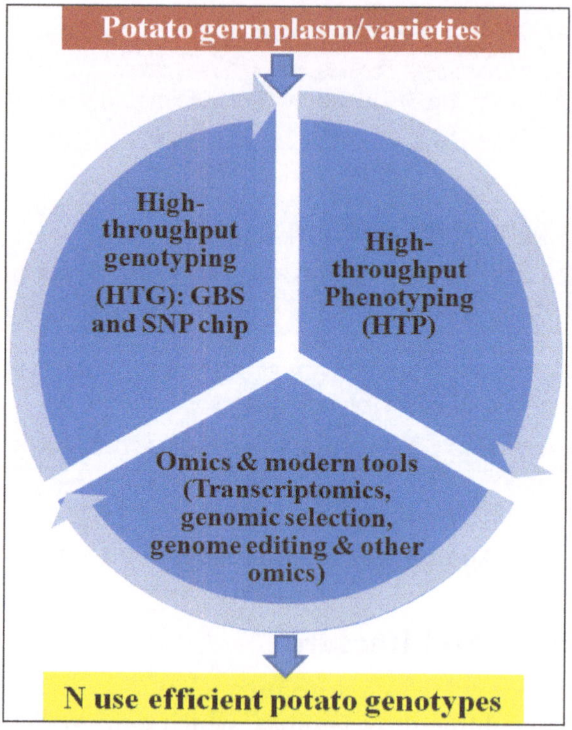

Figure 25.3: Genomics and Phenomics Approaches to Study NUE.

Figure 25.4: Phenotype of Potato Plants under Low and High N in Aeroponics.

6. Genomics Approaches

Knowledge about genes involved in N metabolism is an essential requirement for improving plant NUE via genetic interventions applying modern genomics tools. Target genes for transgenic manipulation of N uptake and metabolism include nitrate transporter (NRT) or ammonium transporter (AMT), nitrate reductase (NR), nitrite reductase (NiR), glutamine synthetase (GS), glutamate dehydrogenase (GDH), aminotransferases (alanine and asparagine) and signalling molecules/regulatory elements (transcription factors and miRNAs *etc.*). Transcriptomics approaches like RNA sequencing is potential technology to identify genes involved in N stress and N sufficient conditions, which could be further manipulated to develop N use efficient genotypes. Further validation of genes by real-time PCR or other technology could be attempted. Besides, genome-wise association mapping (GWAS) is also important to discover genes/QTL in a large population and Genomic Selection by GBS and/or SNP chip methods are cutting edge technologies to unravel NUE in potato.

7. High-throughput Phenotyping

High-throughput phenotyping platform is a non-destructive method of plant phenotyping based on sensor, chlorophyll fluorescence, thermal and hyperspectral imaging (*e.g.* LemnaTech GmbH). It has capacity to measure detailed growth parameters of plants. Since, underground part root and stolon/ tuber are most economically important in potato, there is need to develop HTP platform to measure them.

8. Genome Editing

Genome editing is the targeted alteration in a genome, which creates new allelic variation. The CRISPR/Cas9 technology is the most widely used today for genome editing. The CRISPR/Cas9 is a RNA-guided method to target DNA sequence, which is used widely due to its simplicity, multiplexing capability, cost-effectiveness, and high efficiency. Genome editing is an important tool to create new variants with desirable gene combination. Unlike GM technology which creates stable integration in line by cisgene or transgene. In potato TALENs and CRISPR/Cas9 has been used to some extent for traits like cold induced sweetening, glycoalkaloids, acetochalactate synthase, granule bound starch synthase *etc.* Thus. there is potential to apply this technology for improving NUE in potato.

Chapter 26

Good Agricultural Practices (GAP) for Potato Cultivation

Sanjay Rawal and Pooja Mankar

ICAR-CPRI, Regional Station, Modipuram, Meerut, Uttar Pradesh

1. Introduction

World market is witnessing an increasing demand for healthy and hygienic food. Local as well as global consumers are getting serious about healthy, hygienic food and cautious about chemical residue in food and its consequences on health. Also, agriculture production is facing complex problems like climate change, increase in biotic and abiotic stresses, deterioration of natural resources, rising cost of cultivation *etc.* which has put forward many challenges to produce quality food. This changing scenario has led to the genesis of Good Agricultural Practices (GAP) with the objective to develop common standards and procedures for food production and marketing. The concept of GAP is based on fundamental principles of safe and quality food production, conservation of environment and natural resources, sustainable farming and economic and social sustainability. Food and Agriculture Organization (FAO) describe GAP as collection of principles to apply for on-farm production and post-production processes, resulting in safe and healthy food and non-food agricultural products, while taking into account economic, social and environmental sustainability (FAO, 2003).

2. IndGAP Certification Scheme

Quality Council of India (QCI) has launched IndGAP certification scheme in September, 2014 with the objective of production of safe and hygienic food at the Indian farms and providing the opportunity of farm certification to a local standard at low cost. IndGAP is inclusive of Global GAP standards, best practices, prevalent industry standards and related ISO standards and relevant guidelines. The IndGAP standards comprises of control points and compliance criteria as per national farming practices and regulatory requirements. The scheme is framed as per the growing need of quality and safe food in domestic and international market. It takes into account not only the large farms but also address the small and marginal farmers. The focus of IndGAP standards is to address hygienic and quality food production taking into consideration workers' health and safety, conservation of the environment and farm biodiversity, waste management and recycling *etc.* The IndGAP Basic is meant for small or marginal farmers or for those who are interested to introduce quality farm production for fresh consumption or for processing. Whereas, IndGAP Premium is targeted for large farms or farmer's group interested to seek certification for their farm produce, either for fresh consumption or for processing purpose.

2.1 Advantages of IndGAP Certification

- ☆ Enables farm produce to be globally accepted by following Global GAP norms.
- ☆ Ensures better price realization to the produce in domestic as well as international market.
- ☆ Secure and strengthen the livelihood of small and marginal farmers.
- ☆ Ensures safe, hygienic and quality food production.
- ☆ Create new market opportunities for farmers and exporters.
- ☆ Facilitates capturing new market advantages by modifying management in supply chain.
- ☆ Helps to safeguard the worker's health and safety at working places.
- ☆ Minimize wastage of natural resources.
- ☆ Creating trust among producers, suppliers, traders and consumers.

3. Scope of Certification

3.1 IndGAP Basic

This covers the basic requirement of GAP and is applicable for all farming practices in sustainable manner for maintaining quality and food safety of agriculture of agriculture produce. This standard covers control points and compliance criteria mainly for farms that are either small or marginal or the ones that aspire to introduce quality in their farms for the farm produce in fresh unprocessed form or for further processing.

3.2 IndGAP Premium

It encompasses control points and compliance criteria for any farm wanting to be assessed for global standards for farm produce in fresh unprocessed form or for further processing. The main components of this standards are all farm base module and crop-based module:

a) All farm based module
b) Crop based module
   - ☆ Fruits and Vegetables
   - ☆ Combinable crops
   - ☆ Tea
   - ☆ Green coffee

4. Key Components of IndGAP

4.1 Site Selection

- ☆ Risk assessment of new sites for vicinity to industries, brick kilns, toxic waste, heavy metals, industrial and urban effluents, river, canals all possible contamination, land reclamation, irrigation water quality and drainage *etc.* should be done. It helps in appropriate crop management.
- ☆ If irrigation is required, the farm should have access to reliable source of water.
- ☆ Risk management plan should be prepared to minimize all identified risks.
- ☆ Previous years' meteorological data should be considered for judging the suitability of the site for a particular crop production system.

4.2 Soil Conditions/Management

- ☆ Soil testing for precise nutrient application, soil amendments, site-specific nutrient.
- ☆ Physico-chemical analysis of soil, mapping of field's nutrient status to understand the suitability of the land for a particular crop and cropping system.
- ☆ Follow region specific recommended package of practices for cultivation.
- ☆ If the soil is prone to erosion, then adoption of soil conservation (cultural or mechanical) techniques is essential.

4.3 Seed Material

- ☆ Grow most suitable variety recommended by central and state agencies as per region/agro ecology.
- ☆ Adopt recommended agro-technologies for the selected variety.

- ☆ Quality of planting material *viz.*, seed source, seed health, purity, seed treatment to be critically followed.
- ☆ Keep all required records in compliance with national/state legislation and GAP standards.

4.4 Field Management

- ☆ Precision in variety specific planting for optimum seed rate, planting depth and geometry, plant population *etc.*
- ☆ Practice of crop rotation.
- ☆ Keep record of previous years cropping sequences.
- ☆ Optimum soil tilth and judicious soil preparation.

4.5 Nutrient Management

- ☆ Soil test based, variety specific and mandatorily integration of organic and inorganic sources.
- ☆ Ensure supply of micro-nutrient requirements of a crop.
- ☆ Organics, green manure and recycled material should be the first choice.
- ☆ Risk assessment of well treated/composted material is necessary for disease transmission, weed seed dispersal and heavy metal content.
- ☆ The sources of organic manures, their nutrient content must be known.
- ☆ Farmyard manure, sheep/poultry manure, oil cakes or slaughterhouse wastes must be fully decomposed prior to their application.
- ☆ Use of raw human sewage, municipal and industrial sludge is completely banned.
- ☆ Bio-fertilizer (specifically microbial) should only be applied after recommendation from technically competent source.
- ☆ Extra requirement of a crop should be met by judicious use of inorganic products only after utilization of organic sources. Their source and nutrient contents should also be known.

4.6 Water Management

- ☆ Water quality should be tested and assured for its use in irrigation, post-harvest washing, treatment *etc.*
- ☆ Region and variety specific water requirement should be known and accordingly a plan can be prepared for judicious and efficient use of water.
- ☆ Risk assessment to be done for probable microbial, chemical and physical pollutants in irrigation water.

- ☆ Implement adequate measures to prevent the flow of contaminated water into fields.
- ☆ Proper drainage, water harvesting and conservation methods should be followed at farm site.

4.7 Weed Management

- ☆ Adopt recommended, crop specific integrated weed control strategy.
- ☆ Initial flush of weeds to be effectively controlled to ensure weed free environment for crop.
- ☆ Cultural (crop rotation, hot weather cultivation, green manuring, proper seedbed preparation, precise planting and placement of manures and fertilizers, mulching *etc.*) and mechanical (manual or mechanical intercultural operations) weed control measures to be prioritized.
- ☆ At last, use of only recommended herbicides with prescribed doses applied at right crop growth stage.

4.8 Disease and Pest Management

- ☆ Prepare a comprehensive crop protection plan.
- ☆ Follow recommended integrated pest and disease management practices.
- ☆ Cultural, bio-agents and biological control measures should be preferred.
- ☆ Use of chemicals should be restricted as a last resort with technical advice and recommendations.
- ☆ Only registered plant protection products recommended for specific crop should be used.
- ☆ Observe pre-harvest intervals for specific plant protection products.
- ☆ Do not repeat the spray schedule without any recommendations.
- ☆ When chemical pesticides are used, the residue analysis of final product should be carried out through certified laboratory.

4.9 Harvest and Post-harvest Management

- ☆ Physiological maturity period of the cultivated variety should be known and harvesting to be done accordingly.
- ☆ Harvesting, handling, packaging and storage must be done as recommended.
- ☆ Storage and packing space should be covered, ventilated, maintained clean and contamination free.

- Storage area must be kept free from animals and rodents and follow proper measures for their control.
- Follow pest control procedures for storage/packing space.
- Different varieties must be stored separately with proper labels and signs to avoid their mixing.
- Method for sorting, washing, grading and drying should be well defined and equipment should be well synchronized and calibrated.
- Standards of grading, purity of produce must be strictly followed.
- Water used in washing should be of required standards and quality.
- Proper norms should be followed for packaging material, weight per packet, handling of farm produce during temporary storage and container loading.

4.10 Storage of Agro-inputs

- Fertilizer should always be stored in a separate, covered, clean, dry place and away from plant protection chemicals and farm produce.
- Similarly, agro-chemicals should be stored in separate, sound storage under cool, dry and secured condition away from farm produce.
- Storage space should be well protected from spillage, leakage and seepage to avoid surrounding contamination.
- Provide facilities for accurate measurement of chemicals.
- Mixing of chemicals can be done in storage facility.
- Spillage can be dealt with by equipping mixing area with containers of absorbent inert material like sand.
- Surplus mixture or tank washing should be disposed-off at designated fallow area in the farm.
- Empty containers of plant protection products can only be used for the same product at farm, otherwise safe mechanism is to be followed for their disposal.
- Similarly, expired chemicals should either be disposed-off by following proper procedure or securely maintained in storage.

4.11 Personnel

- Key resource person at farm (farmer and supervisor) must be familiar with all aspects of crop production.
- Workers should also have training on specific skills, hygiene and safety aspect.
- Persons responsible for plant protection measures at farm must have training for integrated pest and disease management.

- Workers must be provided with proper gears/safety cloths during application of plant protection products for their personal safety and health.
- Train the supervisor and workers to maintain hygiene during handling, sorting, washing, grading and packing of produce.
- All workers should have proper knowledge and training for their respective work such as handling and application of chemicals and fertilizers, operating farm machines and equipment, health and safety measures, wearing of protective gear, clothing and shoes *etc.*
- Minimum one worker is to be trained in first aid assistance and must be present during all the farm operations.

4.12 Equipment

- All machinery used in fertilizer and chemicals application, harvesting and post-harvest handling should be calibrated as prescribed by the manufacturers.
- Machinery and equipment should be in good condition and maintained at a regular time frame.
- Regular and cleaning of equipment parts is important.
- The material (specifically metal) coming into direct contact should be of good quality to avoid any contamination. Equipment that pose a risk of hazardous metallic contamination of the harvested crop should be avoided.
- Records of their regular maintenance and calibration should be maintained.

4.13 Workers Health, Safety and Welfare

- A farm should have a written risk assessment document for various operations which should be updated periodically.
- There should be a written policy on health, safety and hygiene. This will include risk assessment issues *viz.*, accident, emergency, hygiene *etc.*
- Adequate precautions to be taken to prevent on farm accidents of farm equipment/machinery.
- A comprehensive procedure for handling emergency situations should be developed in local language and in pictorial format at farm. It should include farm and building map, farm address, contact person responsible for worker's welfare, nearest communication point of phone, important phone numbers (ambulance, hospital, fire brigade and police), emergency exits, location of fire extinguishers, emergency cut off for electricity, gas, water supply *etc.*

- ☆ Permanent warning signs should be prepared in readable ways to indicate hazardous sites such as fuel tank, plant protection storage, workshops, electricity control panel, disposal area for chemicals, *etc.*
- ☆ Electrical installations, plant protection storage, fuel tanks and risky equipment must have adequate safety measures.
- ☆ Provide designated areas and clean, hygienic facilities for hand washing, eating, and toilets *etc.*

4.14 Record Keeping and Internal Inspection

- ☆ Maintain a minimum record of past two years or more from the date of the first/previous external inspection as advised by the certifying agency.
- ☆ Keep a record of all farm operations with exact date (ploughing, sowing, weeding *etc.*), fertilizers (stock, source, quantity applied and time of application *etc.*), irrigation scheduling, weed control schedule, plant protection product (chemicals, doses and mode of application *etc.*) with exact dates of application.
- ☆ All the IPM measures, various approaches adopted for disease and pest management, chemicals including their storage, crop stages and quantity used *etc.* should be recorded.
- ☆ Maintain a record of harvesting procedures, handling and storage, *etc.*
- ☆ Minimum one internal self-assessment per year to be taken against the required standards.

4.15 Traceability

- ☆ Each consignment of produce must be legibly marked with details of variety, month and year of harvest, name, and address of grower *etc.*, following trade practices/legal requirements.
- ☆ Likewise, there should be documented identification and traceability system of produce so that produce can be traced back to registered farm including particular field and tracked forward to the immediate customer.

4.16 Waste and Pollution Management, Recycling and Reuse

- ☆ Identify and list all possible waste products and source of pollutants.
- ☆ Prepare a comprehensive, documented strategy after review of current practices for minimum wastage during production, recycling and re-use of waste.
- ☆ Disposal procedure for non-recyclable materials should exist at the farm.

- ☆ Facility should be developed for on farm organic waste composting.
- ☆ Farm and premises should be kept free of litter and waste.
- ☆ Farm must have separate identified areas for storing/decomposing different types of litter and waste.

4.17 Environment and Conservation

- ☆ There must be a written action plan for improving habitats and to increase farm biodiversity.
- ☆ A baseline survey of flora and fauna should be done and aftereffects on agriculture should be studied to improve the action plan.
- ☆ Unproductive site (lowlands or unfertile soils) may be identified for conserving natural flora and fauna.
- ☆ Fuel efficient farm machinery and equipment should be selected to minimize the use of non-renewable energy.
- ☆ Renewable sources of energy like solar, wind *etc.*, may be integrated in farm power supply plan.
- ☆ Regular maintenance and servicing of equipment should be done for efficient energy consumption.
- ☆ Maintain a record of all the activities for external audit.

4.18 Complaints

- ☆ Complaint procedure must be in place and made available on request for GAP related issues. This should have an appropriate recorded follow up action.
- ☆ A documented recall procedure for produce should be adopted and person responsible for taking decisions must be identified.
- ☆ Recall procedure must be speedy, transparent and accurate. For this, GlobalGAP traceability and recall standards may be followed.

4.19 Subsidy for this Scheme

- ☆ Agriculture and Processed Food Export Promotion Scheme of Agriculture and Processed food products Export Development Authority (APEDA) for the Medium Term Expenditure Framework (2017-18 to 2019-20). (http://apeda.gov.in/apedawebsite/Announcements/SchemeGuidelinesMTEF27042018.pdf)
- ☆ Ministry of Agriculture for mission on Integrated Development of Horticulture has subsidize the IndGAP certification cost (https://midh.gov.in/PDF/midh(English).pdf).

5. The Potato Crop and GAP

GAP for potato production begins right from the selection of most appropriate site for cultivation, input quality check, soil and water testing, right choice of variety, following of all the recommended package of practices as per the purpose of crop utilization (table, seed or processing) *etc.* Scientific precautions during nutrient application, irrigation, weeding, intercultural operations, spraying, harvesting, grading and storage will ensure product quality. Use of safer chemicals within permissible limits and their integration into management with other alternatives will help in conserving ecology. In fact, GAP would help potato growers in reducing cost of cultivation, achieving maximum marketable tuber yields and fetch higher net farm profits due to ensured market by building trust in long run. Implementation of GAP would enable farmers to produce fine quality potatoes for organized domestic market and export, and high quality raw material for the processing industry.

6. Way Forward

This is an appropriate time to promote good agricultural practices for the production of potato crop to increase its utilization in domestic market and for export at global level. Government is also preparing policies and strategies for next generation agricultural reforms in context of eco-friendly and sustainable agriculture, and improving farm income. ICAR-CPRI has already compiled IndGAP Basic and IndGAP Premium protocols for potato production as per Quality Council of India's guidelines (Technical Bulletin No. 108). This can be downloaded from the institute website. The institute is now working on technical compliance of GlobalGAP for exporting quality potato production.

6.1 Useful Sites for Update

- ☆ Agricultural and Processed Food Products Export Development Authority: www.apeda.gov.in/
- ☆ CODEX Alimentarius International food Standards: www.fao.org/fao-who codexalimentarius/en
- ☆ Directorate of Plant protection, Quarantine and Storage: www.ppqs.gov.in
- ☆ Food and Agriculture Organization: www.fao.org
- ☆ Food Safety and Standards Authority of India: www.fssai.gov.in
- ☆ GlobalGAP: https://www.globalgap.org
- ☆ ICAR-CPRI: http://www.cpri.icar.gov.in
- ☆ International Potato Center, Lima, Peru: www.cipotato.org
- ☆ Ministry of Agriculture and Farmers welfare: www.agriculture.gov.in Quality Council of India: http://qcin.org/india-good-agriculture-practices.php

Chapter 27

Breeding Potato Varieties for Different Agro-Climatic Zones of India

S.K. Luthra

ICAR-CPRI, Regional Station, Modipuram, Meerut, Uttar Pradesh

1. Introduction

The potato (*Solanum tuberosum* L.) is believed to have originated near Lake Titicaca basin in Peru-Bolivian region, where the largest genetic diversity of potato exists. Potato belongs to the family *Solanaceae* and the large genus *Solanum*, which contains about 2000 species, out of which nearly 235 are tuber bearing (Hawkes 1990). These are confined mainly to the Andean region of South America. The basic chromosome number in potato is x = 12 (Howard 1970). The ploidy level of potato species varies from 2x to 6x (Hawkes 1994). The commonly cultivated potato is an autotetraploid (Swaminathan and Howard 1953) and belongs to the species *Solanum tuberosum* L (2n = 4x = 48), which includes two subspecies *viz.*, ssp. *tuberosum* adapted to long days and ssp. *andigena* adapted to short days. Besides these, seven primitive species *viz.*, *Solanum ajanhuiri*, *Solanum goniocalyx*, *Solanum phureja*and *Solanum stenototum* (all diploids); *Solanum chaucha* and *Solanum juzepczukii*(triploid); and *Solanum curtilobum* (pentaploid) are also under cultivation, but are restricted to the high Andes from Central Peru to Central Bolivia. While ssp. *andigena* is confined to Andes of Venezuela, Colombia, Equador, Peru, Bolivia and North-West Argentina; the ssp. *tuberosum* has become most widespread and cultivated in Europe, Asia and else where.

In India, potato breeding programme was initiated in 1935 at the Potato Breeding Station, Shimla. However, a regular breeding programme was started in 1949 with the establishment of the Central Potato Research Institute (CPRI) at Patna, Bihar. Headquarter of the CPRI was later on shifted to Shimla in order to facilitate hybridization and maintenance of seed health. The major breakthrough in potato improvement programme came in 1963 with the development of "Seed Plot Technique", which made it possible to raise, evaluate, select and multiply breeding material under disease free conditions in plains. This led to the development of a system, wherein crossing was attempted in the hills and raising of seedling, evaluation and maintenance of segregating population was done in the plains. This approach yielded rich dividends both for potato improvement and potato seed production in the country. All varieties released by the CPRI carry the prefix 'KUFRI' as a memento to the place of hybridization. India has diverse soil types and agro-climatic conditions. Successful potato cultivation requires maximum and minimum temperatures of 33 °C and 15 °C with sunny days and cool nights. Indian sub-tropical plains offer optimum conditions for potato cultivation, where 85-90 per cent of potatoes are grown during short winter days from October to February. The hills account for less than 5 per cent of the total potato production where the crop is grown during long summer days from April to September/October. The plateau regions of South-eastern, central and peninsular India constitute about 6 percent area where potato is grown mainly as rainfed or irrigated winter crop. On the basis of the diverse soil, climate and other agronomic features, the potato growing areas in India can be divided into eight zones. These zones (Table 27.1) lie in two major potato growing area *i.e.* north Indian hills and north Indian plains, while southern and north Bengal and Sikkim hills and plateaus are three special problem areas.

Table 27.1: Potato Zones of India

Zone	*Varietal Requirements*
North-western plains	Short day adaptation, early bulking, heat tolerance and late blight resistance. Tolerance to frost is an added advantage.
West-central plains	Short-day adapted, early bulking, moderate resistant to late blight and slow degeneration rate
North-eastern plains	Short day adapted, early bulking and late blight resistant. Red skin tubers are preferred in some areas.
Plateau region	Early bulking, able to tuberize under high temperatures and resistant to bacterial wilt, mites and potato tuber moth and slow rate of degeneration.
North-western and central hills	Long day adaptation (14 h days), highly resistant to late blight and bacterial wilt.
North-eastern hills	Long day adaptation (14 h days), highly resistant to late blight and bacterial wilt.
Southern hills	Long day adaptation, early bulking, late blight and cyst nematode resistance.
Sikkim and north Bengal hills	Resistance to late blight and immunity to wart. Red skin potatoes are also preferred.

2. Breeding Strategies

Potato variety is a group of identical plants produced asexually from a single genotype and is devoid of heritable variation during its propagation. Plants of potato variety are homogenous because they are produced asexually, and heterozygous, because they originate from a heterozygous individual. The term cultivar and variety are synonyms to each other. It is perhaps an impossible task to combine all traits to obtain an ideal variety because of complex heterozygous nature of potato. More than 50 traits should be combined in a modern potato variety (Ross 1986) and this complicates the choice of parents for production of such a progeny. It is perhaps an impossible task to combine all traits to obtain an ideal variety because of complex heterozygous nature of potato. The new variety thus, should be superior to existing in at least one important characteristic, without being significantly inferior to it in any other important traits. Heterosis is observed on crossing diverse parents. Hence, breeding of potato involves hybridization between identified parents and selection of superior clones from the segregating population. The genetic constitution of the genotype obtained following hybridization is fixed in seedling stage and due to vegetative propagation the genetic constitution of potato genotype with all its intra- and inter-locus interactions responsible for its phenotypic expression are maintained in the clonal generations. Hence, a clone if perceived desirable can be multiplied for commercial cultivation even though initially it may be present as a single plant. Parents for hybridization programme are selected based on the objectives of the programme. In general, phenotypic performance, genetic divergence, combining ability, heterosis and progeny tests are considered for selection of promising parents, to generate variability for making selection effective. In order to have a reasonable chance of identifying a 'winning' selection, an initial population of 50,000 seedlings of about 25 cross combinations need to be raised. A minimum of 2,000 seedlings per cross are needed to have a reasonable chance of selecting a superior genotype from a cross (Luthra *et al.*, 2006; 2020). The clones from the seedling stage are evaluated in clonal generations. A few promising advance clones are finally included in multi-location testing trials across the country, before a clone is found worth release as a variety for general cultivation.

3. Selection in Early Clonal Generations

In potato breeding programmes, the major constraint in effective evaluation of genotypes in early stages is non-availability of sufficient number of plants per clone. The assessment is based on un-replicated plots of just a single plant in the seedling (F_1) stage, five in F_1C_1stage, 30 in F_1C_2 stage and 120 plants in F_1C_3 stage. Further, the seed tubers of clones used for planting often vary in size. There is also a location and year effect. These factors make assessment of quantitative characters like tuber yield difficult in early generations. Hence rejection of undesirable clones (rather than selection of desirable clones) based on stable characters that have no or very little environmental effect, *viz.*, tuber color, shape, depth of eyes and tuber cracking is practiced in early generations.

Seedlings of poor vigour may be rejected prior to transplanting in the field as they would likely have low yields in later generations. Clones with undesirable tuber color, tuber shape, eye depth and tuber cracking may be rejected from seedling stage onward, as these characters have a high repeatability over generations. No rejection should be done on the basis of tuber yield, average tuber weight or number of tubers in the seedling generation. Negative selection (rejection of poor phenotype) for tuber yield and tuber weight can be initiated from the first clonal generation onwards, whereas, number of tubers can be considered as a basis for the rejection of undesirable types from second clonal generation onward. In subsequent generations, average tuber weight should be used as a selection parameter after fixing a standard for minimum number of tubers required in selected types. In breeding programmes involving resistance to late blight, cyst nematodes, wart, *etc.* it is possible to discard susceptible genotypes in seedling stage itself. The breeding population in such cases is subjected both to resistance tests and to rejections based on unacceptable tuber characters.

3.1 Screening for Late Blight Resistance

The 4-6 week old seedlings are screened under controlled conditions against complex races of *Phytophthora infestans* and resistant seedlings are transplanted in the earthen pots in the glass house/field and retained for further evaluations.

3.2 Screening for PCN Resistance

The selected F_1C_2 stage clones are screened for PCN resistance in the glass house. Five tubers of the selected clone are planted in the pots containing a population of 200-250 cysts of both species of *Globodera* per 100 ml soil. Clones showing 0-5 females/root ball after 60-70 days of planting are selected.

3.3. Screening for Processing

Five tubers of each clone in the in F_1C_1 stage are dipped in brine solution for specific gravity estimation and clones with >1.08 specific gravity are selected. The selected clones in F_1C_2 generation are tested for chips/French fries color. Clones showing poor keeping quality due to high shrinkage, rottage, weight loss and excess sprouting after cold store are rejected.

3.4 Screening for Heat Tolerance

The seedlings are screened by exposing them at >20 °C night temperature and seedlings with ability to form tubers after 4-8 weeks of treatment are selected for further evaluations.

The trait specific screening of other desirable traits (table, nutritional superiority, drought tolerance) is done in initial clonal stages as per objective of the breeding programme.

4. Selection in Advanced Trials

In advanced trials (F_1C_4 onwards), advanced clones are evaluated in replicated trials of adequate plot size (Luthra *et al.*, 2020). Two to three standard varieties are included as checks. With advancement of clonal generations, plot size and number of replications are increased. The plots are dehaulmed at different dates to ascertain the potential of clones at various crop durations. The clones are also evaluated for their suitability of planting on different dates in order to judge their potential for inclusion in different cropping sequences operating in a region. The data is recorded on plant emergence, plant vigour, foliage maturity, tuber traits, tuber yield, tuber number per plant, general impression, tuber dry matter, organoleptic test, disease resistance, processing, nutritional components (anthocyanins, carotenoids, iron zinc *etc.*), shelf life *etc.* The produce of clones in yield trials is graded, weighed and their tuber characters, *viz.*, specific gravity, skin/flesh color and internal defects such as hollow heart and internal necrosis, if any are recorded. Samples are boiled and their texture, degree of sloughing is estimated. The produce is also tested for after-cooking darkening, flavor, suitability for processing and for storage at ambient temperature. For identification of bio fortified potatoes, sampled are assessed for nutritional and mineral components. Cold storage of potatoes tends to promote conversion of starch to reducing sugars, which create dark color in chips and French fries. To identify potential genotypes for processing, tests are conducted after different periods of storage, at more than one temperature and after various periods of reconditioning. In general, two to three years' data from replicated trials is considered adequate to identify superior clones for introducing in multi-location tests under All India Coordinated Research Project on Potato (AICRP-Potato).

5. Multi-location Evaluation through All India Coordinated Research Project on Potato [AICRP (P)]

It is functioning since 1971, conducts multi-location testing of potato varieties and technologies. Presently, the AICRP (P) has 25 centers located at several State Agricultural Universities and regional stations of ICAR-CPRI.The main objectives of AICRP (P) is to conduct multi-location testing of advanced hybrids and recommendation for release of the promising hybrids superior to controls in terms of tuber yield and/or many other desirable attributes. The advanced stage hybrids identified in various breeding programmes of the institute are introduced into AICRP (P) for multi-location evaluation. During first year, about 3000 tubers of each advanced stage hybrid are given to Seed Preparatory Unit (SPU) of AICRP (P) for multiplication and subsequent supply to various centers for conducting trials. The performance of newly developed advanced hybrids is evaluated at different centers across the country under the coordinated project of AICRP (P) for a minimum period of two to three years/

seasons. The hybrids are first tested for yield under the initial varietal trials (IVT); the promising ones are promoted to the advanced varietal trials (AVT) in the second year and AVT 2 in the third year. The hybrids that give good performance in AVT-1 and AVT-2 are selected. Data are recorded on plant emergence, tuber yield (marketable and total), tuber dry matter, reaction to various biotic and abiotic stresses and storage behavior at ambient temperature. The quality and processing attributes of advanced stage hybrids are also estimated.

6. Potato Breeding Methodology

The outline of potato breeding methodology is depicted in Figure 27.1 (Luthra *et al.*, 2020).

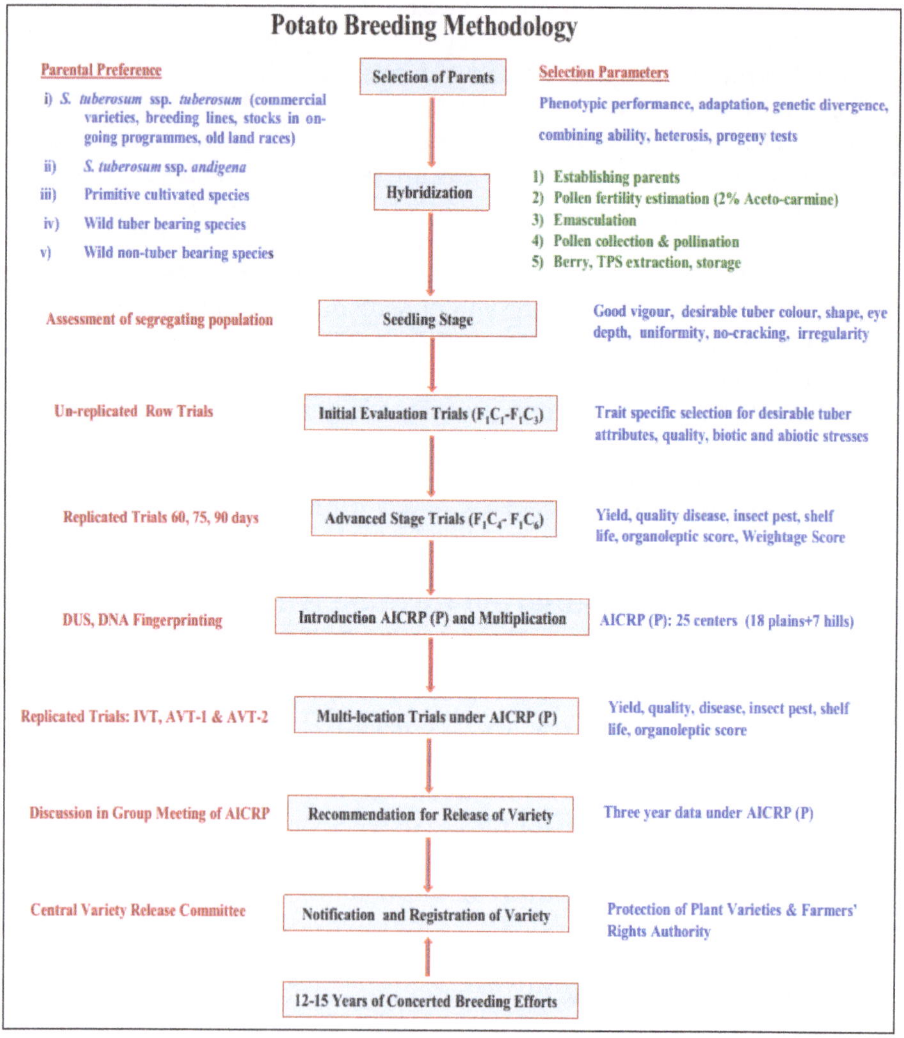

Figure 27.1: The Outline of Potato Breeding Methodology.

6.1 Maintenance of Seed Stock of Advanced Clones

In potato breeding, maintenance and multiplication of selections under disease free conditions, during the course of evaluation over years, is of utmost importance. The seedlings and initial clonal generations are grown in low aphid (virus vector) zone and precaution is taken to minimize introduction and spread of virus diseases. To avoid aphid transmitted viruses, insecticides are used and vines are killed before aphid population reaches a critical level. Plants with visible virus symptoms are rogued out to reduce possibility of secondary infection. The seed production of advanced clones is done in separate seed-maintenance field following seed plot technique.

7. Release and Notification of New Potato Variety

The advanced clones which show good performance for yield and other desirable traits in IVT and AVT are identified and are considered for release in the Annual Group Meeting of AICRP (P). Based on discussion during the group meeting of the workshop, best ones are recommended for release. Crop production and management of newly recommended varieties are also worked. The proposal for release of new varieties is put up in a prescribed proforma to variety release committee. There are two types of variety release committee *viz.*, State Variety Release Committee (SVRC) and Central Variety Release Committee (CVRC). The notification proposal is submitted to the 'Central Sub-Committee on Crop Standards Notification and Release of Varieties for Horticultural Crops', Ministry of Agriculture, Department of Agriculture and Co-operation, Government of India, New Delhi. The Central Seed Committee discusses the results in its meeting and if satisfied, notifies the variety through official Gazette under section 5 of the Seed Act 1966 (54 of 166) for a period of 15 years. It is mandatory to deposit a sample of the variety (to be notified), with the ICAR-National Bureau of Plant Genetic Resources (ICAR-NBPGR) before notification for release of variety. The ICAR-NBPGR issues IC numbers to all the varieties recommended for release. It usually takes 12-15 years of concerted efforts to produce a potato variety. The breeder seed production can be taken up only after notification of new varieties.

8. ICAR-CPRI Varieties Adoption and Impact

Breeding efforts at the ICAR-Central Potato Research Institute have led to development of 66 improved potato varieties and a TPS population (92-PT-27) over years, for cultivation in diverse agro-climatic zones of the country (Tables 27.2 and 27.3, Figure 27.2). Of these, two varieties, *viz.*, Kufri Safed and Kufri Red are clonal selections from indigenous *'desi'* varieties Phulwa and Darjeeling Red Round, respectively, whereas, the remaining 65 are hybrids. The parentage of 65 hybrid varieties shows that 25 varieties have both the parents of Indian origin, 28 have one parent of Indian origin and one of exotic origin and the remaining 12 have both the parents of exotic origin. In total, 99 parents have been used in the breeding programme and most frequently used indigenous parents are EX/A 680-16, Kufri Bahar, Kufri Alankar, Kufri Chipsona-2, Kufri Jyoti, Kufri Kundan, Kufri Pukhraj, Kufri Red, MP/92-35, MS/82-638 and Phulwa. The exotic cultivars that have figured more frequently as parents of Indian varieties are Adina, 27/40 Craig's Defiance, Ekishiraju, Katahdin, Seedling 698-D and parental lines obtained from late Dr. William Black of UK. Recently heat tolerant advanced clone HT/7-1329 has been recommended for release in cultivation in areas of North Indian plains having high temperature during early planting.

At present 44 varieties are under breeder/quality seed production programme of the institute. The statistics of nucleus and breeder seed produced by CPRI during 2017-18 reveals the maximum share of popular potato variety Kufri Bahar (25.8 percent) followed by Kufri Jyoti (14.6 percent), Kufri Khyati (10.6 per cent), Kufri Pukhraj (7.9 percent), Kufri Sindhuri (6.9 per cent), Kufri Chipsona-1 (6.8 percent), Kufri Surya (5.5 percent), Kufri Chipsona-3 (4.0 percent), Kufri Chandramukhi (3.0 percent), Kufri Anand (2.7 percent), Kufri Garima (2.5 percent) and Kufri Frysona (2.4 percent). The CPRI varieties are well adopted by the farmers in different parts of the country. In India, nearly 95 per cent area under potato is grown with indigenous improved varieties developed by the CPRI.

Table 27.2: Potato Varieties Released by CPRI, Shimla

Name	*Year of Release*	*Areas of Adaptation*	*Salient Features*
Kufri Kisan*	1958	North Indian plains	Maturity late; tubers white round, eyes deep and prominent eyebrows and flesh white.
Kufri Kuber*	1958	North Indian plains and plateau region	Maturity medium; tubers white ovoid, tapering towards crown end, eyes medium deep, flesh white; resistant to PLRV and immune to PVY.
Kufri Kumar*	1958	North Indian hills	Maturity late; tubers white ovoid, tapering towards heel end, eyes fleet, flesh white; immune to wart and resistant to charcoal rot.
Kufri Kundan*	1958	North Indian hills	Maturity medium; tubers white round-ovoid, flattened, eyes medium deep, flesh cram; resistant to charcoal rot.
Kufri Red*	1958	North eastern plains	Maturity medium; tubers red round, eyes medium deep and flesh yellow.
Kufri Safed*	1958	North Indian plains	Maturity late; tubers white round, eyes medium deep and red-purple and flesh light yellow.
Kufri Neela*	1963	South IndianHills	Maturity late; tubers white ovoid, eyes medium deep, flesh white; resistant to cyst nematodes.
Kufri Sindhuri	1967	North Indian plains	Maturity late; tubers red round, eyes medium deep, flesh cream; moderately resistant to early blight and tolerant to PLRV.
Kufri Alankar*	1968	North Indian plains	Maturity medium; tubers white ovoid, eyes fleet, flesh white; moderately resistant to late blight.

Name	*Year of Release*	*Areas of Adaptation*	*Salient Features*
Kufri Chamatkar*	1968	North Indian plains	Maturity late; tubers white round, eyes medium deep, flesh yellow; resistant to early blight, charcoal rot and immune to wart.
Kufri Chandramukhi	1968	North Indian plains and plateau region	Maturity early; tubers white ovoid, slightly flattened, eyes fleet and flesh white.
Kufri Jeevan*	1968	North-Indian hills	Maturity late; tubers white ovoid, eyes fleet, flesh white; moderately resistant to late blight.
Kufri Jyoti	1968	Hills, plains and plateau region	Maturity medium; tubers white ovoid, eyes fleet, flesh white; moderately resistant to late blight and immune to wart.
Kufri Khasigaro*	1968	North-eastern hills	Maturity late; tubers white round-ovoid, eyes deep, flesh cream; moderately resistant to late blight and early blight.
Kufri Naveen*	1968	North-eastern hills	Maturity late; tubers white ovoid, eyes fleet, flesh white; moderately resistant to late blight and immune to wart.
Kufri Neelamani*	1968	South Indian hills	Maturity late; tubers white ovoid, flattened, eyes fleet, flesh white; moderately resistant to late blight.
Kufri Sheetman*	1968	North-western plains	Maturity medium; tubers white ovoid, eyes fleet, flesh cream; moderately resistant to charcoal rot and tolerant to frost.
Kufri Muthu*	1971	South Indian hills	Maturity medium; tubers white round-ovoid, eyes medium deep, flesh white; moderately resistant to late blight and immune to wart.
Kufri Lauvkar	1972	Plateau region	Maturity early; tubers white round, eyes fleet, flesh white; tolerant to heat.
Kufri Dewa*	1973	North Indian plains	Maturity late; tubers white round, eyes pink and medium deep, flesh cream; tolerant to frost.
Kufri Badshah	1979	North Indian plains and plateau region	Maturity medium; tubers white ovoid, eyes fleet, flesh white; moderately resistant to late blight, early blight and PVX.
Kufri Bahar	1980	North Indian plains	Maturity medium; tubers white round-ovoid, eyes fleet, flesh white; immune to wart.
Kufri Lalima	1982	North Indian plains	Maturity medium; tubers red round, eyes medium deep, flesh white; moderately resistant to early blight and resistant to PVX.
Kufri Sherpa*	1983	North Bengal hills	Maturity medium; tubers white round flattened, eyes medium deep, flesh cream; moderately resistant to early blight and late blight, and immune to wart.
Kufri Swarna	1985	South Indian hills	Maturity medium; tubers white round-ovoid, eyes fleet, flesh white; moderately resistant to late blight and early blight, immune to wart, highly resistant to cyst nematodes.

Name	Year of Release	Areas of Adaptation	Salient Features
Kufri Megha*	1989	North-eastern hills	Maturity late; tubers white round-ovoid, eyes fleet, flesh white; resistant to late blight.
Kufri Jawahar	1996	North Indian plains and plateau region	Maturity medium; tubers white round-ovoid, eyes fleet, flesh white; moderately resistant to late blight and immune to wart.
Kufri Sutlej	1996	North Indian plains	Maturity medium; tubers white ovoid, eyes fleet, flesh white; moderately resistant to late blight and immune to wart.
Kufri Ashoka	1996	North Indian plains	Maturity early; tubers white ovoid, eyes medium-deep, flesh white.
Kufri Pukhraj	1998	North Indian plains and plateau region	Maturity early; tubers white ovoid, slightly tapered, eyes fleet, flesh yellow; resistant to early blight; moderately resistant to late blight and immune to wart.
Kufri Chipsona-1	1998	North Indian plains	Maturity medium; tubers white ovoid, eyes fleet, flesh white; resistant to late blight; suitable for chips and French fries.
Kufri Chipsona-2	1998	North Indian plains	Maturity medium; tubers white round, eyes fleet, flesh cream; resistant to late blight; tolerant to frost and suitable for chips.
Kufri Giriraj	1998	North Indian hills	Maturity medium; tubers white ovoid, eyes fleet, flesh white; resistant to late blight and immune to wart.
Kufri Anand	1999	North Indian plains	Maturity medium; tubers white ovoid-oblong, eyes fleet, flesh white; moderately resistant to late blight and immune to wart.
Kufri Kanchan	1999	North Bengal hills and Sikkim	Maturity medium; tubers red ovoid-oblong, eyes fleet, flesh cream; moderately resistant to late blight and immune to wart.
Kufri Arun	2005	North Indian plains	Maturity medium; tubers red ovoid, eyes medium deep, flesh cream; moderately resistant to late blight.
Kufri Pushkar	2005	North Indian plains	Maturity medium; tubers white round-ovoid, eyes medium deep, flesh light yellow; resistant to late blight and immune to wart.
Kufri Shailja	2005	North Indian hills	Maturity medium; tubers white, round-ovoid, eyes fleet, flesh white; moderately resistant to late blight.
Kufri Chipsona-3	2006	North Indian plains	Maturity medium; tubers white, round-ovoid, eyes fleet, flesh cream; resistant to late blight; suitable for chips and French fries.
Kufri Surya	2006	North Indian plains and plateau region	Maturity early; tubers yellow oblong, eyes fleet, flesh yellow; immune to wart; resistant to hopper burn and heat tolerant; suitable for early plating.
Kufri Himalini	2006	North Indian hills	Maturity medium; tubers white ovoid-oblong, eyes shallow, flesh cream; moderately resistant to late blight.

Name	Year of Release	Areas of Adaptation	Salient Features
92-PT-27**	2007	Eastern plains	Maturity medium; tubers white-cream-yellow round-ovoid-oblong, eyes shallow to medium deep, flesh white-cream-yellow; field resistant to late blight.
Kufri Himsona	2008	North Indian hills	Maturity medium, tubers white round- ovoid, eyes fleet, flesh cream; resistant to late blight and suitable for chips.
Kufri Sadabahar	2008	Uttar Pradesh and adjoining areas	Maturity medium; tubers white oblong, eyes fleet, flesh white; resistant to late blight.
Kufri Girdhari	2008	North Indian hills	Maturity medium; tubers white ovoid-oblong, eyes fleet, flesh pale yellow; resistant to late blight.
Kufri Khyati	2008	North Indian plains	Maturity early; tubers white ovoid, eyes fleet, flesh cream; moderately resistant to late blight.
Kufri Frysona	2009	North Indian plains	Maturity medium; tubers white-cream long-oblong, shallow eyes, flesh white; resistant to late blight; suitable for French fries.
Kufri Neelima	2010	Nilgiri hills	Maturity medium; tubers white ovoid, eyes shallow, flesh white; highly resistant to cyst nematodes; resistant to late blight.
Kufri Garima	2012	Indo-gangetic plains and plateau	Maturity medium; tubers light yellow ovoid, eyes shallow, flesh light yellow; resistant to late blight.
Kufri Gaurav	2012	North Indian plains	Maturity medium; tubers white-cream ovoid, eyes medium deep, flesh white-cream; nutrient use efficient variety.
Kufri Lalit	2013	Eastern plains	Maturity medium; tubers light red round, eyes medium deep, flesh light yellow; resistant to late blight.
Kufri Mohan	2015	Northern and Eastern plains	Maturity medium; tubers white-cream ovoid, eyes shallow, flesh white; moderately resistant to late blight.
Kufri Kesar@	2017	Eastern plains	Maturity medium; tubers red ovoid, eyes shallow, flesh yellow; moderately resistant to late blight.
Kufri Sukhyati@	2017	North Indian plains	Maturity early; tubers white round ovoid, eyes fleet, flesh cream; moderately resistant to late blight.
Kufri Lima	2018	North Indian plains	Maturity medium-late; tubers white-cream ovoid, shallow eyes, cream flesh; extreme resistance to PVX and PVY and moderately resistant to root knot nematode; resistant to hopper burn and heat tolerant; suitable for early plating.
Kufri Ganga	2018	North Indian plains	Maturity medium; tubers white-cream ovoid, shallow eyes, flesh cream; moderately resistant to late blight.

Name	Year of Release	Areas of Adaptation	Salient Features
Kufri Neelkanth	2018	North Indian plains	Maturity medium; tubers purple ovoid, eyes shallow, flesh yellow; moderately resistant to late blight; speciality potato, rich in anti-oxidants with excellent flavour.
Kufri Sahyadri	2018	Nilgiri hills	Maturity medium; tubers light yellow ovoid, eyes shallow medium, flesh yellow; resistant to late blight and potato cyst nematode.
Kufri Karn	2018	Hills and Plateau region	Maturity late; tubers white-cream ovoid, eyes shallow, flesh white; resistant to late blight, APCLV, PVX, Y, S, A, M, PLRV; moderately resistant to PCN.
Kufri Manik	2018	Eastern plains	Maturity medium; tubers red round, eyes medium deep, flesh yellow; moderately resistant to late blight; possess anthocyanin and carotenoids
Kufri FryoM	2018	Northern and Central plains	Maturity medium; tubers white oblong, eyes shallow, flesh white; field resistance to late blight and potato virus Y; suitable for French fries
Kufri Chipsona 4	2019	Karnataka, West Bengal, Madhya Pradesh and Gujarat	Maturity medium; tubers white-cream round, eyes shallow, flesh white; resistant to late blight; suitable for chips.
Kufri Sangam	2019	Northern (table) and Central (table and processing) plains	Maturity medium, tubers white-cream ovoid, eyes shallow, flesh white; moderately resistant to late blight and excellent storability; Very good taste, aroma, mealy texture; suitable for table purposes and processing to chips and French fries.
Kufri Thar-1	2019	Orissa and Uttar Pradesh	Maturity medium, tubers white-cream round-oval, eyes shallow-medium, flesh cream; moderate resistance to late blight; very good storability; suitable for drought prone areas.
Kufri Thar-2	2019	Uttar Pradesh, Rajasthan, Haryana and Chhatisgarh	Maturity medium, tubers light yellow ovoid, eyes shallow, flesh light yellow; good taste, typical potato flavor, mealy texture, free from discolouration after cooking; Very good tuber appearance and devoid of tuber cracking, very good keeping quality, long tuber dormancy; suitable for drought prone areas.
Kufri Thar-3	2019	Haryana, Uttar Pradesh and Chhatisgarh	Maturity medium; tubers white round-oval, eyes shallow, flesh cream. Slightly resistant to late blight; High water use efficiency.
Kufri Kiran#	2020	North Indian plains and Plateau	Maturity early; tubers white-cream ovoid, eyes shallow, flesh cream; Tolerant to hopper/mite burn and high temperature; suitable for early plating.

*Not under seed production at present, **TPS population; @ Institute release, #Notification awaited

Plains: Early (70-90 days), Medium (90-100 days) and Late (>100 days);

Hills: Early (100-110 days), Medium (110-120 days) and Late (>120 days)

Table 27.3: Potato Varieties Recommended for Cultivation in different Agro-ecological Zones of India

Agro-ecological zones	*Duration**	*Recommended Varieties*
North-Western plains	Early	K. Ashoka, K. Chandramukhi, K. Jawahar, K. Khyati, K. Lima, K. Pukhraj, K. Surya.
	Medium	K. Anand, K. Arun, K. Badshah, K. Bahar, K. Chipsona-1, K. Chipsona-3, K. FryoM, K. Ganga, K. Garima, K. Gaurav K. Jyoti, K. Mohan, K. Neelkanth, K.Pukhraj, K. Pushkar, K. Sadabahar, K. Sangam, K. Sutlej
West-Central plains	Early	K. Chandramukhi, K. Jawahar, K. Khyati, K. Pukhraj, K.Lauvkar, K. Lima, K. Surya
	Medium	K. Arun, K.Anand, K. Badshah, K. Bahar, K. Chipsona-1, K. Chipsona-3, K. Ganga, K. Frysona, K. Garima, K.Jyoti, K. Mohan, K. Neelkanth, K. Pukhraj, K. Pushkar, K. Sadabahar, K. Sangam, K. Sutlej
	Late	K. Sindhuri
North-eastern plains	Early	K. Ashoka, K. Chandramukhi, K. Khyati, K. Lima, K. Pukhraj, K. Surya
	Medium	K. Arun, K. Bahar, K. Chipsona-1, K. Chipsona-3, K. FryoM, K. Frysona, K. Ganga, K. Gaurav, K. Jyoti, K. Manik, K. Mohan, K. Neelkanth, K. Kanchan, K. Lalima, K. Lalit, K. Pukhraj, K. Pushkar, K. Sutlej
	Late	K. Sindhuri
Plateau region	Early	K. Chandramukhi, K.Jawahar, K. Khyati, K. Lauvkar, K. Pukhraj, K. Surya
	Medium	K.Badshah, K. Garima, K. Jyoti, K. Lauvkar, K. Pukhraj, K. Karn
North-western hills	Medium	K.Girdhari, K. Giriraj, K. Himalini, K. Karn, K. Himsona, K.Jyoti, K. Shailja
North eastern hills	Medium	K. Girdhari, K. Giriraj, K. Himalini, K. Jyoti, K. Megha, K. Shailja
North Bengal and Sikkim hills	Medium	K. Girdhari, K. Jyoti, K. Kanchan
Southern hills	Medium	K.Girdhari, K. Giriraj, K.Himalini, K.Jyoti, K. Neelima, K.Shailja, K. Sahyadri, K.Swarna

* **Plains:** Early (70-90 days), Medium (90-100 days) and Late (>100 days);

Hills: Early (100-110 days), Medium (110-120 days) and Late (> 120days).

Three indigenous potato varieties namely, Kufri Pukhraj, Kufri Jyoti and Kufri Bahar are the most popular among the farmers and occupy 33, 21 and 17 percent area in six major potato growing states (Pradel *et al.,* 2019). Kufri Pukhraj, released two decades ago has become farmer's first choice in Bihar, Gujarat and Punjab, occupying 39, 85 and 64 percent cultivated area in these states, respectively. This variety is also popular in Uttar Pradesh and West Bengal where it occupies about 28 and 19 percent of total area under potato cultivation.

Kufri Jyoti is another widely adapted variety and is grown throughout India in hills, plains and plateau. It is the most popular variety in West Bengal and Karnataka, occupying about 56 and 94 percent area under potato cultivation. Kufri Bahar is the farmer's preferred choice in Uttar Pradesh occupying about 45 percent area. These three Indian potato varieties *viz.*, Kufri Pukhraj, Kufri Jyoti and Kufri Bahar occupy 2nd, 5th and 6th place, respectively in South East Asian potato growing countries too. Farmers' preferences for red skinned potatoes keep the demand high for Kufri Sindhuri in Bihar. Kufri Kanchan, a wart immune variety, is popular among the farmers of North Bengal and Sikkim hills where wart is an endemic problem and there is a distinct preference for red skin tubers. Similarly, Kufri Swarna and Kufri Neelima having resistance to cyst nematodes is popular in Nilgiri hills. These varieties released specifically to contain the endemic problems of wart and cyst nematodes, respectively, have helped in reducing the incidence of these diseases in the respective areas. Late blight has been a perpetual challenge for potato production throughout the world including India. However, the Indian potato breeding programmes have been fairly successful in releasing improved varieties having high resistance to late blight at regular intervals. The late blight resistant varieties developed by the institute, Kufri Himalini initially released for Northern hills is also becoming popular in plains and plateau areas, where late blight is a production constraint.

Figure 27.2: Important Indian Potato Varieties.

The institute has developed new improved varieties for the farmers of plains. These include varieties, Kufri Mohan, Kufri Gaurav and Kufri Ganga with high yield, Kufri Lima with extreme resistance to PVX and PVY, early bulker Kufri Sukhyati, speciality potato variety Kufri Neelkanth-rich in anti-oxidants with excellent flavour and red skin variety Kufri Manik. These varieties are likely to pick up and become popular among farmers. The institute has also developed seven varieties exclusively for processing purpose namely Kufri Chipsona-1, Kufri Chipsona-2, Kufri Chipsona-3, Kufri Frysona, Kufri Chipsona-4, Kufri FryoM and Kufri Sangam for cultivation in plains and Kufri Himsona for cultivation in hills. Among these varieties, Kufri Frysona and Kufri FryoM is suitable for French fries, while all other are chipping varieties. These varieties were developed keeping in mind the demand of the processing industry and also to deal with the frequent glut situation in the market. These varieties besides being suitable for processing also have high yield and resistance to late blight. Kufri Chipsona 2 and Kufri Chipsona 3, are notable varieties for processing. Kufri Chipsona 1 is primarily found in Uttar Pradesh and is suitable for the industry due to its low sugar content, high dry matter, and good tuber size and shape (Pradel *et al.*, 2019). Recently three drought tolerant/water use efficient varieties namely Kufri Thar-1, Kufri Thar-2 and Kufri Thar-3 has been released in cultivation in drought prone areas of North Indian plains.

Availability of indigenous varieties, good quality seeds and right package of agronomic practices triggered a revolution in potato productivity causing very fast growth in area, production and productivity. The potato production in India during 2017-18 was 51.31 million t from 2.14 million ha area with a productivity of 23.97 t/ha as compared to 1949 when India used to produce 1.54 million t potatoes from 0.23 million ha area at an average productivity level of 6.6 t/ha. It is the hard work of potato farmers, scientists and policy makers that potato production, area and productivity increased over more than seven decades by 33.31, 9.30 and 3.63 times, respectively. India is the second largest annual producer of potato after China. Considering all other factors involved in this quantum jump in potato productivity, we can safely assume that at least 50 percent contribution is due to release of improved, high yielding varieties possessing resistance/tolerance to various biotic and abiotic stresses.

Chapter 28

Potato Breeding for Abiotic Stresses

Devender Kumar

ICAR-CPRI, Regional Station, Modipuram, Meerut, Uttar Pradesh

1. Introduction

Although cultivated potatoes originated in highlands of South America, it is in the temperate countries where the full potential of the crop was first exploited as a result of organized breeding work. From these countries, potato was introduced to the tropical and subtropical areas of the world. However, the yield of potatoes in the tropical and subtropical parts of the world is less than one third compared to the temperate zones. Although poor cultural practices and high incidence of diseases and pests are responsible for the low yields, yet abiotic factors like high temperatures and limited moisture availability are other major factors affecting the crop growth and yield adversely. Recently, salinity stress has also been included to introduce potato into salt affected areas including coastal regions of the country. Though, Incorporation of resistance genes in cultivars is the most effective, economical and eco-friendly method of management of these stresses. Yet, It is difficult to breed cultivars with resistance against all stresses. The stresses are, therefore, dealt with on the basis of importance in various agro-climatic zones and trait specific potato breeding schemes *i.e.* heat tolerance, drought and salinity tolerance are elaborated and discussed in this chapter.

2. Breeding Varieties for Heat Tolerance

In potato (*Solanum tuberosum* L.), the optimum temperature for growth of haulms is about 25 °C, whereas, for rapid tuberization moderate temperatures (about 18-20 °C) are beneficial. Optimum temperatures for tuber bulking are between 15-20 °C. A progressive reduction in tuberization is encountered with increasing temperature up to 30 °C (Khanna, 1966). Minimum night temperature plays a crucial role during tuberization in potato. The tuberization is reduced at night temperatures above 20 °C with complete inhibition of tuberization above 25 °C. Higher temperature during haulm growth causes the plant to lose its capacity to initiate tubers or to accumulate starch in them. Exposure of potato plants to heat stress alters the hormonal balance in plants and results in partitioning of most of the assimilated carbon in favor of above ground vegetative parts at the cost of tubers. In India, the potatoes with high productivity per unit area, short duration and excellent nutritive value has the potential to significantly augment food availability. The crop is grown in northern plains during winters and in northern hills during summers. High temperatures, during crop growth restrict its planting early in the crop season in North western plains, and in peninsular India, which are characterized by high night temperatures (above 20 °C). It is imperative to develop potato varieties which could germinate, grow and tuberize well under high night temperatures to mitigate the likely effect of climate change and to introduce potato cultivation in non-traditional areas/ seasons, with relatively higher temperature during crop growth period.

3. State of the Developments

Heat stress in potato is influenced by genotype as well as environment (Midmore 1992). Selection of promising parents and development of superior crosses is important for obtaining heat tolerant high yielding progenies. To locate sources of resistance to heat stress, several efforts have been made. At CIP, 6000 potato genotypes were evaluated for yield at high temperatures, only 34 gave reasonably good yields. Studies with these genotypes showed that a higher level of heterozygosity conferred a greater capacity to adapt to divergent environment. The best performing genotypes had genes from several species like subsp. *tuberosum*, subsp.*andigena* and *S. phureja*. Genotypes with early tuberization and fast bulking rate performed better under such conditions. In India, the work on response of potato plants to heat stress was started in 1984 with screening of germplasm collection. No relationship of tuber yield could be established with various traits (plant height, internode length, number of auxiliary shoots and rate of leaf expansion) at high day/night temperatures (35/23 °C) in glass house experiments. In 1985, six genotypes were evaluated in glass-house at high temperatures and in the net house at ambient temperatures (27/16 °C). It was observed that the per cent increase in inter-node length at high temperature was highly correlated with tuber yield under high temperature (Nagarajan and Minhas 1995). The ability of leaf bud cuttings to tuberize at 25 °C night temperature was also associated with heat tolerance (Minhas *et al.*, 2001). Basu and Minhas (1991) reported that in heat tolerant genotypes, chlorophyll stability index was higher, GA: ABA ratio was lower, photo-respiration was lower, sucrose phosphate synthatase activity was higher and 20-25 percent of the photosynthesis products accumulated as sucrose (compared to 4-6 per cent for the susceptible genotypes). Singh *et al.* (2015) attempted gene expression analysis in two Indian cultivars with contrasting tuberization behavior at night temperatures of 24 °C. Gene expression analysis using a customized microarray chip representing 39,031 potato genes showed that a total of 2500 genes were differentially expressed on 21 days and 4096 genes on 14 days after stress. Gene ontology and pathway analysis provided insights into the probable biological processes and pathways governing tuberization at elevated temperature. Heat stress condition is associated with mite/hopper burn in northern plains and with bacterial wilt in peninsular India. In early crop of potatoes mite and leafhopper causes significant reduction in yield (Malik and Luthra 2007). Efforts, therefore, need to be made to identify genetic resources having heat tolerance with bacterial wilt resistance or with mites/hopper burn tolerance/resistance. *Solanum chacoense* and *Solanum microdontum* are good sources of resistance to bacterial wilt, whereas, *Solanum chacoense* and *Solanum commersonii* are known to be good sources of tolerance to heat stress. Moderate tolerance to mites/hopper burn damage under field conditions in early planting has been observed in germplasm accessions CP1588, CP1927, CP2058 and CP2123 (CPRI, 1998-99). Potato varieties Katahdin, Desiree and Kufri Lauvkar and genotypes LT-l, LT-2, LT-5, LT-7, LT-8, LT-9, DTO-28, DTO-33 from CIP have also been identified as heat tolerant (Minhas *et al.*, 2001).

4. Parameters of Heat Tolerance

4.1 Small Leaves

The potato plant with relatively smaller leaves exposes smaller part of the plant to high temperature leading to less use of photosynthates for respiration activity, thus making photosynthates available for translocation to the developing tubers.

4.2 Internode Length

Heat tolerant genotypes show higher inter-node elongation on exposure to high temperatures as compared to heat susceptible genotypes (Nagarajan and Minhas 1995). Percent increase in inter-node length at high temperature (as compared to normal temperature) is associated with tuber yield. Hence, potato genotypes with long inter-nodal length at high temperatures may be considered heat tolerant.

4.3 Leaf Bud Cuttings

It is a model system to study the effect of temperature on tuberization in potato as well as for screening for heat tolerance. Single leaf cuttings are placed in the soil/sand medium so that the leaf bud is buried in the soil. These cuttings are exposed to 22 °C night temperature and leaf buds are examined for tuberization after three weeks. The presence of tubers, stolon's, or tubers on long stolon's in the leaf bud provide fair indication of the heat tolerance of the plant from which leaf bud cuttings were taken. This technique is used to identify promising parents and for selection of progeny from the segregating populations.

4.4 Yield under High Temperature

Potato genotypes show variability with respect to their performance when grown under heat stress conditions and screening can be done accordingly.

4.5 Gibberellin Content in the Leaves

Exposure of potato plant to heat stress alters the hormonal balance between roots and shoots and affects tuberization and bulking. Higher temperature is known to increase the gibberellin content in the leaves thus promoting haulm growth and inhibition of tuberization (Menzel, 1983). Thus, level of gibberellins content in the leaves can be considered a criterion for selection.

4.6 Abscisic Acid Concentrations in Shoots

Low ratio of gibberellins and Abscisic acid concentrations in shoots is an indication of tolerance of genotype to heat (Basu and Minhas 1991). Tuberization at high temperature may be related with high levels of ABA like inhibitors in shoots during tuber induction.

4.7 Calcium Application

Stomatal conductance remains higher in Calcium treated plants under heat stress conditions and it helps to reduce the heat stress via enhanced cooling through increased transpiration (Kumar *et al.*, 2003). Calcium also helps in maintaining the structural and functional integrity of the cell membrane at high temperatures and minimizing heat induced leaf injury.

4.8 Photosynthetic Rate

The optimum temperature for photosynthesis is about 25 °C and higher temperature decreases rate of photosynthesis due to non-stomatal factors (Ku *et al.*, 1977). Low night temperature may result in less respiratory loss of photosynthates, thus making photosynthates available for translocation to the developing tubers.

4.9 Chlorophyll Fluorescence

At temp. higher than 38 °C, activity of PS-II was irreversibly reduced due to chloroplast membrane disorganization, however, its activity was intact up to 38 °C

4.10 Thermostability of Membrane

This is another reliable method to measure the heat tolerance of a genotype. When thermo-stability of the cell membrane is measured by electrolyte leakage after exposing the leaf discs to 50 °C for 30 minutes, heat tolerant genotypes show lower electrical conductivity compared to susceptible genotypes.

4.11 Chlorophyll Stability

Chlorophyll of heat tolerant genotypes exhibits higher stability after short term exposure to high temperature (Nagarajan and Bansal 1990)

5. Breeding Plan

For developing potato varieties possessing tolerance to heat, it is important to identify suitable parents/progeny from segregating population by glass house/field evaluation. Kumar *et al.* (2003) suggested multiple selection criteria for abiotic stress-resistance to achieve high yielding populations. Simultaneous selection for higher values of number of tubers (0.68**) and tuber dry weight (0.89**) and lower values of plant height (-0.18*) and shoot/root ratio (-0.68**) is required for achieving high tuberization efficiency for identifying heat stress tolerant genotypes (Bhardwaj *et al.*, 2003). Screening of genotypes with tolerance to leafhopper and mite may be helpful to obtain good yield under heat stress conditions. Further, high temperature stress during vegetative stage and tuberization is known to trigger accumulation of glyco-alkaloids in tubers. Therefore, screening of parental lines and advanced stage hybrids to identify genotype with low glyco-alkaloids content (>15 mg/100 g fresh tuber weight) is desirable for consumption of tuber. During selection, care is to be exercised for i) ability of the plants to tuberize at night temperature of 22 °C and above, ii) low shoot/root ratio at high temperature, and iii) early maturity of the crop.

5.1 Hybridization

Promising parents are selected based on their tuberization efficiency and inter-nodal length and other parameters of heat tolerance. The selected parents are crossed with well adapted genotypes of the region.

5.2 Screening

The seedlings are raised in pots in glass house at ICAR-CPRI, Shimla. These are exposed to 22 °C night temperature, created using heat convectors. Observations on plant height, number of leaves, leaf area, number of tubers, tuber dry weight, shoot/root ratio and tuberization efficiency are recorded for identifying heat tolerant genotypes. Observations on tuberization are recorded after 4 and 8 weeks of planting. The seedlings that tuberize under these conditions are early and heat tolerant genotypes.

5.3 Evaluation in Plains

Normally 5-6 tubers of each seedling are available after screening in seedling stage. These tubers are first multiplied for two years under main planting conditions in the plains to obtain enough tubers for evaluation. Subsequently, evaluations are done in early planting and multiplication in main planting.

Outline of breeding procedure employed for developing heat tolerant varieties is given in CPRI Tech Bulletin No.74 (Revised)

6. Current Status

ICAR-CPRI has developed first heat tolerant variety Kufri Surya (Minhas *et al.*, 2006). Kufri Surya (Figure 28.1) is an early maturing, heat tolerant and hopper/mite-burn resistant potato variety produces yellow, oblong tubers with shallow eyes and yellow flesh. This variety is meant for early (September) planting in NW plains and for rabi and kharif plantings in peninsular India. It can also be grown in the main crop season in north-western and west-central plains. Further potential areas for the deployment of heat tolerant variety were identified by weather simulation studies (Minhas *et al.*, 2011). Sixteen locations in non-traditional potato growing areas of peninsular India were selected and an estimate of the available number of days in a year suitable for potato crop at different night temperature scenarios, *viz.*, 18, 20 and 22 °C were simulated. The results showed that heat tolerant varieties having 2-4 °C advantage for tuberization were suitable for more number of locations and could grow for a longer duration as compared to existing high yielding varieties. Kufri Lima is another released variety suitable for growing in warmer areas. K. Lima (CP4054) is an early season variety suitable for planting in north Indian plains. It produces attractive white-cream ovoid uniform tubers with shallow eyes and white-cream flesh. It possesses tolerance to hopper and mite burn, good keeping and culinary quality. Kufri Kiran is the latest heat tolerant potato variety released in 2021 for UP, Bihar, Haryana, Punjab, Uttarakhand, Gujarat, Odisha and Maharashtra and also in West Godavari District of Andhra. The variety is early-medium maturing (85-90 days), with yield potential (30-35 t/ha) and excellent storability under ambient conditions. Many Advanced stage hybrids namely HT/7-620, HT/7-1105, and HT/11-3 for heat tolerance has been introduced in AICRP (P) for multi-location testing. Some new locations *viz.*, Thiruvananthapuram, West Godavari (Andhra Pradesh), Bhuvneshwar, Ladol (Gujarat) and Pune (Maharashtra) *etc.* are some of the new bench marks/ locations in this segment.

(A) Kufri Surya **(B) Kufri Lima** **(C) Kufri Kiran**

Figure 28.1: Heat Tolerant Potato Varieties.

7. Breeding Varieties for Drought Tolerance

Potato is susceptible to drought (Yuan *et al.*, 2003; Monneveux *et al.*, 2013) due to shallow root system, low capacity of recuperation (Iwama and Yamaguchi 2006) and poor soil water extraction (Weitz *et al.*, 1994). Drought is considered to be the major limiting factor for potato production influencing negatively not only yield but also tuber quality. The magnitude of drought effects depends on timing, duration and severity of the stress. Insufficient water supply may occur almost anywhere, where potatoes are grown. In sub-tropics where potato production is only possible with irrigation, short periods of drought often arise because of inadequate irrigation techniques or shortage of water. Taking into account production conditions and the present yield levels, it is estimated that the average potato yield in the world could be increased by at least 50 percent if the water supply to the crop could be optimized. Therefore, identification of drought tolerant genotypes for yield maintenance and breeding purposes is now a priority for improving drought tolerance of potato crop, saving irrigation water, and ensuring yield and food security in changing scenario of global climate and growing demand of water (Luthra *et al.*2018). With the spread of potato cultivation to subtropical and tropical areas, the crop is likely to be exposed to increasing incidences of drought stress. Therefore, to cope with it, the first line of defense is the development of drought tolerant varieties. Genetic variability exists in the potato germplasm, varieties and wild species for tuber yield under drought stress (Schafleitner *et al.*, 2007). Newer techniques like heavy carbon isotope (13C) discrimination based on Δ13C values indicate water use efficiency of the plant. The technique is based on the fact that the ratio of 13C/12C is fixed in the atmospheric air. The heavier isotope 13C is slower to diffuse through stomata and RuBP carboxylase prefers 12C or fixation over heavier isotope 13C. As a result of both these factors, the ratio of 13C/12C (Δ13C) is lower in the plant tissue and more so if the water use efficiency if high. These ratios can be easily measured by isotope ratio mass spectrometer (IRMS). Screening of potato germplasm for Δ13C showed good genetic variability for this character and can be utilized for breeding drought tolerant varieties (Minhas *et al.*, 2003).

8. State of the Developments

Drought may affect potato growth and production in three ways:1) by reducing the amount of productive foliage, 2) by decreasing photosynthesis per unit of leaf area and 3) by shortening the vegetative period. Drought after planting may delay or even inhibit germination. Drought after planting is generally experienced by the potato crop under rain-fed conditions. Bansal and Nagarajan (1987) found that water stress caused reduction in leaf growth in all the eight cultivars tested by them, although the extent of reduction varied within the varieties. Even mild water stress of -3 to -5 bars greatly reduced leaf extension in potatoes. Similar results on reduction of leaf growth under water stress were obtained by other workers (Kumar and Minhas 1994). Insufficient water supply in the period between emergence and the beginning of tuber bulking may therefore lead to a limited growth rate of foliage characterized by smaller leaves and smaller plants (Kumar and Minhas 1999, and Kumar *et al.*, 2017b). As a consequence, soil cover with green foliage will often be incomplete and yields will be below optimum (Kumar *et al.*, 2017c). Decline in the rate of photosynthesis is fast and substantial even at relatively low water potentials of -3 to -5 bars. Even in the irrigated crop, plants experience water stress during the mid-day as the roots are not able to fulfill the transpirational demands of the plants fully. Mid-day depression of about 40 percent in the rate of photosynthesis in well irrigated crop was reported by Minhas and Sukumaran (1988). Plants respond to water stress by closing their stomata, thus shutting out the supply of CO_2. Sugar concentration within the leaf tissue increases which enhance the osmotic potential of the plant, thus leading to feedback inhibition of photosynthesis.

Sensitivity of the potato crop to water stress varies with the developmental stage of the crop. Various workers have defined these stages as per their convenience, however, the stage between stolon initiation and early tuber development is the most sensitive and stress at this stage causes maximum reduction in tuber yield (Minhas and Bansal, 1991). Water stress during the tuber bulking stage caused a reduction in the leaf expansion rate, but to a lesser extent, as compared to plants before tuber initiation. Presence of the tubers probably increased the water capacitance of the plants (Kumar and Minhas, 1999) leading to reduced effect of water stress. Apart from reduction in leaf growth, water stress during tuber bulking stage accelerates plant senescence resulting in decreased leaf area index (LAI). At first, lower leaves start to wilt and drop off, simultaneously drought inhibits the development of new leaves.

Water stress affects tuber quality characteristics like shape, dry matter and reducing sugars content (Kumar *et al.*, 2006; Kumar and Minhas 2013; Kumar *et al.*, 2015). Tuber shape defects like dumb-bell shaped, knobby or pointed end tubers are caused by short periods of drought during the tuber bulking stage. In dry soils, where temperature can go high, there is a possibility of occurrence of misshapen tubers due to secondary growth. This phenomenon may also result in poor cooking quality (glassiness) in some of the tubers or jelly end or translucent end tubers. All these tubers have high content of reducing sugars, which makes them unfit for the processing industry. Tubers of water stressed plants often have higher content of total sugars than well irrigated plants (Levy, 1983). Some studies have shown that the dry matter and starch content of potato tubers grown under low soil moisture was higher than the well-watered plants thus improving the quality (Kumar and Minhas, 2013, 2015). Other studies have shown that dry matter declines upon exposure to drought, however, there were varietal differences for this character. Tuber starch content has been reported to increase under drought stress (Ghosh *et al.*, 2000).

9. Indicators of Drought Tolerance

The plants water status can be described by leaf water potential (LWP) indicating the energy status of water in the plant and the relative water content (RWC). Since these parameters are related to leaf growth and photosynthesis rates, they can provide information on drought tolerance. Further, direct or indirect indications of drought tolerance can be obtained by measuring photosynthesis rate, water retention of excised leaves, anatomical structure of the leaf, depth and extension of the root system, yield under dry growing conditions in the field, water use efficiency and transpiration rate, the extent of wilting and recovery after severe drought stress.

9.1 Relation among Water Deficit Stress Parameters

When plants are exposed to water stress, reduced water uptake leads to reduction in RWC and LWP, increased stomatal resistance due to closure of stomata and consequently reduction in the rate of photosynthesis. It has been found that under controlled conditions the photosynthesis rate of potatoes was reduced by 17 percent at an LWP of –4 bars and by 50 percent at –11bars. Both passive and active osmotic adjustment of the cell sap allows the plant to maintain the photosynthetic rate at lower LWP. When RWC of a leaf decreases; the osmotic value of the cell sap will increase passively. However, it is not clear whether active osmotic adjustment takes place in potato plant or not.

9.2 Varietal differences in Sensitivity to Water Stress

It is well known that varieties of many field crops including potato respond differently to water stress with respect to various growth parameters and yield. The varietal differences in response to stress can be due to variation in one or more of the factors like anatomical structure of the leaf, stomatal behavior, leaf water potential, osmotic potential and relative water content, water retention by the leaves, water use efficiency and transpiration rate, leaf proline content, leaf longevity, chlorophyll fluorescence kinetics, root factors (depth and volume), cell membrane stability, wilting and recovery after a long water stress period, yield determinations on a soil profile with varying depths, photosynthesis rate, secondary metabolites in leaf and tuber tissues, water use efficiency and harvest index. These parameters can serve as a basis for screening clones for tolerance against drought but most of them are not easy to apply or are not sufficiently reliable, so that it could be used to screen a large number of clones. The most promising method seems to be estimating water use efficiency and transpiration rate based on isotope ratio of carbon and oxygen by isotope ratio mass spectrometer.

This instrument can measure time averaged rate of water use efficiency based on ratio of carbon isotopes and transpiration based on ratio of oxygen isotopes.

10. Screening Techniques for Drought Tolerance

Various techniques have also been attempted by different workers. Using these techniques drought tolerant varieties have already been developed/ screened in potato, sorghum, wheat, maize *etc.* Some of the screening techniques for tolerance to water deficit in potato are as under-

- Leaf extension rates are drastically reduced during stress. When stress is relieved, leaves start growing again and the rate of growth has been found to be related to degree of drought tolerance (Bansal and Nagarajan 1987).
- Isotope discrimination ratio (Δ13C), as described by Udaya Kumar *et al.*, 1988.
- Field screening using line source principle, where test population is planted in radial lines starting from the sprinkler source, the amount of water received by the plants is reduced as we move away from the sprinkler and the resultant yield at various distances gives a good indication of drought tolerance (Steyn *et al.*, 1998).
- A higher total root mass in in-vitro plantlets is an indication of drought tolerance in potato (Gopal and Iwama 2007)

11. Breeding Plan

For developing potato varieties possessing tolerance to drought, itis important to identify suitable parents/progeny from segregating population by evaluating them in lab/field condition. Kumar *et al.* (2003) suggested multiple selection criteria for abiotic stress-resistance to achieve high yielding populations. During selection, care is to be exercised for i) ability of the plants to maintain high relative water content (RWC) in its leaf tissues ii) low shoot/root ratio under water stress, iii) stay green trait or higher leaf longevity particularly during post stress period of the crop and iv) higher water use efficiency.

11.1 Hybridization

The promising parents are selected based on their water use efficiency and other parameters of drought tolerance. The selected parents are crossed with well adapted genotypes of the regions female parent.

11.2 Screening

The seedlings raised in seed beds are transplanted in the field. The selected clones are multiplied for two years under main planting conditions in the sub-tropical plains to obtain enough tubers for evaluation. Subsequently, the evaluations are done under water stress conditions (attained through delayed frequency of irrigation) and multiplication in normal irrigation. Watering is regulated (using Parshall Flume) to be in equal quantity in each plot. Total irrigation is five in normal irrigation and three in water stress conditions. Field phenotyping traits *viz.*, percent reduction in leaf area, relative water content (RWC), plant height, photosynthesis rate, stomatal conductance, transpiration rate, leaf temp are recorded at peak stress levels along with tuber shape, appearance and yield at harvest. The outline of breeding procedure employed for developing drought tolerant varieties is given in CPRI Tech Bulletin No.74 (Revised)

12. Current Status

Considering future water scarcity, a programme to breed drought tolerant varieties was started in 2005. Besides field screening of existing varieties, hybrids and exotic clones (Kumar *et al.*, 2005 and Kumar and Minhas 2013); carbon/oxygen isotopes discrimination, proline profile *etc.* were considered in the selection of parents. Ninety-eight germplasm lines were screened for water use efficiency and transpiration rate by isotope ratio and good variation was found in the potato germplasm (Minhas and Kumar 2012). Highly water use efficient and high transpiration rate lines were selected and used in breeding for combining both these characters to develop drought tolerant potato varieties. Established protocols were followed for breeding and selection of desired tubers traits in subsequent generations. Promising hybrids were evaluated for leaf area pattern, photosynthesis rate and associated parameters, antioxidant defense mechanism, leaf senescence, tuber yield and tuber defects under reduced water regimes (25 percent less).Based on yield performance under mild and severe (Modipuram) and severe (Jodhpur) drought conditions, Luthra *et al.*, 2018c, reported high yield and high drought tolerance index in advanced stage hybrids MS/6-1947, which has been released as Kufri Ganga for cultivation in north Indian during main crop season. Further, breeding efforts for identification and release of drought tolerant/water use efficient varieties led to identification and release of three drought tolerant varieties namely Kufri Thar-1 (WS/05-146), Kufri Thar-2 (CP4175) and Kufri Thar-3 (J/92-167) (Figure 28.2) for drought prone areas in different agro climatic zones of the country recently in 2020 (Kumar *et al.*, 2020).

(A) Kufri Thar-1 **(B) Kufri Thar-2** **(C) Kufri Thar-3**

Figure 28.2: New Potato Varieties (with high water use efficiency) for Drought Prone Areas.

13. Future Thrusts

High temperature and water scarcity are the most crucial factors affecting potato yield. Night temperature beyond 20-22 °C drastically reduces tuber formation, constraining potato cultivation in tropics and subtropics. Similarly, owing to shallow root system, potato is extremely sensitive to water deficit stress. Identification of genes and pathways affected by high temperature is crucial for developing thermo tolerant cultivars for these regions. There is urgent need to improve water use efficiency (WUE) and associated traits in potato. Therefore, gene expression analysis and pathway maps need to be constructed to precisely understand the gene expression patterns under these two prominent abiotic factors (heat and drought). Genes associated with photosynthesis, hormonal activity; sugar transporters and transcription factors are also known to be differentially expressed. The effect of high temperature on expression of genes controlling tuberization, photosynthesis and also secondary metabolites, particularly glycoalkaloids needs to be emphasized to make available a framework for further investigations in to heat stress in potato. Besides, in the present scenario of climate change potato respond moderately sensitive to soil salinity and needs freshwater irrigation for tuber production. Salinity is a major problem for potato cultivation in many parts of the world where majorly potato is grown extensively. The recent trends showed that it might be a major problem in India in the coming future due to its expansion in non-conventional parts, particularly in coastal regions. To select salt-tolerant cultivars attempts are under way to screen potato cultivars/germplasm using standard protocol involving tissue culture technology. Identification of salt tolerant genotypes/germplasm is first step and the identified material shall be exploited to develop varieties suitable for salt affected non-traditional areas.

Chapter 29

Economics, Marketing and Export of Potato in India

Nalini Ranjan Kumar and N.K. Pandey

ICAR-National Institute of Agricultural Economics and Policy Research (NIAP), New Delhi

1. Introduction

Fruits and vegetables are more difficult to market than to produce. Perishable and bulky nature of the produce makes marketing process more difficult and vulnerable to uncertainties to a higher extent. India has experienced silent potato revolution during last 65 years. In 1949 (when CPRI was established), India produced 1.54 million tonnes potato out of 0.234 million ha with an average yield of 6.58 t/ha. However, the country produced 45 million tonnes of potatoes from 1.9 million ha with an average yield of 23.68 t/ha showing an increase of 8.1, 29 and 3.6 times in area, production and yield of potato, respectively. At present India is facing a problem of plenty as far as potato is concerned. However, CPRI believes to take this production to 125 million tonnes by the year 2050 (Singh *et al.*, 2014).

Potato production is not a problem in India and planners need to think from marketing end for further improvement. The reported post-harvest losses for potato are 16 percent of the annual production (FAOSTAT). The export performance of Indian potato is quite dismal as we have rarely been able to export more than 0.5 percent of our domestic production (Kumar 2005). Potato processing in India is growing fast, yet the proportion of processed potato is still low *i.e.* nearly 7.5 percent of the production against up to 60 percent in developed countries (Rana 2011 and Singh *et al.*, 2014). Other marketing functions such as grading, packing, storage and transportation *etc.* also need transformations and special attentions as for as potato is concerned (Pandit *et al.*, 2003).

2. Potato Marketing System

2.1 Conventional (APMC Act)

Conventionally potatoes are taken to wholesale markets through different marketing channels and then go to consuming markets or consuming centres from these markets. A chain of market intermediaries such as field procurement/ assembling agents, forwarding agents, commission agents, wholesalers, and retailers are operating to complete the marketing process. However, some of these commission agents and wholesalers may act at more than one point. Due to lesser quantity of produce flowing through a particular marketing chain, the marketing efficiency of the process is generally low. For example, most of the small retailers have a target of earning Rs 400 to 800 a day and they accordingly adjust their marketing margins which may be up to 100 percent, or even more, of their purchase price.

2.2 Model Act

With the promulgation of Model Act in 2003 a new era of agricultural marketing was promised. The provision of creating private markets in various parts of the country resulted in varied performances in different states. Some of the states were quick in creating state of the art private markets having all modern amenities while the others showed lack-lustre interest. With the results big corporate houses to came and joined the business of procurement and retailing in agriculture. Since marketing margins under conventional marketing system were huge the new business model was assessed to be profitable for big companies.

2.3 Retail Chains

At present a large number of big and reputed corporate houses have shown interest in this sector. Very high level of apprehensions was built among the conventional retailers regarding the dangers to their existence. However, Dr P Chengappa, present Vice Chancellor at the University of Agricultural Sciences, Bangalore opined very early that fresh fruits and vegetables are a small part of this retail juggernaut and even if retail chains do very well, they will capture only 10 percent of the market in another 10 years. Actually the retail chains have failed to corner a significant share of sales of fruits and vegetables in India. It is in spite of strong theoretical explanation that the new marketing system will eliminates market intermediaries by directly purchasing from the farmers and selling to the retailers, providing cheaper and better quality produce. Now the estimated 1.2 crore street vendors and small retailers in our country, out of which nearly 30 lac retailers and vendors sell fruit and vegetables, are much assured about their livelihood sustenance even in every part of every urban area in India.

2.4 Future Trading

Futures trading is an agreement between a buyer and a seller obligating the seller to deliver a specified asset of specified quality and quantity to the buyer on a specified date at a specified place and the buyer, in turn, is obligated to pay to the seller a pre-negotiated price in exchange of the delivery. However, for speculators futures trading is a process under which sellers make promises to deliver something they don't have; and buyers promise to accept delivery of something they don't want- and both legally break their promises. Here profit maximization is the prime objective of the speculator while loss minimization is hedger's aim. Both speculators and hedgers seldom allow future contracts to mature, by nullifying the contract with reverse sale or purchase of contracts. In India MCX, NCDEX and NMCE are the three major commodity future exchanges while the first two trade in potato. Major part of potato future trades take place in MCX and some details of the potato contacts are given in Table 29.1. Price discovery, hedging and speculation are the main functions of future markets. For further details on the subject the readers may refer to Rana *et al.* (2008).

Table 29.1: Contract Details (MCX Potato)

Lot size	30 MT
Margin	6 percent of value
Leverage	16.67 times (Pay for 1.8 MT to buy a contract of 30 MT)

2.5 Contract Farming

Contract farming is a written or verbal contract between buyer and seller of purchase/sale of an agricultural commodity at pre-defined terms and conditions. These terms and conditions are generally related to price, quantity and quality of the produce. Contract farming has been an important tool to cut risk in the procurement process of processing quality potatoes, seed potatoes and specialty potatoes. There are many contract farming models prevalent in various Indian states and some of them are exploitative too. The concept is still maturing in India and its popularization is required because from farmers' point of view contract farming can reduce price risk associated with excessive price volatility of farm produce. Potato is a preferred crop for the industrial contractors due to the importance of potatoes as suitable raw material in the processing industry. It is worth mentioning here that industry needs potato tubers having high dry matter and low reducing sugars contents. Seed supply chain of specialized processing varieties is another important issue in this business. Hence, the processing industry opts contract farming in order to exercise better control over quality and quantity of the required raw material. For further details on the subject the readers may refer to Singh *et al.* (2011).

3. Important Potato Marketing Functions

3.1 Grading

Grading is very important function and offers a convenience of choosing a product according to customers' taste and preference(s). In Indian potato markets grades differ from market to market. Since very high proportion of grading is done by human beings hence size under the same grade may differ from person to person. CPRI has developed mechanical graders, which are used by large farmers only. Much needs to be done on popularizing grading practices adopted in international potato markets. Due to non-adoption of such grading practices our potato exporters are facing a lot of problems. Now Indian markets have started rewarding well graded and sorted produce with the higher prices.

3.2 Transportation

The transportation cost in India is very high due to very slow effective speed of goods movement. Rail transport is quite cheap but railway's very low priority for potato transport creates problems for our domestic as well as export marketing. Frequent thefts in rail transport add to the problems. Bad condition of existing roads, absence of roads in rural areas and scarcity of express highways make our cargo movement very slow while frequent interruptions of cargo on interstate borders and traffic corruption aggravate the situation. Potatoes transported both using rail and road are liable to large scale spoilage due to hot weather and prolonged exposure to the sun. It creates wide price gap between potato producing and consuming area.

3.3 Cold Storage

Shortage of cold storage space in many states of India, erratic electricity supply, lack of cold stores on and near ports, inadequate cold chain and expensive cold storage or cold chain facilities create problems for potato marketing in India. We know that about 90 percent of potato in India is produced in Rabi season hence storage plays an important role in distributing supply over the remaining months of the year. Cold store operators also indulge in different malpractices such as changing/stealing stored material. Farmers do not get any compensation even if the produce is spoiled due to the negligence of cold store operators. Except for few states where electricity supply is too much interrupted of obstructed, the cold storage facilities have been adequate. However, India needs to do a lot on the front of modernization of existing cold storage facilities. In order to cater to the needs of processing industry and consumers who dislike sweetness of conventionally cold stored (2-4 °C) potatoes, the popularization of potato storage at elevated temperature (10 to 12 °C) after CIPC treatment has proved to be a boon.

3.4 Processing

Food processing is the sunrise sector of India. The importance of this industry is not limited only to the contribution to the GDP, but it also provides many other desirable socio- economic benefits such as increased employment opportunities; improvement in income and lifestyle of the rural people leading to reduction of migration of rural masses to cities; and mitigation of huge post-harvest and storage losses, specially, in fruits and vegetables. Tremendous potential of Indian potato processing industry and its rapid growth during recent past has been widely discussed (Chengappa 2004; Rana *et al.*, 2004; Pandey *et al.*, 2006; Rana and Pandey 2007; Singh 2010; Rana 2011; Singh and Rana 2012). India possesses wide agro- climatic conditions and areas suitable for adequate and round the year supply of processing quality potatoes. In addition, now we have potato varieties that are bred according to the requirements of the processing industry. The increasing proportion of potato processing in India shall not only avoid glut like situations but also carry forward the potato revolution in India. However, estimation of the requirement of raw material (processing quality potatoes) for potato processing industry is of utmost importance. McCain Foods is popularizing French-fries in India in anticipation of a strong demand for convenience and variety in food products. Changing eating habits, growing number of working women and acceptance of western life style, especially in large cities has encouraged setting up of a large number of fast food restaurants in India that serve French-fries too. However, McCain Foods experienced constraints in the availability of raw material for French-fries in addition to infrastructural bottlenecks in the country. The company set up 30000 tonnes French fries plant in Gujarat during 2006 which has now been upgraded to more than double its original capacity. French fries are the fastest growing potato product in our country. With the results other French fries' unit have started processing in Gujarat during the recent years. Potato chips, potato sticks, and potato 'bhujia' are one of the most common other processed products available in markets. Besides our traditional potato products like dehydrated chips, papads, waries and Laccha *etc.* are also finding increasing popularity not only in India but among NRIs abroad also. Potato powder/flakes is another rapidly growing segment of potato processing industry in India. New players are establishing facilities and the existing ones are expanding aggressively. The raw material requirement of potato processing industry and the corresponding demand for processed potato products can be viewed from Tables 29.2 and 29.3, respectively.

Table 29.2: Raw Material Demand of Potato Processing Industry (million t)

Product(s)	*2010*	*ACGRs*	*2050*
Potato chips	2.45	4.5	14.22
Potato flakes/powder	0.29	7.6	5.44
Frozen potato products	0.06	11.6	5.40
Total	**2.80**	**5.61**	**25.06**

Table 29.3: Demand of Potato Products (thousand t)

Product(s)	*2010*		*2050*	
	Thousand t	*Per capita g*	*Thousand t*	*Per capita g*
Potato chips	612	506	3555	2194
Potato flakes/powder	52	43	979	604
Frozen potato products	34	28	2808	1733
Total	698	577	7342	4532

3.5 Exports

Although, Indian potato is free from major prohibitive diseases like wart, ring rot, black leg, and nematodes *etc.*, yet due to the want of appropriate certification our seed is badly affected under sanitary and phyto-sanitary sanctions. Opinion on genetically modified potatoes and anti-dumping duties are potential trade barriers of the future for Indian potato exports. Very high levels of farm subsidies by developed countries distort international potato trade too. Unfair means adopted by Dutch seed-potato multinationals who have been reported giving expensive gifts and bribes to control their business in Bangladesh and Sri Lanka. Lack of long term and consistent potato exports policy is one of lacking initiatives at our part. Absence of required market intelligence *i.e.* information on international prices and preference and non-seriousness on quality assurance are costing us a lot. Time of product arrival in the international market plays a crucial role in earning profits. Variety is a most important consideration in the mind of farmers going for buying seed-potato. A number of Indian potato varieties and hybrids developed by the CPRI are already grown in different countries including Afghanistan, Nepal, Bhutan, Bangladesh, Sri Lanka, Philippines, Madagascar, Bolivia and Vietnam either under Indian names or modified local names *e.g.* Kufri Chandramukhi (in Afghanistan), Kufri Jyoti (in Nepal and Bhutan) and Kufri Sindhuri (in Bangladesh and Nepal). Five Indian hybrids, I-654 (as CCM-69.1 in Mexico); I-822 (as cv. Krushi in Sri Lanka); I-1035 (as cvs. Montanosa in Philippines and Mailaka in Madagaskar); I-1039 (as cvs. India in Bolivia and Red Skin in Vietnam); and I-1085 (as cvs. Sita in Sri Lanka and BSUP-04 in Philippines) are being officially grown. Many countries are growing Indian potato varieties/ hybrids unofficially too.

Since many countries out of these are our actual or potential seed-potato importers, we have an important opportunity of increasing our exports to them. In most of our neighbouring countries potato in grown under short day conditions like India and the varieties as well as physiological age of seed produced under long day (temperate) conditions of Europe and Canada *etc.* are not suitable here. India is the only country in South Asia that has its own potato seed production programme, which further increases our seed potato export potential to these countries. The annual seed potato market in SAARC countries alone is 616,000 tonnes. Our exporters are getting inquiries even from the countries like Egypt and Turkey. Bulk of Indian potato is harvested during January to March when fresh potato is not available in most parts of the Northern Hemisphere. It creates an important export opportunity for Indian table potato to many EU and other countries. Sri Lanka, Bangladesh, UAE, Nepal, Mauritius, Pakistan, Maldives, Singapore, Japan, Malaysia and Saudi Arabia are our existing or potential table potato importers. Our table potato of Kufri Sindhuri and Kufri Jyoti varieties is in great demand in Nepal and Bangladesh while Kufri Chandarmukhi has high preference in Afghanistan (Singh and Rana 2005). Iran, although an important exporter of table potato in one part of year, imports sizeable potato during the other part. India should target this potential market too.

India is gradually becoming an important supplier of French-Fries to countries like Mauritius, UAE, Sri Lanka, Malaysia, Bangladesh, Nepal and Singapore *etc.* MNCs are showing increasing interest to make India an outsourcing destination for their global French-Fries need. McCain Foods, a leading MNC in French-Fries has started large scale contractual potato growing for French fries with the Indian farmers specially in Deesa area of Gujarat. This year a lot of such potato has been exported by McDonald's, another MNC in instant potato processing, after buying from the McCain Foods. This is an example that needs to be emulated by other corporates in India. Besides this India has been exporting potato chips to Bangladesh, Oman, Sri Lanka, Maldives, Canada, Kuwait, UAE, UK and USA *etc.* Dried potato products like potato flour, starch, dehydrated chips and flakes *etc.* are exported from India to Sri Lanka, UAE Maldives, South Africa and UK. This is an area where Indian exporters should also concentrate. There is an untapped potential of exporting canned or baby potatoes from India. Presently we are canning potatoes for the requirements of our military and para-military forces. Processing quality potato from India is getting demand from Middle East, especially Dubai. Countries like Japan, China, Korea, Malaysia and Singapore *etc.* have great demand for processing potato. We have three varieties suitable for processing namely Kufri Jyoti, Kufri Chandramukhi and Kufri Lauvkar and four specially bred for processing, namely Kufri Chipsona -1, Kufri Chipsona-2 and Kufri Chipsona -3 for chipping purpose and Kufri Surya for making French fries. In India we have a lot of scope for growing potato suitable of processing under warmer conditions. This is a great opportunity we should exploit for the betterment of our national economy and potato farming.

3.6 Gluts

Glut is a market situation where supply of a commodity exceeds its demand and with the result prices of that commodity crash. This term is more commonly associated with the agricultural goods and perhaps the most with the potato. The years 1975, 1979, 1982, 1985, 1987, 1988, 1997, 2000 and 2003 were the glut years for Indian potato. Although now potato is not thrown on the roads yet farmers face low prices after every 3-4 years. Acreage adjustment to the last year's potato prices; seasonal and regional production of potatoes; release of many short duration and high yielding potato varieties; inelastic price and income elasticities of potato demand; inadequate cold storage capacity; under-developed potato processing industry; very low exports of potato as well as its products; seed being the most important cost factor of potato cultivation, farmers use last year's unsold (cold stored) potato as seed; market manipulations; transportation bottlenecks; farmers' financial commitments against the crop to be harvested; majority of small farmers going for early potato crop with the objective of taking three crops a year; unawareness of general masses on potato nutrition; and non-adoption of low-cost/input technology are the main reasons for potato gluts in India.

Large scale technical expertise and huge investments in mechanization restricts regular potato growers from switching over to other crops however, higher profits in some years tempt new farmers to jump into potato business increasing crop area significantly. There are some other biotic and abiotic factors that contribute their share to make it very difficult to avoid potato gluts. As a result of easy availability of loans in agriculture sector over the recent past, more and more farmers are falling under the burden of debts. Many potato farmers fear loss of their land and occurrence of suicide stories like cotton growers under the influence of more frequent and violent price fluctuations of the concerned produce. However, prudent behavior at farmers' level with the adequate support at government and research level can go a long way in mitigating the impact of gluts on farmers and their families. Potato being capital intensive crop, gluts have far reaching socio-economic adverse impact on potato farming community and it is very important to identify control measures for such gluts as well as to take concerted steps towards their proper implementation. Fortunately, the current market conditions indicate lower or even no incidence of gluts in agricultural commodities in general and potato in particular.

4. Projected Scenario

Potato has been found assuming an important role of food security option in India (Singh and Rana 2013). Rapid urbanization, rising number of working women, elevated tendency of eating outside and increasing number of fast food restaurants in India is increasing per capita demand of potato in India at an annual compound growth rate of 2.04 percent. The population of India is estimated to rise at an annual compound growth rate of 0.78 percent. The component wise analysis of this huge rise in estimated demand (including potato demand in processing and seed sectors in addition to the wastage on account of post-harvest losses) has been given in Table 29.4. The marketing of fresh, processing quality and seed potatoes will bring new challenges in the future.

Table 29.4: Component-wise Potato Demand (million t)

Year	*Demand$*	*Fresh*	*Processing*	*Seed*	*Waste*
2010	35.10 (100.00)	23.94 (68.21)	2.67 (7.61)	2.96 (8.43)	5.53 (15.75)
2050	121.81 (100.00)	78.47 (64.42)	25.06 (20.57)	6.10 (5.01)	12.18 (10.00)

Note: Figures in parenthesis are percentages; $: net domestic demand Source: Singh *et al.* (2014)

5. Conclusion

India is the second largest producer of potatoes after China however, cultivation of majority of potatoes in India under sub-tropical conditions pose highly area specific problems for us. More than 90 percent of our potato production faces hot summer soon after its harvest creating very strong need for cold storage in the country. Similarly, more than 85 percent potato production takes place in North India states which emphasize the need of efficient transportation. Majority of potato growers being small and marginal are subjected to exploitation in the markets. Due to smaller volumes flowing in many marketing channels the marketing efficiency in India is low. Smaller holdings and marketable surpluses create conditions adverse for proper marketing especially grading, sorting and selling in the markets at the right time. The huge estimated rise in potato production in the future will further make the marketing process complicated and challenging.

Chapter 30

Insect Pest Management in Potato

Anuj Bhatnagar, S. Subhash, Subhash Katare, Kailash C. Naga and Aarti Bairwa

ICAR-CPRI, Regional Station, Modipuram, Meerut, Uttar Pradesh

1. Introduction

Potato crop is attacked by a number of insect pests which are one of the important limiting factor in raising productivity and production of potato in the country due to transfer of number of viral diseases, resulting in quick degeneration. Insect vectors like whitefly and thrips have given new dimension and challenges to IPM of potato in the country. Adoption of IPM is more relevant in potato because potato is often consumed directly as a vegetable. Still, there is a gap in available technology and the technology being adopted by the potato farmers. Beside other factors, foliar diseases and soil and tuber borne diseases, viruses, soil pests, foliage feeders, sap feeders and storage pests cause considerable damage at various stages of the crop and their severity and incidence varies from region to region. Amelioration in crop losses due to diseases and insect pests through scientific interventions is very critical and forms sound means of their management. In this chapter, insect vectors in potato seed production and their management (IPM) have been dealt comprehensively. Its interventions should be suitably modified and validated according to the region before its implementation.

2. Insect Pest Scenario in Potato

Insect pests cause direct damages to the crop by feeding on potato tubers in soil, feeding on foliage and on potato tubers in stores, while indirectly by sucking the sap and transmission of viral diseases in the crop resulting quick degeneration of seed stocks. The number of insect pest species causing serious damage to potato crop has been few in 70's to 10-15 in the last 5 years. Among the major vector pest, aphid *Myzus persicae* Sulzer is the most destructive. They inflict severe losses in potato particularly by transmitting number of viruses like PLRV, PVY, PVA and PVM. White fly, *Bemisia t abaci* (Gennadius) now assumed significant importance in North Central India as; they are main vector of Gemini virus, which causes apical leaf curl disease in potato. Similarly, leafhoppers and mite are now serious problem in early potato crop in North-Western plains and Plateau region. These vectors are considered more important because they are responsible for significant decline in potato yields directly or indirectly across the country.

2.1 Sucking Pests (Sap Feeders)

The major sap feeders include aphids, whitefly, thrips, leafhoppers and mites. Among these group, aphids are of the greatest economic importance as a vector of virus diseases in potato. Whitefly is assumed significant status as a vector in North Central India but high populations of these sap feeders can also cause direct damage to crop. Thrips, mites and leafhoppers are sporadic and restricted to certain potato growing areas. These sap feeders not only cause direct damage to potato crop but also transmit viruses, resulting degeneration of potato seed.

2.2 Aphids

Aphids are one of the most important insect vectors in the production of healthy seed tubers. Among the dozen species, *Myzus persicae* (Sulzer) and *Aphis gossypi* have been reported widely on potato crop in India. As a result of feeding the leaves curl downward, turn yellow, become wrinkled and ultimately die. Aphids transmit potato virus in two ways *i.e.* non persistent and persistent manner in potato crop. Potato leaf roll virus (PLRV) and potato virus (PVY) are estimated to 20 to 50 percent and 40 to 85 percent yield losses, respectively. In the North-Western and Central Indo-gangatic plains the population of aphids remained low from October to December. Aphids are crossed the critical limit (20 aphids/100 compound leaves) in these regions only after last week of December in North-Western plains and second week of January in North-Eastern plains. The maximum population is generally observed in February and March. The seed potatoes can be grown successfully by adjusting the planting time and cutting of haulms as per the activity of aphids. Reproduction of aphids is usually parthenogenetic and viviparous. The *M. persicae* feeds on several ornamental plants, weeds and vegetables crops.

2.2.1 Management

- ☆ Potato seed production area should be selected on the basis of population studies of aphids for the production of disease free potato seeds.

- ✰ Eliminate alternate host plants or infected diseased plants; use balanced fertilizer and good water management in potato fields.
- ✰ Haulm killing and harvesting should be completed as per the activity of aphids or before the population attaining 20 aphids/100 compound leaves.
- ✰ Monitoring of aphid population through aphid count on 100 compound leaves or use of water and yellow sticky traps (20 traps/ha).
- ✰ Treatment of chitted tubers with Imidacloprid 17.8 SL @ 0.05 percent (5ml/10 lit.) for 10 min dip or spraying on whole tubers
- ✰ Foliar first spray of pymetrozine 50 WG @ 0.06 per cent followed by second spray ofThiamethoxam 25 WG @ 0.05 percent, (5 g/10 lit.) 10-15 days after last spray are recommended for controlling aphids and other sucking insects on seed crop. Soil application of cartap hydrochloride 4G @ 20 kg/ha insecticide applied at earthing up has also been found effective. If required 3rd spray with Flonicamid 50 WG @ 0.04 percent (4 g/10 lit.) 15 days after last spray especially for aphids
- ✰ Conservation of natural enemies like *Coccinell aseptempunctata* L., *Chrysopa.*

2.3 Whitefly

Bemisia tabaci is tiny, soft, white winged insect, found mostly on the lower surface of potato leaves and sucking sap from succulent leaves. Whitefly is a newly emerging vector and transmit Gemini virus in potato, resulting apical leaf curl disease in North-Central India. Management is followed as per aphids.

2.4 Thrips

Thrips is very small insect of 1mm with fringed wings, adhered on the apical portion of the foliage. *Thrips palmi* Karny is one of the predominant species responsible for transmission of disease in early potato in Central India. It will acquire tospo virus only at nymph stage and transmit throughout their life. The wide range of disease symptoms include leaf droop and hanging, blackening and cracking of stem and concentric ring spots in affected stems. Potato yield losses due to the disease vary greatly from place to place and year to year and may range from 15 to 30 percent.

2.4.1 Management

- ✰ Avoid early potato planting.
- ✰ Grow tolerant potato varieties for early planting like Kufri Sutlej, Kufri Badshah and Kufri Jawahar.

- ☆ Seed treatment Imidacloprid 17.5 SL @ 5ml/10 l of water for 10min. or spray with same insecticide just after the emergence of crop @ 4ml/10 l of water.
- ☆ Frequent irrigation reduced the activity of thrips on crop.

2.5 Mites

Mites cause severe damage (20-60 percent) to early potato crop in Indo-gangetic plains, part of Himachal Pradesh, Maharashtra, Karnataka, Gujarat, Madhya Pradesh and Western Uttar Pradesh. Mites suck the sap from leaves and in severe damage resulting wither and waxy lower surface. Mite is very minute insect and difficult to see through necked eyes.

2.5.1 Management

- ☆ Delay in potato planting.
- ☆ Spraying of Abamectin benzonate 1.8 EC 0.5 ml/lit.) As and when required on potato crop.

2.6 Leafhoppers

Amrasca biguttula and *Empoasca devastans* cause direct as well as indirect damage to potato crop. Both the nymphs and adults suck the sap from tender parts and in severe condition plants showed hopper burn, which is very common in case of early planted potato crop in northern plains. The nymphs are light yellow to pale yellow in colour with well-developed head capsule Management as per aphids.

2.7 White Grubs

These are polyphagous beetle whose larvae live in soil and these larvae are called 'white grubs'. White grubs are widely distributed world over hence they are cosmopolitan in nature. The number of grubs can be more in light soil with high organic matter content. This had been most destructive insect threatening potato production in hilly regions, now it is becoming a problem in potato producing areas in plains too. In plains, white grub has long been associated with sugarcane crop and now causing the damage to potato crop. Out of many, *Brahmina coriacea, Holotrichia seticollis, H. Longipennis, Anomala dimidiata* and *Melolontha indica* are most destructive in north western hilly region and many others are found in different parts of the country. In India, adult beetles emerge from the soil during April to June in response to pre monsoon showers. The second fortnight of June observed as the peak period of emergence of beetles, adults mate in the evening and at dawn. Females return to the ground to deposit the eggs in the soil depending upon the softness of the soil. The beetles are strong fliers, in the evening of warm sunny days they come out of the soil and fly to nearby trees/shrubs at dusk. The damage is done by second and third instar grubs (larval stage) which feed on underground part of the plant by making large shallow and circular holes in the tubers. The grub stage which can be easily identified is creamy white in colour with dark brown head and attains 'C' shape of English letter when disturbed.

2.7.1 Management

Light traps can also be useful to catch the adults as soon as they emerge and kill them in water mixed with kerosene. For chemical control, damage can be minimized by the application of soil insecticide at the time of planting or at the time of earthing-up. Spray of chlorpyriphos 20 EC (0.1 percent) immediately after first monsoon showers on weeds and bunds around the field will reduce the number of grubs emerging out of eggs. Spray the crop (ridge portion) with chlorpyriphos 20 EC@ 2.5 l/ha after earthing-up to kill the larvae.

Grub management: Deep ploughing after harvest of potato is the best way to expose white grubs to high temperature and natural predatory birds. Similar ploughing should be done at the time of planting also. Only well rotten Farm Yard Manure should be applied to the fields as this acts as attractant for grubs to feed on if not fully rotten. Entomopathogenic nematodes (EPN)are effective in management of white grubs in potato crop in mid and higher hills of Himachal Pradesh.

2.8 Cutworms

Cutworms are polyphagous and most destructive insects. *Agrotis segetum, A. Ipsilon and A. flammatra* are reported to occur on potato. In India, cutworms are more serious in northern region than south. Peak activity occurs during May-June in Shimla hills, in August in peninsular India and in March April in Bihar and Punjab. In plains of country, the population of cutworm *A. ipsilon* is able to attain the first peak in mid of December, second peak will be observed during third to fourth week of March, and type of the soil has a large influence on the rate of infestation. Cutworms tend to be more frequent in soil with plenty of decaying organic material. Damage is worse where cutworms are present in and around the field in large number before planting.

The eggs are small (about 0.5 mm in diameter). The freshly laid eggs are cream coloured and turn reddish-yellow to blackish before hatching. They are laid singly or in clusters. Eggs are laid on vegetation, on moist ground around plants, or in cracks in the soil. Eggs hatch in 10 to 28 days. Young caterpillars are smooth, stout, cylindrical with blackish brown in colour dorsally. Fully-grown caterpillars are 4 to 5 cm long. The larvae curl into a C shape when disturbed and remain motionless for a short period. They commonly hide just under the soil or litter by day. They are nocturnal and hide in the soil or under stones and plant debris during the day. At night they move up to the soil surface to feed. Caterpillars construct burrows or tunnels in the soil near the host plant. They pupate in an earthen cell in the soil. Adults are dark grey, black or brown coloured medium sized moths with markings on the front wings. The life cycle can be completed in 6 weeks under warm conditions.

Crop damage is caused by caterpillar (larva) stage only. They cause damage by- (i) young larvae feeding on leaves (ii) mature larvae by cutting the stem of the plant just near the ground and (iii) making irregular holes in the tubers. Larvae also feed on roots also. Smooth, grayish-brown, greasy and plump looking caterpillars are found hiding in the soil near to the stem of the plant during day time. After tuber formation, they start feeding on tubers and roots resulting in a variety of holes, ranging from small and superficial to very large deep ones so it directly affects the tuber yield and its market value. Tuber damage can be from 9-16 percent.

2.8.1 Management

Deep summer ploughing in potato fields to exposes immature stages to predators and sun. Fields should be prepared and vegetation and weeds destroyed 10 to 14 days before planting the crop in the field. Flooding of the field for a few days before planting can help to kill cutworm caterpillars in the soil. Exposing the larvae to bird predators is the best way. Light traps installed in/around potato fields attract the adult moths.

For effective chemical control, the chemical should be sprayed at the appropriate time. The best time is when caterpillars are small and still feeding on the haulms. Chemical control would be more effective when soil is dry and weather is warm. For efficient chemical control thorough coverage of foliage with good amount of water is needed. Chlorpyriphos 2.5lit/ha and flubendiamide 20 WG@ 0.06 percent are used to control cutworms on potato.

2.9 Potato Tuber Moth (*Phythorimea operculella*)

The potato tuber moth also referred as PTM, is the most significant insect pest of the potato and three species are most important, common potato tuber moth, *P. Operculella*. The common PTM today it has become a cosmopolitan pest. It has been reported from more than 90 countries worldwide. The moth occurs in all tropical and subtropical potato producing systems. The damage has been reported from Maharastra, Bihar, Madhya Pradesh, Uttar Pradesh, Kangra valley (Himachal Pradesh), Tamil Nadu, North Eastern hill states and plateau region and Karnataka. The range of infestation can be 30-70 percent in stored potato.

The adult moth prefers to lay eggs on green foliage than on tubers. Eggs are laid on underside of the leaf in the field the young larva feed inside the tunnel between two layers of leaf tissue and later on to leaves. The larvae destroy the crop by injuring the leaves and boring into petioles and terminal shoots causing wilting. After tuberization, the eggs are laid in the eyes of tubers through cracked soil or if tubers are exposed. The larvae feed on them causing mines. Because of the heat produced by the activity of larvae in the heaps the losses become huge due to tuber rottage also. In country store the larvae bore into stored potatoes causing 18-83 percent tuber damage in NEH hills. Life cycle of PTM is completed in 21-30 days at 27-35 °C. Upper and lower threshold temperatures for PTM are 40 °C and 5.0 °C.

2.9.1 Management

Storage of healthy, un-infested potato tuber is the best way to control potato tuber moth. Cultural practices can contribute significantly to reduce PTM infestation at harvest.

2.9.1.1 Use of Botanicals/Biopesticides

Covering of potato heaps with 2.5 cm thick layer of chopped dried leaves of Lantana/Eucalyptus can prevent tuber infestation. Use of Granulosis Virus (GV) is extremely effective in reducing PTM damage. Use of sex-pheromones can be made by mass trapping PTM male adults.

2.9.1.2 Chemicals/Semiochemicals and Attract and Kill Approach

Dusting of the tubers with 5 percent Malathion or 1.5 percent quinalphos (125g dust/100 kg potato) will result in good control of PTM but these potatoes should not be consumed. Sex pheromones are available for all the PTM species. Attract and kill product (attracticide) consisted of pure pheromone, cyfluthrin as contact insecticide, formulated with plant oils and ultraviolet screens was successful in both laboratory and field experiments.

2.10 Termites

Common termites are-*Microtermes obesi, Odontotermes obesus,* and *Eromotermes* spp. The damage done by termite is more in rain fed crops than to frequently irrigated crops. They can be easily identified with brown head and dirty white soft body. It is the worker caste of the termite which damages the crops by damaging the roots as a result the leaves of the plant turn yellow and plant starts wilting and ultimately dries. Tunnels can be made into the tubers. The tubers become hollow and filled with soil and that is the typical symptom of termite. These insects are social insects. Most of the individuals perish but the remaining few will mate, shed their wings and burrow in the soil to form new colony of which they become king and queen, the queen lives 5-10 years and lays large number of eggs 70,000-80,000 per day. Crop residues should not be left in the field as they provide food to the termites.

2.11 Red Ants

Red ants, *Dorylus orientalis* and *D. Labiatus* are reported as a pest on potato, cauliflower, cabbage, groundnut, sugarcane, coconut seedlings. They also have a habit of attacking the underground parts of the plant but do not avoid light. They can be seen in large numbers in the field. The ants seen in the fields are workers. It is a pest of warmer climates (> 26 °C), high temperature and dry weather favours the population build up. Workers of red ant feed on tubers by nibbling and making small but deep circulars holes. The damaged plants wilt in sunlight and eventually dry up.

2.11.1 Management

The pest is very difficult to eradicate. Spraying of the crop and drenching of the ridges with chlorpyriphos 20EC @ 2.5 l/ha helps.

Chapter 31

An Overview of Virus-Vector Interaction with Reference to Whitefly-Begomoviruses System

Kailash C. Naga

ICAR-Central Potato Research Institute, Shimla, Himachal Pradesh

1. Introduction

Potato is attacked by many insect vectors during its growth stages. The insects which also causes indirect losses by transmitting the plants viruses are, whitefly, aphids, thrips, and leaf hoppers. The whitefly, *Bemisia tabaci* (Gennadius) (*Hemiptera: Aleyrodidae*), is one such highly notorious insect vector and a polyphagous pest which is responsible for significant yield losses in many crops including potato. Whitefly may cause enormous direct losses by sucking plant sap and reducing plant vigor or indirectly by transmitting several plant viruses. *B. tabaci* has ability to ingest viruses belonging to at least five genera: *Begomovirus, Ipomovirus, Crinivirus, Carlavirus, Torradovirus* and can transmit more than hundred plant viruses. The transmission of several plant viruses along with its ability to develop resistance to pesticides makes it a most devastating pest in agriculture. Begomoviruses and *B. tabaci* have been considered as one the most economically important virus-vector complexes during the last two decades. This virus-vector complex poses a serious threat to major vegetable and field crops across the developing world where availability of food remains scare. The plant viruses are transmitted in three manners non-persistance, semi-persistance and persistence. The whitefly transmits begomoviruses in persistence manner. The persistence transmission is also called as circulative transmission and in this type the transmission the virus gets into the insect gut along with juice and crosses the midgut and reaches to haemolymph and finally reaches to salivary glands. When such whitefly feed on healthy plants it transmits the virus along with the saliva. This process of acquiring to becoming the whitefly viruliferous require minimum period called as latent period. Once the whitefly become viruliferous remains viruliferous for rest of its life time hence such type of virus transmission is called as persistence type of transmission.

During the translocation of viruses inside vector body will encounter various organs and interaction of viruses with these organs decides the fate of the virus transmission. In this chapter we will see the virus and vector interactions specially whitefly and begomoviruses virus-vector system.

2. The Circulative Pathway of Begomoviruses in the Whitefly Vector

2.1 Minimum Time Needed for Efficient Acquisition and Inoculation of ToLCNDV by *B. tabaci*

To observe the effects of different Acquisition Access Period (AAP) on the rate of virus transmission *B. tabaci* was allowed to feed for five, 15, 30 minutes and one, three, six, 12hrs and 24hrs on the *ToLCNDV-potato* affected potato plants separately. After the prescribed AAP, the *B. tabaciwere* transferred on the 6 days old healthy potato plants at the rate of 15 whiteflies a constant 24hrs Inoculation Access Period were given. To determine the influence of different Inoculation Access Period (IAP) on transmission of virus, *B. tabaci* were allowed for 24hrs as AAP on *ToLCNDV-potato* affected potato plants. The viruliferous whiteflies were then transferred to 6 days old healthy potato plants at the rate of 15 whiteflies per plant, separately and allowed IAP for five, 15, 30 and one, three, six, 12 and 24hrs. Whiteflies were then killed by insecticide spraying as described earlier and seedlings were kept in insect proof cages for symptom development.

The whitefly is capable of transmitting the virus with a minimum time of 30 minutes of each AAP and IAP and the transmission of virus increased with the increasing in the AAP and IAPs. The whitefly, *B. tabaci* is capable of transmitting virus with 100 percent efficiencies with minimum period of 12 and 6 hrs of AAP and IAP, respectively.

2.2 Penetration of the Stylet into the Phloem of Infected Plants

2.2.1 Virus Source

Most begomoviruses are phloem-limited and are restricted to the vascular system, *B. tabaci* ingests begomoviruses by inserting its stylets in the vascular system of infected leaves. Virus amounts in an infected plant vary with the age of the tissues; therefore, the amounts of virus ingested and the kinetics of virus accumulation may depend on the virus concentration in the target leaves. The amount of virus uptaken by *B. tabaci* B was dependent on the virus concentration in the plant source and not on the virus itself.

2.2.2 Virus Ingurgitation

B. tabaci feeds on phloem sap by inserting its stylets into the vascular tissue. *B. tabaci* maxillary stylet is paired and doubly grovved when joined from two channels, first is food canal (through which phloem sap is ingested) and second salivary canal (through which saliva is injected into the plant). The mouth parts also have a pair of mandibular stylets which help in piercing the plant tissues. Before plant penetration with their stylets and during stylet movement through the apoplast, like aphid's, whiteflies also form a salivary sheath around the stylet by secreting a gel like substance along with saliva. After penetrating a sieve tube, whiteflies secrete watery saliva prior to ingestion. Both saliva types are produced in the salivary glands. Ingestion is facilitated by hydrostatic pressure in the sieve tubes and stylet. Ingestion of *tomato yellow leaf curl virus* is quite rapid. The virus is detectable by PCR in the head of whiteflies 5 min after the beginning of the AAP. This correlates with the time it takes until stylets reach the sieve elements and start phloem ingestion, as determined by the electrical penetration graph (EPG) technology.

2.2.3 Entry of Virus into the Gut

The esophagus is a narrow tube connects the cibarium to the filter chamber. The virus starts to be detected in the thorax of some insects after 10 min and in the abdomen after 25 min. The cuticular intima of esophagus makes it an unlikely site of virus entry into the haemolymph.

2.2.4 Guts, Midgut and Filter Chamber

The midgut, is the site for the digestion and absorption of the food material. The filter chamber of mid gut is the site of numerous events that affect the translocation and transmission of begomoviruses. Also determine the amount of virus translocation, replication and expression. It is also the site where processes and proteins, which are beginning to be uncovered, affect the amount of virus associated with the insect and impact transmission to plants.

2.3 Accumulation of Begomovirus in the Midgut and Filter Chamber

It has been shown that begomoviruses accumulate in vesicle-like structures in whitefly midgut cells and that clathrin-mediated endocytosis is responsible for their internalization. However, the process by which begomoviruses are trafficked within whitefly midgut cells remains largely unknown. Disrupting vesicle trafficking using RNA silencing and inhibitors Vesicle trafficking is responsible for the transport of TYLCV, within midgut cells of whitefly (*Bemisia tabaci*). It was also observed that the early steps of endosomal trafficking are important for the intracellular transport of TYLCV in the whitefly midgut. Like animal viruses, TYCLV is trafficked within cells in a manner independent of recycling endosomes, late endosomes, lysosomes, the Golgi apparatus and the endoplasmic reticulum. Instead, that TYLCV might be transported directly from early endosomes to the basal plasma membrane and released into the hemolymph. Silencing of the sorting nexin Snx12, which may be involved in membrane tubulation, resulted in fewer viral particles in hemolymph; this suggests that the tubular endosomal network may be involved in the transport of TYLCV. Virions are transported through the cytoplasm of the filter chamber epithelial cells in vesicles that fuse with the basal plasma membrane, releasing particles between the membrane and the basal lamina. Similar vesicles were described in the TYLCV non-vector whitefly *T. vaporariorum*; however, they were devoid of virus.

2.3.1 Interaction of Begomovirus Proteins with Whitefly Guts Proteins

Once the virus particle get entry in the gut then interact with several proteins and have been shown recently to bind to TYLCV in the midgut and thereby to influence TYLCV transmission. TYLCV induces protein aggregation in infected tomatoes and in its vector *B. tabaci*. The interactions between TYLCV and HSP70 and HSP90 in plants and vectors are essential for virus infection to proceed. Upon TYLCV infected HSP70 and HSP90 are redistributed from a soluble to an aggregated state in the host cells. These aggregates contain, together with viral DNA/proteins and virions, HSPs and components of the protein quality control system such as ubiquitin, 26S proteasome subunits, and the autophagy protein ATG8. In the infected host cell, HSP70, participates in the translocation of CP from the cytoplasm into the nucleus. Viral amounts decrease when HSP70 is inhibited, but increase when HSP90 is down regulated. In the whitefly vector, HSP70 impairs the circulative transmission of TYLCV; its inhibition increases transmission. Hence, the efficiency of virus acquisition by whiteflies depends on the functionality of both plant chaperones and their cross-talk with other protein mechanisms controlling virus-induced aggregation. A midgut protein (coined MGP) was identified using the CP of the begomoviruses Tomato leafcurl New Delhi virus (ToLCV) and Cotton leaf curl Rajasthan virus (CLCuV) as bait in a yeast two hybrid screen. MGP was localized mainly in the gut epithelial layer. Oral ingestion of anti-MGP antibodies resulted in a 2/3 decrease in transmission of ToLCV or CLCuV, thus indicating the importance of MG Pin the virus-vector interaction process.

2.3.2 Gut-haemolymph Barrier

For successfully transmission, begomoviruses need to cross the gut haemolymph barrier. This checkpoint was demonstrated to be genetically determined by selecting good and poorly TYLCV transmitting lines from *B. tabaci B* populations. In the poor transmitter whitefly isolines has a considerably delayed and reduced virus translocation from the midgut into the haemolymph, indicating to a block at the gut/haemolymph barrier. TYLCV is unable to cross the midgut barrier of *T. vaporariorum* to enter the haemolymph and from there the salivary glands hence make this species as non-vector of TYLCV.

Immuno-localization of TYLCV-Mld in the vector *B. tabaci* B and the non-vector *T. vaporariorum* showed that the virus was present at the brush borders and in the surrounding luminal surface of the midgut epithelial cells of both whitefly species. However, it was exclusively found in the cytoplasm of the midgut cells of *B. tabaci*, not in that of *T. vaporariorum*. Endocytosis was the major process serving to transport TYLCV across the *B. tabaci* midgut barrier. Indeed, endocytosis is the most common path exploited by virus to enterits host. TYLCV accumulated in vesicle-like structures in the midgut epithelial cells of the vector whitefly *B. tabaci*, but not in those of the non-vector whitefly *T. vaporariorum*. Hence, in order to cross the gut/haemolymph barrier, TYLCV must first enter the midgut cells, likely by endocytosis. This route was confirmed by disrupting components involved in clathrin-mediated assembly at the cell surface and disassembly at the endosome such as clathrin heavy and light chain (CHC and CHL), dynamin (DYN), rab5 and rab7 associated with early and late endosome, respectively.

2.3.3 Haemolymph

Once begomoviruses cross the gut barrier, they are prone to destruction by the proteases and nucleases conspicuous in the haemolymph. To protect the virus in this hostile environment, the CP of begomoviruses interacts with a 63-kDa GroEL chaperone produced by the facultative endosymbiotic bacteria of whitefly and released into the haemolymph. The role of the whitefly GroEL in ensuring TYLCV circulative transmission is similar to that first demonstrated for the transmission of luteoviruses by their aphid vector. The release of virions protected by GroEL occurs adjacent to the PSGs. The whitefly houses many facultative endosymbionts by GroEL produced by *Hamiltonella* is mainly shown strong interaction with TYLCV CP and protect from viral degradation in the haemolymph. The poor TYLCV transmitter whitefly biotype Q does not contain *Hamiltonella*. But the *B. tabaci* biotypes do not house the *Hamiltonella* are still good transmitter of other begomoviruses. *B. tabaci* Asia II isolates from Rajasthan, India, contain the primary endosymbiont *Portiera*, but only the secondary endosymbiont, *Arsenophonus*. Hence, GroEL from *Arsenophonus* ensures thetransmission of cotton leaf curl virus (CLCuV). So specific virus can interact with the specific endosymbiont present in the *B. tabaci* and ensure the successful transmission of the virus.

2.3.4 Primary Salivary Glands

The Salivary glands of *B. tabaci* has a pair of primary salivary glands (PSGs), the accessory salivary glands. The PSGs are involved in begomovirus circulation. The cells around the secretory region of the PSGs play a central role in determining specificity of begomovirus transmission. The *B. tabaci* contains species-specific receptor-like elements that are able to discriminate between begomoviruses, even as similar as TYLCV and TYLCCNV. While *B. tabaci* B efficiently transmits TYLCV and TYLCCNV, *B. tabaci* Q ingests and retains both viruses, but is unable to transmit TYLCCNV. The selective transmission of TYLCV/TYLCCNV by Q whiteflies was investigated by following the time course of virus penetration in the PSGs, using CP immunofluorescence of dissected glands. The viruses followed a specific path in the glands. Following a 4–6-hrs AAP, TYLCV CP spread to the entire PSG; at 12–24 hrs, virions gradually accumulated to the central region of PSGs; at 36–72 hrs, the viral CP was observed only in the few cells along the salivary duct. In comparison, at 4–12 hrs AAPs, TYLCCNV was distributed in cells in the boundary of Q PSGs and disappeared thereafter. Therefore, secretory cells in the central region of whitefly PSGs govern the transmission specificity of begomoviruses.

3. Conclusion

The virus interacts at many places in the whitefly body for its successful transmission. There are mainly two barriers in the vector body crucial for successful transmission of the virus. The first barrier is Gut-haemolymph and second is haemolymph-salivary gland barrier. Whitefly endosymbionts also releases GroEL like proteins that protect the virus in the haemolymph. Such understanding of the virus and vector is important for breaking of the virus and vector interaction at its weakest points.

Chapter 32

Protocols for Analysis Physio-Biochemical Parameters in Potato

Dharmendra Kumar, Som Dutt, Sushil K. Changan, Milan K. Lal, Asha Thakur, Vandana and Brajesh Singh

ICAR-Central Potato Research Institute, Shimla, Himachal Pradesh

1. Introduction

Potato is considered as the most productive vegetable and provides a major source of nutrition and income to many populations and communities. Its content of dry matter, edible energy and edible protein make it a good choice for nutrient availability. Potato is known to everyone as a supplier of energy but its ability to supply vital nutrients is vastly underestimated. Potato is an excellent source of complex carbohydrates, dietary fibres and vitamin C. It also contains a variety of health-promoting compounds, such as phytonutrients that have antioxidant activity. Among these, important health-promoting compounds are carotenoids, flavonoids, and caffeic acid, as well as unique tuber storage proteins, such as patatin, which exhibit activity against free radicals. Potato is a substantial source of ascorbic acid, thiamine, niacin, pantothenic acid and riboflavin. Potato is highly desirable in the human diet as many of the compounds present in potatoes are important because of their beneficial effects on human health.

There are many misconceptions prevalent in society concerning the nutritional value of potatoes. The most common misconception is that potatoes are fattening and can cause or worsen diabetes. To remove such misconceptions it is pertinent to establish the nutritional value of potatoes. Therefore, for this purpose we have developed/optimized protocols for estimations of health-promoting compounds from potatoes. This information will be useful for students, researchers and potato processors. The protocols are descriebd here for the nutritional analysis in potato.

- Resistant Starch
- Total Soluble Proteins
- Total Phenolics
- Individual Phenols
- Ascorbic Acid
- Total Carotenoids
- Anthocyanin
- β-Carotene
- Biguanide and Related Compounds
- Metformin

2. Resistant Starch

Starch is the most abundant storage polysaccharide in plants and is the major component of the diet. Digestibility of starch improves during cooking and not all of the starch present in the food is digestible. A part of starch present in the diet escapes digestion and absorption in the small intestine and is fermented in the large intestine of humans, with the production of short-chain fatty acids (SCFA). This is termed as 'Resistant starch'. Resistant starch (RS) is a form of dietary fibre and is naturally present in many of starchy foods. Raw potatoes contain RS type 2, whereas cooked and cooled potatoes contain RS type 3.

2.1 Reagents

- KCl-HCl Buffer, pH 1.5
- 0.1M Tris-Maleate Buffer, pH 6.9
- 0.1M Potassium Hydroxide
- 0.4M acetate buffer, pH 4.75
- 2M HCl
- Pepsin Solution: 400mg pepsin in 10mL KCl-HCl buffer, pH 1.5
- Pancreatic α-amylase Solution: 800mg α-amylase in 10 mL 0.1M Tris-Maleate buffer, pH 6.9
- Amyloglucosidase
- GOD-POD assay kit (glucose oxidase peroxidase)
- Glucose standard (provided with kit)

2.2 Extraction Procedure

- Take 100 mg of dry sample in centrifuge tube and add 5mL pepsin solution.
- Incubate the samples at 40 °C for 1 h in water bath with constant shaking.
- Take out the samples and bring to room temperature.
- Add 5 mL of pancreatic α-amylase solution.
- Mix well and incubate at 37 °C for 16 h with constant shaking.
- Centrifuge (30 min, 10000 g) and discard the supernatant. Wash the residue with DW and centrifuge. Discard the supernatant.
- Add 3 mL of DW and 3 mL of 4 M potassium hydroxide to the residue.
- Leave for 30 min at room temperature with constant shaking.
- Add 5.5 mL of 2 M HCl and 3mL of 0.4 M acetate buffer, pH 4.75.
- Add 600 µl of amyloglucosidase to the sample and incubate at 60 °C for 45 min with constant shaking.
- Centrifuge (15 min, 3000g) and retain supernatant.

2.3 Analysis

- Take 0.3 mL of aliquot and 2.7 mL of GOD-POD reagent.
- Mix well and incubate at 37 °C for 5 min.
- Measure the absorbance at 450 nm.

3. Total Soluble Proteins

Potato is a very good source of high-quality protein. The average protein content in potatoe is 2 percent on fresh weight basis and about 10 percent on dry weight basis. The total nitrogen of potato tubers can be divided into soluble protein, insoluble protein and soluble non-protein nitrogen. The insoluble protein fraction is mainly present in the peel. Soluble potato protein contains substantial level the essential amino acids. Free amino acids present in potatoes are totally available for absorption. Potato protein has an adequate ratio of total essential amino acids to total amino acids and a balance among individual essential amino acid concentrations to meet the needs of infants and small children. Potato protein has a very high biological value since all the essential amino acids are present in it in a good proportion. With its high lysine content, potato can supplement diets which are limiting in lysine. Potato is a superb food that has correct balance between net protein calories and total calories adequate for all age groups.

3.1 Reagents

- 0.1 N sodium hydroxide

- ☆ 15 percent trichloroacetic acid
- ☆ Solution A: 2 percent sodium carbonate in 0.1 N sodium hydroxide
- ☆ Solution B: 0.5 percent $CuSO_4.5H_2O$ in 1 percent sodium potassium tartarate
- ☆ Solution C: Prepared by mixing 50 mL of reagent A with 1 mL of solution B. This reagent is freshly prepared at the time of use.
- ☆ Solution D: Folin-Ciocalteu phenol reagent and distilled water are added in equal quantity at time of use
- ☆ Standard: 20 mg Bovine serum albumin/100 mL DW

3.2 Extraction Procedure

- ☆ Take 1 g dry sample and add 5 mL of 0.1 N sodium hydroxide.
- ☆ Vortex and centrifuge (10000 rpm, 10 min).
- ☆ Collect supernatant and raise the volume to 5mL with 0.1N sodium hydroxide.
- ☆ Add 1mL of 15 percent trichloroacetic acid.
- ☆ Incubate the samples at 4 °C for 24 h.
- ☆ Centrifuge the samples at 5000 rpm for 20 min.
- ☆ Discard supernatant.
- ☆ Dissolve the residue in 5 mL of 0.1N sodium hydroxide.

3.3 Analysis

- ☆ Take 0.5 mL of aliquot and add 5 mL of solution C to it.
- ☆ Shake the samples and keep it for 5 min.
- ☆ Add 0.5 mL of solution D. Incubate the samples at room temperature for 60 min.
- ☆ Measure the absorbance at 570 nm.

4. Total Phenolics

Potatoes are a good source of phenols. Because of high per capita consumption of potatoes, they are able to contribute towards high intake of phenols. Phenols are present in high concentration in potato peel as compared to flesh. Phenolic compounds present in potatoes are mainly phenolic acids and flavonoids including flavonols, flavanols, and anthocyanins.

4.1 Reagents

- ☆ 80 percent ethanol
- ☆ Folin-Ciocalteu reagent

- ☆ 35 percent sodium carbonate
- ☆ Standard

4.2 Extraction Procedure

- ☆ Grind 1g tissue in 10 mL of 80 percent ethanol.
- ☆ Centrifuge at 10,000 rpm for 20 min.
- ☆ Collect supernatant and re-extract the residue with 5 mL 80 percent ethanol.
- ☆ Centrifuge and pool supernatants.
- ☆ Evaporate the supernatant to dryness.
- ☆ Dissolve the residue in 5mL distilled water.

4.3 Analysis

- ☆ Mix 100 µl aliquot with 0.5 mL Folin-Ciocalteu phenol reagent.
- ☆ Vortex mix and raise the volume to 10 mL with distilled water.
- ☆ Incubate for 3 min and add 1 mL of 35 percent sodium carbonate.
- ☆ Vortex and incubate the samples for 1h.
- ☆ Measure the absorbance at 620 nm and calculate the phenolic content after comparing with standard curve.

4.4 Total Phenols through ELISA Plate Reader

4.4.1 Reagents

- ☆ 80 percent ethanol
- ☆ Folin-Ciocalteu reagent
- ☆ 5 percent Sodium Carbonate
- ☆ Catechol Standard

4.4.2 Extraction Procedure

- ☆ Extract 100 mg dry sample in 5 mL of 80 percent ethanol.
- ☆ Incubate for 2 h at 40 °C in shaking water bath.
- ☆ Cool to room temperature, centrifuge and collect the residue.
- ☆ In 96 well ELISA plate, add 25 µl of extract along with 25 µl of 80 percent ethanol.
- ☆ Add 25 µl of Folin-Ciocalteu reagent and 200 µl of distilled water.
- ☆ Add 30 µl of 5 percent sodium carbonate.
- ☆ Incubate the plate for 1 hrs at room temperature.
- ☆ Measure the absorbance at 620 nm and calculate the phenolic content after comparing with standard curve.

5. Individual Phenols

Potato phenolics are present in both soluble (free and soluble esters and soluble glycosides) and insoluble-bound forms. Most common potato phenolics are chlorogenic acid, caffeic acid, ferulic acid, gallic acid, rutin and protocatechuic acid. These phenolics confer several health benefits.

5.1 Reagents

- ☆ Extraction solution: Mix Methanol:Water:Acetic acid in 80:19.5:0.5 ratio.
- ☆ Water: Methanol: Mix methanol: water in 90:10 ratio
- ☆ Mobile phase A: 1 percent Formic acid in water
- ☆ Mobile Phase B: Methanol
- ☆ Standards: Chlorogenic acid, Catechin, Quercetin, Caffeic acid, p-Coumaric acid, Sinapic acid, Ferulic acid, Gallic acid, p-Catechuic acid and Rutin

5.2 Extraction Procedure

- ☆ Extract finely chopped 1 g sample with extraction solution.
- ☆ Sonicate samples for 15 minutes in sonicator.
- ☆ Centrifuge at 9000 g for 10 minutes at 4 °C.
- ☆ Collect supernatant and re-extract residue with 2 ml of water: methanol solution (90:10 ratio).
- ☆ Centrifuge and pool the supernatants.
- ☆ Lyophilize the samples.
- ☆ Reconstitute samples in 1ml of methanol and filter through 0.45 µm membrane.
- ☆ Store at -80 °C until analysis.

5.3 Analytical Procedure

5.3.1 Chromatographic Conditions in HPLC

- ☆ Column: 250 x 4.6 mm, 5µm Purospher RP-18e
- ☆ Pump: LaChrom L-7100 (Merck-Hitachi, Germany)
- ☆ Oven: Lachrom L-7350 (Merck-Hitachi)
- ☆ Column temp: 35 °C
- ☆ Mobile phase A: 1 percent Formic acid in water
- ☆ Mobile Phase B: Methanol
- ☆ Detector: UV detection, LaChrom L-7420 (MerckHitachi)
- ☆ Injection: 20 µl in rheodyne injector

- ☆ Flow rate: 0.95mL/min
- ☆ Detector wavelength and absorbance: 324nm and 270nm, 0.04 AUFS

5.3.2 Gradient Used

Time (minutes)	*Mobile Phase A (Percent)*	*Mobile Phase B (Percent)*
0	95	5
2.5	84	16
5	73	27
7.5	62	38
10	50	50
12.5	40	60
15	30	70
17.5	25	75
20	20	80
22.5	10	90
25	95	5

5.3.3 Analysis

Inject 20 µl of stored sample in rheodyne injector.

6. Ascorbic Acid

Vitamin C is a water-soluble vitamin that acts as an antioxidant and stabilizes or eliminates free radicals, thus helping to prevent cellular damage. Potato as major source of Vitamin C provides protection against scurvy. Vitamin C enhances the absorption of iron in human body and may help support the body's immune system. The ascorbic acid content of potato declines when potatoes are stored, cooked or processed. Though potato loses some of its vitamin C during storage, substantial amounts are retained until it sprouts. In Indian scenario, the consumption of fresh fruits and vegetables is much less and for poor it is almost negligible. Potato being in reach of all income groups may serve as the main source of vitamin C, especially for poor people.

6.1 Reagents

- ☆ Bromine water: Add 1-2 drops of liquor bromine in 100 mL cool water
- ☆ 2 percent 2,4 Dinitrophenyl hydrazine (DNPH) reagent
- ☆ 4 percent Oxalic acid
- ☆ 0.5N sulphuric acid
- ☆ 80 percent Sulphuric acid
- ☆ Standard (10 mg Ascorbic acid/100 mL Distilled water)

6.2 Extraction Procedure

- ☆ Grind 10 g of fresh tuber tissue presence of 25 mL of 4 percent oxalic acid
- ☆ Filter and collect the aliquot in conical flask.
- ☆ Add Bromine water drop wise with constant mixing.
- ☆ When the extract turns orange yellow due to excess bromine, expel it by blowing in air.
- ☆ Make volume up to 50 mL with 4 percent oxalic acid.

6.3 Analysis

- ☆ Pipette out 2 mL of brominated sample extract in a test tube and make up the volume in each tube to 3 mL by adding distilled water.
- ☆ Add 1 mL of DNPH reagent followed by 1-2 drops of thiourea to each tube.
- ☆ Thoroughly mix the contents of the tube and incubate at 37 °C for 4-5 hrs in water bath.
- ☆ Add 7 mL of 80 percent sulphuric acid and dissolve the orange-red osazone crystals formed due to reaction with sulphuric acid.
- ☆ Measure optical density at 540 nm.
- ☆ Calculate the amount of ascorbic acid in the sample using a standard curve prepared from pure ascorbic acid.

6.4 Ascorbic Acid Estimation through ELISA Plate Reader

6.4.1 Reagents

- ☆ 2 per cent and 6 per cent metaphosphoric acid
- ☆ Dye solution: Dissolve 100 mg of 2,6-dichlorophenol indophenol dye and 84 mg of sodium bicarbonate in hot (85-95 °C) distilled water, cool and make up to 100 mL. Filter and dilute 25 ml to 500 ml with distilled water.
- ☆ Ascorbic acid standard solution: Accurately weigh 10mg of ascorbic acid and make upto 10 ml with 2 percent metaphosphoric acid. Dilute 4mL of this solution to 100 mL with 2 percent metaphosphoric acid (1 ml = 40 μg of ascorbic acid).

6.4.2 Procedure

1. Extract 5 g of sample with 20 mL of 6 percent metaphosphoric acid
2. Crushed the sample and Filter through muslin cloth.
3. Centrifuge it (at 18 °C, 10,000 rpm for 10 min) to settle down the tissue and cell debris, if any.

4. Load 50 µl of clear supernatant on ELISA plate.
5. Mix 100 µl of 2,6-dichlorophenol indophenol dye just before taking O.D. using 518 nm.
6. Same procedure should be replicated with standard solutions by eliminating the sample aliquot of step 4.0 with serial dilutions of standard solution.
7. The amount of ascorbic acid in the samples can be calculated by comparing the O. D. with the standard graph.
8. Run blank using 150 µl metaphosphoric acid.

   Note: The procedure of taking O. D. should be completed within 10-15 sec after adding dye solution in sample aliquot.

6.5 Ascorbic Acid through HPLC

6.5.1 Reagents

1. 8 percent Acetic acid
2. Mobile Phase: Acetic acid: MilliQ water: Methanol in 1:69:30 ratio. Add 10 g/L sodium chloride
3. Ascorbic acid standard

6.5.2 Extraction Procedure

- ☆ Cut peeled tuber to small pieces and weigh 5 g per replication.
- ☆ Extract the sample with 5 ml of 8 percent acetic acid.
- ☆ Sonicate the sample for 30 minutes.
- ☆ Centrifuge and collect supernatant.
- ☆ Re-extract the residue with 3 ml of 8 percent acetic and sonicate again.
- ☆ Centrifuge and pool the supernatants.
- ☆ Make the volume to 10 ml with 8 percent acetic acid.
- ☆ Filter the sample through 0.45 µm membranes and store at -80 °C until analysis.

6.5.3 Analytical Procedure

6.5.3.1 Chromatographic Conditions in HPLC

- ☆ Column: 250 x 4.6 mm, 5µm Purospher RP-18e
- ☆ Pump: LaChrom L-7100 (Merck-Hitachi, Germany)
- ☆ Oven: Lachrom L-7350 (Merck-Hitachi)
- ☆ Column temp: 35 °C
- ☆ Mobile phase: Acetic acid: MilliQ water: Methanol in 1:69:30 ratio. Add 10 g/L sodium chloride

- ☆ Detector: UV detection, LaChrom L-7420 (MerckHitachi)
- ☆ Injection: 20 µl in rheodyne injector
- ☆ Flow rate: 1 mL/min
- ☆ Detector wavelength and absorbance: 273 nm, 0.04 AUFS
- ☆ Retention time: ≈ 2.9 min

6.5.3.2 Analysis

- ☆ Inject 20 µl of stored sample in rheodyne injector.

7. Total Carotenoids

Yellow pigmented potatoes are known to have high carotenoids content such as lutein, zeaxanthin, violaxanthin and antheraxathin. Potato carotenoids are primarily oxygenated carotenoids which are also known as xanthophylls. Purple pigmented potatoes have health benefit against cardiovascular disease while consumption of yellow pigmented potatoes enhances immune response. Lutein, Zeaxanthin, violaxanthin and neoxanthin are the major carotenoids present in potatoes and β-carotene has been detected only in trace amounts. The orange colour of the tuber flesh is due to zeaxanthin, whereas the yellow colour is due to lutein.

7.1 Reagents

- ☆ Sodium sulfate anhydrous
- ☆ Acetone
- ☆ n-hexane
- ☆ β carotene solution

7.2 Extraction Procedure

- ☆ Cut five tubers into small pieces and mix thoroughly.
- ☆ Macerate 20 g tissue and extract in 100 mL of cold acetone: n-hexane (2:3) twice or till the filtrate become colorless.
- ☆ Take the filtrate in separating funnel and add cold water.
- ☆ Shake carefully to make it free of acetone.
- ☆ Separate the volume of the hexane layer containing the carotenoids.
- ☆ Raise the volume to 100 mL with hexane.
- ☆ The absorbance is measured at 436 nm in spectrophotometer.

8. Anthocyanin

Anthocyanins are present in dark colored potatoes. But white/cream fleshed potatoes also contain this compound to some extent. Colored flesh potatoes, either red or purple pigmented are important due to presence of anthocyanins. Indigenous potato varieties contain varying concentrations of anthocyanins.

8.1 Reagents

- ☆ Ethanol
- ☆ 0.5 N HCL in 80-85 percent methanol
- ☆ Anthocyanin reagent: Mix 1 mL of 30 percent H_2O_2 with 9mL of methanolic HCL
- ☆ Standard cyanin hydrochloride (10 mg/100 mL DW)

8.2 Extraction

1. Grind 10 g of tissue in 50 mL of alcohol and methanolic HCL (1:1).
2. Filter and collect the filtrate.
3. Take 4 mL of filtrate and add 2 mL of anthocyanin reagent.
4. Incubate in dark for 20 min.
5. Take the absorbance at 525 nm against blank.
6. Calculate the amount of anthocyanin present in the samples from standard curve prepared with cyanin hydrochloride.

   Note: If standard is not available, then use the following o.d and concentration for calculation:

   10 µg of cyanin hydrochloride/mL in methanolic HCl = absorbance of 0.405 in a 1.0 cm cell at A_{525}

9. β-Carotene

β-carotene is a red-orange colored pigment present abundantly in plants. β-carotene is distinguished from other carotenes due to presence of beta-rings at both end of the molecule. It is present in low concentration in potatoes compared to other carotenes. Although around 600 different carotenoids have been identified and characterized, those known as dietary compounds like β-carotene, lutein, and lycopene were maily focussed.

9.1 Reagents

- ☆ Absolute ethanol containing 0.5 percent Butylated Hydroxytoluene (BHT)
- ☆ Hexane: Acetone (2: 1 ratio)
- ☆ Methanol containing 15 percent potassium hydroxide and 0.5 percent BHT
- ☆ Petroleum ether: Diethylether (2:1 ratio) containing 0.5 percent BHT
- ☆ Mobile phase A: Methanol containing 0.5 percent triethylamine: Acetonitrile: Chloroform in 50:40:10 ratio

- Mobile phase B: Methanol containing 0.5 percent triethylamine: Acetonitrile: Chloroform in 35:35:30 ratio
- β-carotene standard

9.2 Extraction Procedure

- Weight 20 g of finely chopped sample and extract it with pre-warmed absolute ethanol containing 0.5 percent BHT.
- Add 2 ml of hexane: acetone (2:1)
- Incubate for 10 minutes in dark at room temperature.
- Centrifuge at 5000 rpm for 10 minutes at 4 °C.
- Collect supernatant and add equal volume of methanol containing potassium hydroxide and BHT.
- Incubate at 80 °C for 15 minutes and then chill on ice for 10 minutes.
- Add 4ml of distilled water, 3 ml of petroleum ether: diethylether containing BHT.
- Vortex and centrifuge at 5000 rpm for 10 minutes at 4 °C.
- Transfer upper organic phase to fresh tube.
- Extract the residue twice with 4ml of petroleum ether: diethylether containing BHT.
- Collect upper phase and pool.
- Lyophilize the sample and reconstitute the residue in Mobile phase 'A'
- Store at -20 °C until analysis.

9.3 Analytical Procedure

9.3.1 Chromatographic Conditions in HPLC

- Column: 250 x 4.6 mm, 5µm Purospher RP-18e
- Pump: LaChrom L-7100 (Merck-Hitachi, Germany)
- Oven: Lachrom L-7350 (Merck-Hitachi)
- Column temp: 35 °C
- Mobile phase A: Methanol containing 0.5 percent triethylamine: Acetonitrile: Chloroform in 50:40:10 ratio
- Mobile Phase B: Methanol containing 0.5 percent triethylamine: Acetonitrile: Chloroform in 35:35:30 ratio
- Detector: UV detection, LaChrom L-7420 (MerckHitachi)
- Injection: 20 µl in rheodyne injector
- Flow rate: 1.2 mL/min

- ☆ Detector wavelength and absorbance: nm, 0.04 AUFS
- ☆ Retention time: ≈ 21-23 min

9.3.2 Analysis

- ☆ Inject 20 µl of stored sample in rheodyne injector.

9.4 Gradient used

Time (minutes)	Solution A (Percent)	Solution B (Percent)
0	100	0
2	100	0
2.1	85	15
3	85	15
3.1	70	30
4	70	30
4.1	55	45
5	55	45
5.1	40	60
6	40	60
6.1	25	75
7	25	75
7.1	10	90
8	10	90
9	100	0
25	100	0

10. Biguanide and Related Compounds

Among the antidiabetic agents, biguanides and related compounds reduce hyperglycaemia by increasing insulin sensitivity and decrease glucose absorption. As arginine is precursor for biguanides and related compounds, there is some evidence that biguanides and related compounds are present in plants. In potato metformin and galegine analogue were identified and quantified.

10.1 Reagents

- ☆ Mobile phase: Mix 0.05 M potassium dihydrogen phosphate and 100 percent acetonitrile in 70:30 ratio, adjust pH to 3.5 with orthophosphoric acid
- ☆ Standards: Metformin-HCl, galegine analogue (2-methylthio-2-imidazoline hydroiodide)

10.2 Extraction Procedure

- ☆ Extract finely chopped sample with 10 ml of mobile phase.
- ☆ Centrifuge samples at 4000 rpm for 10 minutes.
- ☆ Collect supernatant and filter through 0.45 μm filter.
- ☆ Store at -80 °C until analysis.

10.3 Analytical Procedure

10.3.1 Chromatographic Conditions in HPLC

- ☆ Column: 125 x 4 mm Purospher RP-18
- ☆ Pump: LaChrom L-7100 (Merck-Hitachi, Germany)
- ☆ Oven: Lachrom L-7350 (Merck-Hitachi)
- ☆ Column temp: 35 °C
- ☆ Mobile phase: Mix 0.05M potassium dihydrogen phosphate and 100 percent acetonitrile in 70:30 ratio
- ☆ Detector: UV detection, LaChrom L-7420 (MerckHitachi)
- ☆ Injection: 20 μl in rheodyne injector
- ☆ Flow rate: 1.2 mL/min
- ☆ Detector wavelength and absorbance: 215 nm, 0.04 AUFS
- ☆ Retention time: ≈ 1 min

10.3.2 Analysis

- ☆ Inject 20 μl of stored sample in rheodyne injector

11. Metformin

Metformin is used as a first line of defense in diabetic patients. Presence of metformin in potato has been confirmed using LC-MS. Presence of this compound in potato may help to remove the misconceptions linked to potato consumption.

11.1 Reagents

- ☆ Ammonium acetate containing Formic acid: Mix 0.5M ammonium acetate and formic acid in 9:1 ratio.
- ☆ Stock standard (1mg/ml) was prepared by dissolving metformin-HCl in Ammonium acetate solution containing formic acid.
- ☆ Solution A: MilliQ water containing 0.1 percent formic acid
- ☆ Solution B: Methanol containing 0.1 percent formic acid

11.2 Extraction Procedure

- ☆ Finely chopped 5 g sample was extracted with 5ml of Ammonium acetate solution containing formic acid.
- ☆ Samples were vortexed and 5 ml of chloroform was added.
- ☆ Samples were vortexed and sonicated for 15 min.
- ☆ Samples were shaken for 3 min and centrifuged for 5 min at 3600 rpm.
- ☆ Upper layer was collected (aqueous) in new tube.
- ☆ SPE-MCAX column (300 mg) was conditioned with 3ml of methanol slowly. Thereafter, followed by 2x3 ml of milliQ water twice (i.e 6 ml).
- ☆ Upper layer of sample was passed through conditioned MCAX column slowly.
- ☆ Filtrate was discarded.
- ☆ Metformin was eluted with 2x1.5 ml of methanol *i.e.* twice (HPLC grade) for complete elution of metformin from column.

11.3 LC-MS Conditions

- ☆ Flow rate: 0.3 ml/min
- ☆ Retention time: ~ 0.95 min

11.4 Gradient Used

Time (minutes)	*Solution A (Percent)*	*Solution B (Percent)*
0	98	2
1.2	98	2
2	20	80
3	20	80
3.2	98	2
6	98	2

Chapter 33

Ploidy Estimation in Potato

Hemant Kardile and Rahul

ICAR-Central Potato Research Institute, Shimla, Himachal Pradesh

1. Introduction

Ploidy estimation is one of the most frequent applications of flow cytometry. The methodology consists of three major steps namely preparation of suspensions of intact nuclei, labeling with the fluorochrome and analysis of nuclear DNA content either with external or internal standards. External standardization involves the individual analysis of a plant with known ploidy (reference standard) followed by the analysis of an unknown specific plant with the same instrument settings. It results in a histogram of DNA content who's relative G_1 peak position indicates the plant ploidy. It is suitable for large-scale screening. Internal Standardization involves simultaneous analysis of a reference and unknown plant of different ploidy results in a histogram having two G_1 peaks. The ratio of the G_1 peak positions reflects the ratio of ploidy levels. It is more precise since it eliminates most of the sources of variation.

2. Ploidy Analysis with the External Standard Method

The test samples and reference standard (the reference standard for the ploidy determination must be from the same species, whose ploidy is already known) were analysed separately on the flow cytometer with same instrument gain settings. The diploid, *S. chacoense* is used as the reference standard for potato clones. The suspensions of intact nuclei were prepared from the young leaf using the modified HPI buffer. As, buffer itself contains the DNA fluorochrome, Propidium iodide (PI), the nuclei get stain during the isolation process itself. The reference samples were first analysed in the flow cytometer and positioned the G_1 peak on the absicca. Similarly, stained nuclear suspension from the potato samples introduced in the flow cytometer and positioned the G_1 peak at the required position on the abscissa with same instrument gain settings. The ploidy was calculated as per the formula:

Sample ploidy (integer) = Reference ploidy × mean position of the G_1 sample peak/Mean position of the G_1 reference peak

3. Ploidy Analysis with the Internal Standard Method

In internal standard method, the test samples and reference standard (the reference standard for the ploidy determination must be from the same species, whose ploidy is already known) were analysed separately on the flow cytometer with same instrument gain settings. The rest of the methodology remains same as that of external method.

4. Preparation of Fluorochrome Labelled Intact Nuclear Suspension

- ✰ Take an approximately of 20 mg young leaf from the apical meristem of potato plants
- ✰ Chop the leaf samples in 200 µl ice-cold modified HPI buffer with new razor blade and add another 600 µl modified HPI buffer to the chopped samples. Mix the homogenate by pipetting up and down for several times.
- ✰ Filter the homogenate through a 40-micron cell strainer into a labelled sample tube
- ✰ Incubate the sample tube on ice for 5-10 min, with occasional shaking in the dark.

5. Ploidy Analysis with External Standard

- ✰ Run a sample prepared from a reference standard, *S. Chacoense* (diploid potato) by introducing the samples into the flow cytometer. Analyze the sample at low flow rate (the sample rate usually does not exceed 20–50 nuclei/s). The DNA peak(s) of G_1 and G_2 cells were positioned to the required position on the abscissa by adjusting the instrument gain settings. About 10,000 populations were measured and the data were saved for the analysis.
- ✰ Analyse unknown samples with same instrument gain settings by following step 5
- ✰ Calculate the DNA ploidy of the unknown sample as follows:

  Sample ploidy (integer) = 2 × mean position of the G_1 sample peak/ Mean position of the G_1 reference peak

6. Ploidy Analysis with Internal Standard

- ☆ Prepare the sample by simultaneous chopping and staining tissues from a reference plant with known ploidy (chromosome number) and from the unknown plant sample as described earlier.
- ☆ Perform sample analysis following Step 5
- ☆ Determine the DNA ploidy of the unknown sample following Step 7

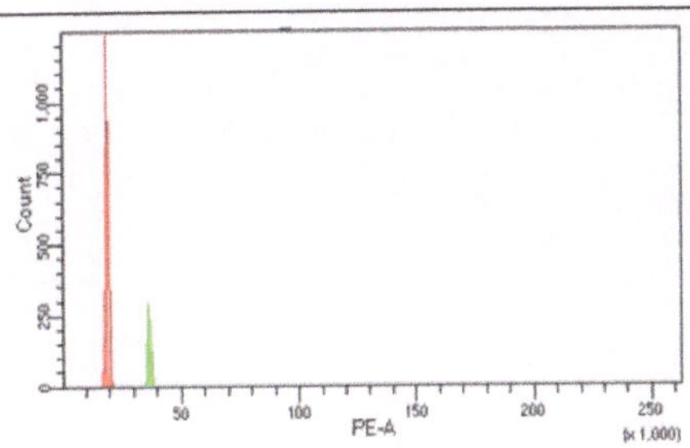

S. Chacoense **(diploid reference standard)**

Experiment Name:	DNA QC 29072017
Specimen Name:	CPRI_BUFFER STD
Tube Name:	C13
Record Date:	Aug 7, 2017 2:41:14 PM
$OP:	Administrator
GUID:	59dc776b-5611-4e74-9a4c-06a...

Population	#Events	%Parent	PE-A Mean	PE-A %CV
All Events	3,197	####	21,736	91.5
G1	1,532	47.9	17,871	2.2
G2	416	13.0	35,620	1.4

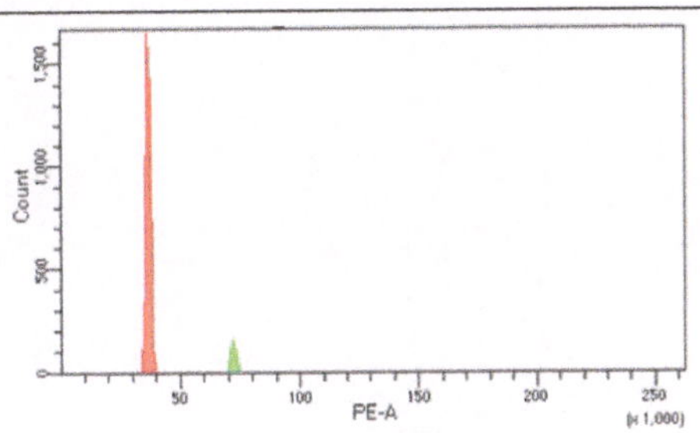

Potato (K. Girdhari)

Experiment Name:	DNA QC 29072017
Specimen Name:	CPRI_BUFFER STD
Tube Name:	DB03
Record Date:	Aug 7, 2017 2:55:27 PM
$OP:	Administrator
GUID:	989d3ab2-ef28-487e-bb03-e59...

Population	#Events	%Parent	PE-A Mean	PE-A %CV
All Events	10,000	####	33,958	89.1
G1	3,682	36.8	35,834	2.1
G2	434	4.3	71,176	1.3

Sample ploidy (integer) = 2X 35,834/17, 871=4.01=Tetraploid

Figure 33.1: Representative Results: Estimation of DNA Ploidy from Potato.

Chapter 34

Micropropagation in Potato

Tanuja Buckseth, Sumita Sharma, R.K. Singh and J.K. Tiwari

ICAR-Central Potato Research Institute, Shimla, Himachal Pradesh

1. Introduction

Potato is a vegetatively propagated crop which is a good host for a large number of viruses besides other pathogens. Infection of planting material by pathogens mainly potato viruses cause severe reduction in yield. Therefore, disease-free/healthy in vitro multiplied plantlets are used in germplasm conservation and conventional seed production. ICAR-CPRI, Shimla is designated as the National Active Germplasm Repository for conservation of potato germplasm in India. Presently more than 4500 accessions of cultivated and wild potato species, obtained from 30 countries, are conserved at the institute. Today nearly 2700 Tuberosum accessions of potato germplasm are maintained through tissue culture (in vitro) at the institute. Micropropagation allows large-scale multiplication of virus-free potato microplants. Nodal segments of virus-free potato microplants are cultured on semisolid or liquid medium under aseptic conditions for obtaining new microplants. Murashige and Skoog's (MS) medium supplemented with 2.0 mg/L D-calcium pantothenate, 0.1 mg/L GA3, 0.01 mg/L NAA and 30 g/L sucrose is best suited for propagation of potato microplants. Cultures are usually incubated under a 16-h photoperiod (50-60 μ mol $m^{-2}s^{-1}$ light intensity) at 24 °C. Usually, two-three nodal cuttings (1.0-1.5 cm) are inoculated per culture tube (25 x 150 mm), and the tubes are closed with cotton plugs. Within 3 weeks the axillary/apical buds of these cuttings grow into full plants. These plants can be further sub-cultured on fresh medium.

At an interval of every 25 days of subculturing, theoretically 3^{15} (14.3 million) microplants can be obtained from a single virus-free microplant in a year.

Virus-free micro plants can be used for direct transplanting after hardening, in the fields or nursery beds for production of normal tubers or minitubers, respectively. Alternatively these plants can also be used for the production of microtubers in the laboratory. Microtubers are miniature tubers produced under tuber inducing conditions in vitro. These small dormant tubers are particularly convenient for handling, storage and distribution. Many protocols have been developed for induction of microtubers in vitro. Most of the published work on potato microtuberization is focused on the use of cytokinins, especially N6-benzyladenine (BA). Other substances like abscisic acid, chlorocholine chloride (CCC), NAA, triazoles, coumarine, acetic acid and jasmonic acid have also been used for induction of microtubers in potato. MS basal nutrient mixtures are universally used for potato microtuberization. Sucrose is the most effective carbon source, and an increase in its concentration to 8 percent induces early tuberization, whereas concentrations above 8 percent are inhibitory. Temperature and photoperiod are two important physical factors that affect potato microtuber induction in vitro. The optimum temperature for in vitro tuberization is 20 °C with a constant temperature being more effective than alternating day-night temperatures. Temperatures below 12 °C and above 28 °C have been found to be inhibitory to potato microtuber production. In general, optimum microtuberization occurs under continuous darkness during cytokinin-induced tuberization, but a longer photoperiod with higher light intensity is required when cytokinin is not used.

At ICAR-CPRI, Shimla microtubers are also induced in MS medium supplemented with 10 mg/L BA plus 80 g/L sucrose, and the cultures are incubated under complete darkness at 20 °C. Microtubers begin to develop epigeally 1-2 weeks after incubation depending on the genotype, and are harvested after 60-75 days of incubation. In general, 15-20 microtubers with an average weight of about 100-150 mg can be obtained from each flask/magenta box. Before harvesting, the magenta boxes are shifted under diffused or artificial light at 20-30 °C for 10-15 days for greening the microtubers. Thereafter, green microtubers are treated with 0.2 percent Bavistin, dried at 20 °C, packed in perforated polythene bags, and stored under dark at 5-6 °C till dormancy release. These microtubers are planted on nursery beds under aphid-proof net houses (@ 50 microtubers/m^2) in seed producing areas of the Indian plains. The microtuber crop is allowed to mature in the nursery beds to produce minitubers. The virus-free minitubers thus produced constitute the pre-basic seed material for production of breeders' seed after 3-4 field multiplications.

In countries, where disease-free seed potatoes cannot be produced for want of vector-free production areas, micropropagation can play an important role to maintain and multiply commercial cutivars under disease-free conditions. In such countries, both microplants and microtubers can be used effectively for producing initial pre-basic (healthy) material in the form of minitubers. The ease with which microtubers are produced and handled also makes them more suitable for long distance transportation and use in potato seed production schemes. Minituber production is affected by genotype and crop husbandry including soil type, type of net house and planting density.

2. Preparation of Stock Solutions

Plant tissue culture media is prepared mainly using five different stock solutions. A general protocol for the preparation of nutrient stock solutions based on MS medium is given below:

2.1 Preparation of Macronutrients Stock Murashige and Skoog (MS-I) in 1000 ml

Sl.No.	*Name of the Chemical*	*Molecular Formula*	*Quantity (g)*	
			(10X)	*(40X)*
1.	Potassium nitrate	KNO_3	19.0	76.0
2.	Ammonium nitrate	NH_4NO_3	16.5	66.0
3.	Potassium di-hydrogen phosphate	KH_2PO_4	1.7	6.8
4.	Magnesium sulphate hepta-hydrate	$MgSO_4.7H_20$	3.7	14.8

Take 500 ml double distilled water in a 2.0 liter beaker, weigh, add and keep on dissolving each salt sequentially in a descending order as listed above (dissolve by magnetic stirring), and finally make up the volume to 1000 ml by distilled water. Store at 4 °C. Always check for precipitation in macro stock before using it for medium preparation.

2.2 Preparation of Calcium Stock Murashige and Skoog (MS-II) in 1000 ml

Sl.No.	*Name of the Chemical*	*Molecular Formula*	*Quantity (g)*	
			(10X)	*(40X)*
1.	Calcium chloride dihydrate	$CaC_{12}.2H_2O$	4.4	17.6

Take 500 ml double distilled water in a 2.0 liter beaker, weigh and add CaCl2.2H20, dissolve by magnetic stirring, and finally make up the volume to 1000 ml by distilled water. Store at 4 °C.

2.3 Preparation of Micronutrients Stock Murashige and Skoog (MS-III) in 1000 ml

Sl.No.	*Name of the Chemical*	*Molecular Formula*	*Quantity (g)*	
			(10X)	*(40X)*
1.	Boric acid	H_3BO_3	62.0	248.0
2.	Manganese sulphate mono-Hydrate	$MnSO_4.H_3O$	169.0	676.0
3.	Zinc sulphate hepta-hydrate	$ZnSO_4.7H_2 0$	86.0	344.0
4.	Potassium iodide	KI	8.3	33.2
5.	*Sodium molybdate di-hydrate	$Na_2MoO_4.2H_2O$	2.5	10.0
6.	**Copper sulphate penta-Hydrate	$CuSO_4.5H_2 0$	0.25	1.0
7.	**Cobalt chloride hexa-hydrate	$CoCl_2.6H_2 0$	0.25	1.0

Note: *$Na_2MoO_4.2H_2 0$ should be dissolved separately in approximately 20 ml of distilled water and added to the stock and $MnSO_4.H_2O$ takes time to dissolve.

**Since it is difficult to weigh small quantities, prepare separate stocks of these salts @1.0 mg ml^{-l} and then add required quantity. Take 500 ml of double distilled water in a 2.0 liter beaker, weigh, add and keep on dissolving each salt sequentially in descending order as listed above (dissolve by magnetic stirring), and finally make up the volume to 1000 ml by distilled water. Store at 4 °C.

2.4 Preparation of Iron-EDTA Stock Murashige and Skoog (MS-IV) in 1000 ml

Sl.No.	*Name of the Chemical*	*Molecular Formula*	*Quantity (g)*	
			(10X)	*(40X)*
1.	Sodium EDTA di-hydarte	Na2EDTA.2H20	373.0	1492.0
2.	Ferrous sulphate hepta-Hydrate	FeSO4.7H20	278.0	1112.0

Take 1000 ml double distilled water in a 2.0 liter amber colored bottle, and warm the water near boiling. Now weigh and add $Na_2EDTA.2H_2 0$ while stirring under a magnetic stirrer; after $Na_2EDTA.2H_2O$ has been dissolved, add gradually $FeSO_4.7H_2O$ while still under mild magnetic stirring. This will yield a clear yellow solution. Immediately after adding $FeSO_4.7H_2O$ close the bottle, keep on stirring at least for an hour and store at 4 °C.

2.5 Preparation of Vitamin Stock Murashige and Skoog (MS-V) in 1000ml

Sl.No.	*Name of the Chemical*	*Quantity (mg)*	
		(10X)	*(40X)*
1.	myo-Inositol	1000.0	4000.0
2.	Glycine	20.0	80.0

Sl.No.	Name of the Chemical	Quantity (mg)	
		(10X)	(40X)
3.	*Thiamine HCl	1.0	4.0
4.	*Nicotinic acid	5.0	20.0
5.	*Pyridoxine HCl	5.0	20.0

Note: *Since it is difficult to weigh small quantities, prepare separate stocks of these salts @1.0 mg ml^{-1} and then add required quantity.

Take 500 ml double distilled water in a 2.0 liter beaker, keep on adding and dissolving the vitamins sequentially in descending order as listed above, and finally make up the volume to 1000 ml by adding distilled water. Store at 0-4 °C. Vitamin stock is very prone to microbial contamination in storage. Therefore, always check the stock before using.

3. MS Media Preparation

Routine preparation of culture media using stock solutions is simple and involves following steps. The method has exemplified for preparing 1.0 liter of MS basal medium.

3.1 Murashige and Skoog (MS) media for 1000ml

Sl.No.	Stock Solution	Quantity (ml)	
		10X	40X
1.	MS-I	100.0	25.0
2.	MS-II	100.0	25.0
3.	MS-III	100.0	25.0
4.	MS-IV	100.0	25.0
5.	MS-V	100.0	25.0

3.2 Mixing of Stock Solutions

For the preparation of 1.0 liter MS medium, the above stock solutions should be added sequentially in about 500 ml of double distilled water at the following rate:

- ✰ In general, for potato micropropagation the Murashige and Skoog medium is supplemented with 0.1 M sucrose/sugar (20 or 30 g) and 4.19 µM (1.0 mg/ml) D-calcium pantothenate. Weigh and add required quantities of sucrose and dissolve by magnetic stirring.
- ✰ According to the purpose of the medium, other medium conjugates/additives are added, and the volume of the medium is made up to 1000 ml by distilled water.

- Adjust the pH of the medium to 5.8 using 0.1 N NaOH or 0.1 N HCl before autoclaving. Note that the pH meter should be calibrated by standard buffers (pH 4.0 and 7.0) immediately before adjusting the medium pH.
- For preparing semisolid medium, heat the medium until near boiling in a microwave oven or gas oven with intermittent stirring and add agar at the rate of 6.0-8.0 gl^{-1} or gelrite at the rate of 2.0 gl^{-1}.
- Mix thoroughly and dispense measured volume into culture tubes/ containers/vessel using automatic media dispenser.
- For preparing liquid medium, pH adjusted media are directly poured in suitable containers.
- Plant tissue culture media are usually autoclaved at 121 °C for 20 min (15 Ib in^2 or 1.05 kg cm^2).
- Autoclaving is generally done in a horizontal or vertical autoclave.
- Minimum time necessary for steam sterilization of media is dependent on volume of medium per vessel
- Autoclaved media are kept in ambient temperature for a day and then transferred in a dust- free closed cabinet for subsequent use. Semisolid medium starts drying up, and therefore should be used within a fortnight after its preparation.

3.3 Sterilization Techniques

3.3.1 Steam Sterilization

- Plant tissue culture media are generally sterilized by autoclaving at 121 °C (15 psi). The time required for sterilization depends on the volume of the medium in the vessel (Figure 34.1).
- Always dispense medium in small aliquots whenever possible because many media components are broken down on prolonged exposure to heat and pressure.
- Media exposed to temperatures in excess of 121 °C may not properly gel or may result in poor culture growth.
- Thermolabile components are prepared and filter-sterilized through a 0.2 µ filter in a sterile container.
- The filtered chemicals is aseptically added to the culture medium, which has been autoclaved and allowed to cool to approximately 35-45 °C.
- The medium is then dispensed under aseptic conditions in the laminar flow.

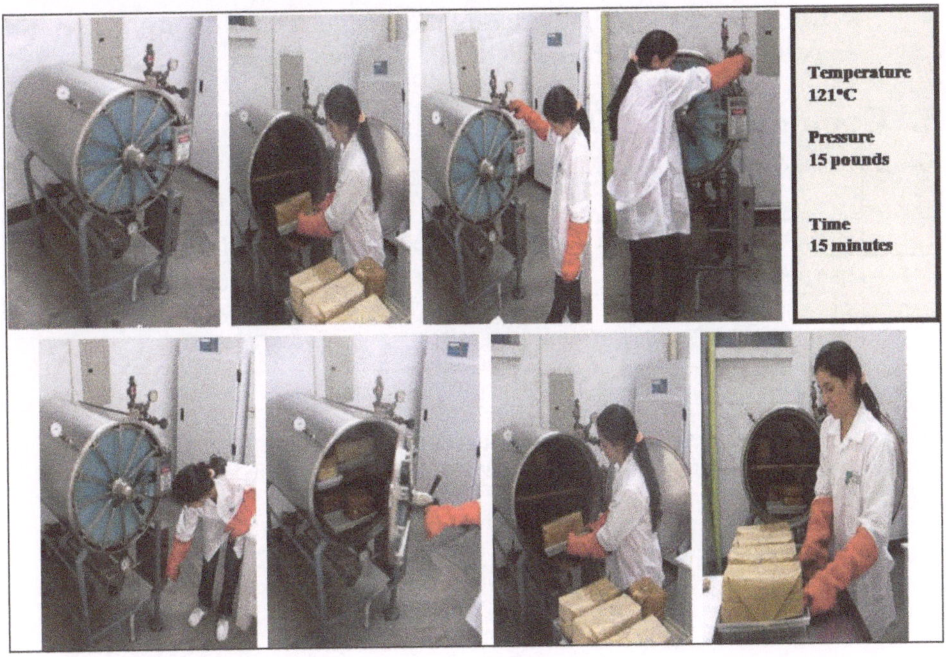

Figure 34.1: Process for the Sterilization of Culture Media.

3.3.2 Filter Sterilization

For sterilization of small volume (50 ml), use syringe filters. Generally syringe filters are available in 4, 13 and 25 mm diameters. Readymade PTFE or cellulose ester membranes (0.2 μ) are convenient to use since they are packed sterile, and manufactured with female luer lok™ inlet and male outlet as a standard; the filter housing can easily be connected to a syringe.

4. Surface Sterilization of Explants

To prevent bacterial and fungal growth, tissue explants are surface sterilized before they are used to establish axenic/*in vitro* cultures. The most common disinfectants are listed below with the concentration and exposure. There is no universal working procedure for disinfecting plant tissue explants; the procedure varies from tissue to tissue and species to species. It is recommended that the experimenter standardize his/her own protocol based on the following guidelines:

- ☆ Wash the tissue explants with mild detergent (Tween-20) before treatment with the disinfectant solution.
- ☆ Rinse the explants thoroughly under running tap water for 10-30 min.
- ☆ Submerge the explants into the disinfectant solution, seal the bottle and gently agitate.

☆ Under sterile conditions, decant the disinfectant solution and rinse the explants several times with sterile distilled water.

5. Commonly Used Disinfectants for Plan t Tissue Culture

Sl.No.	*Name of the Disinfectants*	*Concentration (Percent)*	*Exposure Time (min)*
1.	Hypochlorite	0.5-5.0	5-30
2.	gen peroxide	3-12	5-15
3.	Alcohol	70-95	1-5.0
4.	Nitrate	1.0	5-30
5.	ric chloride	0.1-1.0	1-5

* Commercial bleach contains about 5 percent sodium hypochlorite and thus may be used at a concentration of 10-20 percent which is equivalent to 0.5-1.0 percent sodium hypochlorite.

Chapter 35

Soil and Plant Sample Test

Sushil Kumar and Vineeta Sharma

ICAR-Central Potato Research Institute, Shimla, Himachal Pradesh

1. Introduction

India has about 195 m ha area under cultivation out of which about 63 percent area is rainfed. The everincreasing population and reducing cultivable land holdings due to roads and industrialization are posing tserious hreat on food security in India. In modern agriculture, soil testing is the most important practice to manage fertilizer application and crop production. Soil testing ensures the right amount of the fertilizer to be applied. Soil test- based fertility management is one of the most important approaches for sustainable agriculture production. It provides information regarding nutrient availability in soils, which provides the basis for the fertilizer recommendation to get maximum production. On the other hand modern plant analysis is used primarily as a source of information on plant nutrient status and, ultimately as a tool to help in nutrient management decisions. For the nutrient management of the crops, analytical inputs are used to diagnose existing nutrient problems, predict nutrient problems likely to affect crop production between sampling and harvest and to monitor crop nutrient status for optimal crop production.

The judicious use of chemical fertilizers and organic manures on the basis of soil and plant test values is very important. The soil test helps in selecting the proper manuring schedule for different crop sequence by observing the nutrient build up and depletion from the soil profile. Due to high crop intensity, the removal of nutrients (macro, secondary and micro nutrients) is far higher than their addition in the form of manure and fertilizer.This scenario may be different when we add nutrients more than the requirement of the crop leads to wastage of nutrients.

2. Soil Analysis

In agriculture sector soil analysis is considered as a tool for making fertilizer recommendations for different crops and crop sequences as well as for reclamation of problem soils. The main objectives of soil testing are:

- ✰ Soil fertility evaluation for making fertilizer recommendations
- ✰ Prediction of likely crop response to applied nutrients.
- ✰ Classification of soil into different fertility groups for preparing soil fertility maps.
- ✰ Identification of the type and degree of soil problems like salinity, alkalinity, acidity, soil erosion *etc.*, and their appropriate reclamation procedure.

2.1 Collection and Preparation of Soil Samples for Analysis

Only a minute fraction of huge soil mass of the field is used for the analysis in the laboratory to find out the relevant physical and chemical characteristics. Therefore, for collecting soil samples the following aspects should be considered carefully.

The soil sample collected should be representative of the area sampled. For the collection of composite soil samples variation in slope, color, texture, crop growth and management should be taken into account and separate sets of should be collected from each of such area. During sampling recently fertilized plots, bunds, channels, marshy tracts and spots near trees, wells, compost, or FYM/piles or other non-representation locations must be avoided. Soil sampling should be done well in advance and not just before sowing. Ten to fifteen sub samples of approximately equal weight should be drawn in a zigzag pattern from an apparently homogenous plot. However, where the area with in the field looks different in appearance, topography, divided the field into parts and take sample of each portion (constituting a sample unit) separately.

2.2 Sampling Procedure

To obtain a composite sample, small portions of soil are to be collected up to the desired depth. Generally the samples may be drawn to the plough depth (0-15 cm) by means of suitable sampling tools from at least 10-15 spots after scrapping off the surface litter, if any. For sampling loose and moist soil, the tube auger, spade or khurpi is quite satisfactory. If a spade or khurpi is used, a V shape cut may be made up to the plough layer and a uniform 1.5 cm thick slice is taken out. The soil collected in this manner should be thoroughly powdered and mixed by hand on a clean piece of cloth or polythene sheet or thick paper. The bulk is reduced by quartering and about 500 g of the composite sample is retained. The soil must be quickly dried in shade at room temperature and put in cloth or polythene bags with suitable description and identification marks.

2.3 Preparation of Sample for Analysis

2.3.1 Drying

Samples are generally air dried (25–35 °C), relative humidity (20-60 percent) and stored. Results of soil analysis are expressed on oven dry weight basis. This necessitates determination of moisture percentage by drying a small quantity in an oven at 105 °C for 2 hrs.

2.3.2 Sieving

Moist samples prior to drying may be passed through 6 mm sieve by rubbing with fingers. Soils at right moisture condition can be passed through a 2 mm sieve. If gravels content in the soil sample exceeds 2 percent limit over a 2 mm sieve, their exact percentage should be observed.

2.3.3 Mixing

Sample should be thoroughly mixed by rolling procedure. Place the dried and sieved sample on cloth or paper. Hold the opposite corners and then holding one corner down pull the other corner across the sample. Now the process is repeated in the reverse direction. Use the other two corners and roll the soil from one corner to another repeatedly. Continue this until desired amount of soil is remained.

2.3.4 Storing

Store the soil in paper carton using a polythene bag as an inner lining.

3. Physico-chemical Analysis of Soil

Soil pH, EC, texture, structure, bulk density, pore space, water holding capacity exerts a great influence on nutrient availability.

3.1 Estimation of Soil Reaction

Soil pH and EC influence the availability of different nutrients. Its extreme levels hinder the plant growth. Soil pH is estimated in 1:2 soil water extract prepared by taking 20 g of soil and 40 ml distilled water in 100 ml beaker. The suspension is shaken at regular intervals for half an hour.Saline soils are not fit for potato cultivation.The EC indicates the presence of salts in the soil. The Electrical Conductivity (EC) of a solution is a measure of the ability of the solution to conduct electricity. The EC is reported in either millimhos per centimeter or the equivalent decisiemens per meter.

3.2 Soil Texture and Structure

Soil texture and structure are the good index for assessing the nutrient and water supplying capacity of the soil. Sandy loam and loamy soils having high organic matter are considered best for potato production. The light textured alluvial soils from Indo-Gangetic plains with low amount of organic matter and water retention capacity require higher dose of fertilizer and water for getting optimum yield of potato. Similarly, heavy textured soils like clay loam to clayey soils of peninsular India suffer from waterlogging and cracking, and obstruct the tuber growth beside tuber greening and tuber moth infestation. The International pipette method is standard procedure for soil texture estimation.

4. Soil Testing Methodology

Several methods involving use of selective extractant are available for testing each nutrient essential for plant growth. The method is selected by the availability of nutrients in soluble form for its intake by the growing plant.

4.1 Organic Matter

The contribution of organic carbon towards the availability of all essential nutrients is of great importance. Soil organic carbon is oxidized to CO_2 by heat of dilution obtained when a known quantity of soil is treated with excess of standard K_2Cr2O_7 in the presence of concentrated H_2SO_4. The excess of K_2Cr2O_7 not reduced by the organic matter is back titrated with standard ferrous ammonium sulphate in the presence of phosphoric acid and sodium fluoride using diphenyl as indicator. Color change from violet/blue to bright green is considered as end point.

4.2 Available Nitrogen

Major part of soil N is present in organic form and is closely related to organic matter. It has to undergo mineralization process to produce nitrate and ammonium for plant growth. Chemical method for the estimation of soil Available nitrogen is based on oxidation of organic matter. A known quantity of the soil is mixed with excess of alkaline $KMnO_4$ and distilled. Ammonia gas released by distillation is absorbed in a known volume of standard sulphuric acid excess of which is titrated with standard alkali using methyl red as an indicator.

4.3 Available Phosphorus

Several methods are used for the determination of available phosphorus in soils. Generally two methods are common. Olsen's method (Olsen *et al.*, 1954) is used for neutral to alkaline soils whereas the Bray's method (Bray and Kurtz, 1945) is used for acid soils.Phosphorus fixation take place under acidic and alkali conditions. In Olsen's method available phosphorus in soil is extracted by shaking the soil with 0.5 N sodium bicarbonate solution adjusted at pH 8.5. Color intensity of the molybdophospheric blue color is measured.

Bray's P (or Bray No.1) extractant, which involves soil extraction with a solution consisting of 0.03N NH_4F and 0.025 N HCl is widely used under acidic soil conditions. The fluoride complexes Al and Fe in soil, thus releasing some bound P besides the easily acid soluble P (largely Ca-P). This extractant is suitable for soils containing less than 2 percent calcite or dolomite because in calcareous soils, carbonates quickly neutralize the acid, resulting in less extraction of P.

4.4 Available Potassium

Potassium in soil presents as water soluble, exchangeable, non- exchangeable (fixed) and lattice K. For its determination, most of the methods are based on the determination of easily available fractions *i.e.* water soluble and exchangeable K, which are considered to easily available to plants. When these two forms are depleted, a part of the non- exchangeable K moves to the exchange sites and soil solution. Available K is determined by extracting the soil with N neutral ammonium acetate solution. The ammonium ions replace potassium ions absorbed on the soil colloids. The estimation of potassium in the extract is carried out with the help of flame photometer.

4.5 Secondary Nutrients

Soil test for Secondary nutrients became necessary in view of their widespread deficiency resulting from intensive cropping system and use of high analysis fertilizers. Among Secondary nutrients, for the estimation of calcium and magnesium ammonium acetate method is normally employed. Whereas, 0.15 percent $CaCl_2$ method gives fairly good results for sulfur estimation in potato.

4.6 Estimation of available Micronutrients

Micronutrients are needed by plant in very small quantities, soil testing for the evaluation of Zn, Cu, Fe, Mn, B, Cl and Mo was of little concern till few decade back. Only in some cases, the necessity of assessing these micronutrients was felt. Now in certain areas deficiency of Zn, Fe, Mn and B has been appeared due to use of high analysis inorganic fertilizers, little or no use of organic manures and heavy removal by high yielding crop cultivars. A number of soil test methods, mostly colorimetric, have been widely used for the estimation of micronutrients. Neutral ammonium acetate and chelating agents like EDTA, have been used as extractants for Fe, Zn, Mn and Cu estimation. Zn determination by dithizone method (Shaw and Dean, 1952) has been very popular until AAS became available. For a rapid and accurate analysis of Fe, Zn, Mn and Cu, the DTPA method (Lindsay and Norvell, 1978) is widely used. Available soil B is mostly extracted by hot water and Mo by ammonium oxalate method. The B in the soil extract is determined by azomethine-H and Mo by dithiol method.

5. Plant Analysis

5.1 Sampling and Sample Preparation

Concentration of nutrient element in plant tissues indicates accessibility of the concerned elements from the soil to the plant. However, different plant species and even different varieties of the same species may vary in their nutrient requirements and composition. For plant analysis, collection of specific plant part (tissues) at the right stage of growth as per technical specifications is very important. After sample collection, the fresh tissues should be decontaminated from dust and other foreign material by adopting the following procedure. Plastic containers are taken in which the following solutions are added.

- 0.2 percent liquid detergent
- N/10 HCI solution (8 ml cone, HCl/liter).
- Deionized water

The fresh tissue should be washed in sequence in detergent solution, dilute HCI and deionized water. The liquid detergent will remove waxy coating on leaf surface and any soil particles. N/IO HCI will remove metallic contamination and deionized water will wash the previous two solutions. The extra moisture is wiped out; the sample is placed in new paper bags and dried in an oven at 65 + 5 °C. In case the sampling site is far away from laboratory which require longer travel period, the sample may be placed in a polythene bag and transported in an ice box.

Table 35.1: Recommended Sampling Stage of Growth, Plant Part, and Sample Size of Potato Leaves for Diagnostic Plant Tissue Analysis

Crop	*Stage of Growth*	*Plant Part*	*No of Plant Part to Sample*
Potato	Prior to or at initial flowering	4th petiole and leaflet (whole leaves)	40-50
Potato	Tuber bulking	4th petiole and leaflet (whole leaves)	40-50

5.2 Sample Preparation of Potato Tubers

- Wash the collected tubers with running tap water (trying to remove any soil residue), rinse with distilled water and dry the tubers with paper towel
- Potato tuber samples must be put in a clean and labeled bag and store them at 5 °C
- Sort in order and place the samples in white plastic trays, one day before processing

- ☆ wash the tubers again with distilled water, dry using paper towel, Peel the tubers with a high-grade stainless steel or ceramic peeler, and then cut each tuber longitudinally in 4 sections
- ☆ Peeling should be done carefully, with minimum removal of the flesh.
- ☆ Obtain 3 – 4 slices of each of two opposite sections of each tuber to obtain a 50 g weighed sample.

5.3 Methods of Analysis

5.3.1 Determination of Total Nitrogen

Total nitrogen in plant samples is determined commonly by the Kjeldahl method and by colorimetric procedure. This essentially involves (i) digestion of the sample to convert the N compounds in the sample to the $NH4^+$ form and (2) determination of $NH4^+$ in the digest.The plant material can be digested in a di-acid or tri-acid mixture or dry ashed and dissolved in acid for nutrient estimation other than N.

5.3.2 Estimation of Phosphorus in Plant

Orthophosphates obtained in tri-acid digest react with ammonium molybdate ammonium vanadate in HNO_3 medium and gives a yellow color complex. This color develops in about 30 min and remains stable for 2 to 8 weeks. Intensity of color is measured at 470 nm in a Spectrophotometer.

5.3.3 Determination of Potassium in Plant

Potassium content in plant extract is determined by flame photometry *i.e.* when atoms of potassium are excited in flame, they emit a flame of specific wave length, the intensity of emission is proportional to the concentration of K, which is determined in flame photometer using K filter. The concentration of K is measured by comparing the radiation emitted by a known standard with that of the sample. Most flame photometers manufactured in India give a linear range up to 5 ppm and therefore a desired dilution has to be carried out before determination of K in plant samples.

5.3.4 Estimation of Calcium and Magnesium in Plant

Calcium and magnesium can be determined in plant extract with help of Atomic Absorption Spectrophotometer or versene titration with the help of their standard solutions. Calcium can be estimated on a flame photometer also, though the sensitivity is less than that for K and Na. Calcium (Ca) and magnesium (Mg) react with Ethylene tetra acetic acid (versenate) at different pH values and form stable complexes. Sn, Cl, Zn, Mn and Fe may interfere in the determination of Ca and Mg. This interference may be prevented by using carbonate. In Ca estimation a known amount of di-acid aliquot of plant/soil extract is titrated with standard EDTA solution (0.01 N) using murexide indicator (ammonium purpate, Mur) in presence of NaOH at pH 12, the color changes at end point from orange red to purple, while for Ca + Mg Eriochrome black T indicator is used in the presence of $NH_4OH + NH_4CI$ buffer at pH 10, the color changes at end point from wine red to blue.

5.3.5 Estimation of Sulphur in Plant

The plant material is digested with nitric and perchloric acid and the sulphate content of an aliquot of the digest are determined by its turbidity as $BaSO_4$ by a barium chloride gelatine procedure. A common procedure follows wet ashing of plant tissue sample (described under phosphorus) and the sulphate content in the digest is then determined by barium sulphateturbidimetry method. During ashing/wet digestion of the sample, all the plant sulphur is converted to sulphate form, which when treated with $BaCl_2$ is precipitated as white $BaSO_4$. This provides turbidity to the solution which is proportional to the amount of sulphate present. Measurement of this turbidity provides the means for quantitative determination of sulphur. Barium sulphate tends to precipitate if solution is not properly acidified. On the other hand, if the solution is too acidic, the precipitation of $BaSO_4$ is low. Therefore, buffer salt solution is added to prevent the pH of the medium from fluctuating widely. The rate of precipitation should be controlled by the addition of pure $BaCl_2$ crystals of definite size and by its slow dissolution. Gum acacia solution is added to help in stabilizing turbidity.

5.3.6 Determination of Zinc, Copper, Manganese and Iron in Plant

Plant sample is digestedwith di-acid and make the volume in 100 ml volumetric flask. Use this extract for the determination of Zn, Cu, Mn and Fewith the help of Atomic Absorption spectrophotometer (AAS).The AAS is based on the principle that atoms of metallic elements (Fe, Mn, Cu, Zn *etc.*), which normally remain in ground state under flame conditions absorb energy when subjected to radiations of specific wavelength. The absorption of radiation is proportional to the concentration of atoms of that element. The absorption of radiation by the atoms is independent of the wavelength of absorption and temperature of the atoms. These two features provide AAS a distinct advantage over flame emission spectroscopy. It also has greater sensitivity and accuracy.

6. Plant Test for Modifying Fertilizer Recommendations

Tissue test for N and K have been employed in correcting nutrient deficiency by applying corrective dose to the standing crop at early stage of crop growth. Early sampling for leaf petioles *i.e.* 40-45 days in hills and 20-30 days in plains is best for tissue testing and taking corrective measures. For micro-nutrients deficiency, the foliar application is most appropriate method.

7. Plant Test for Assessing Fertilizer Recovery

Plant tests have been used for working out fertilizer use efficiency, nutrient recovery, depletion of nutrients in a cropping system and planning precise fertilization schedule for the crop sequence so as to maintain soil fertility and productivity. For the estimation of nutrient recovery and fertilizer use efficiency, whole plant *i.e.* haulm, tuber and roots are taken from the control and fertilized plots.

Chapter 36

DNA Fingerprinting and Gene Expression Analysis in Potato

C.M. Bist, Rahul, Sundaresha and J.K. Tiwari

ICAR-Central Potato Research Institute, Shimla, Himachal Pradesh

1. Introduction

The analysis is performed by extracting genomic DNA from the leaf material using standard Plant DNA isolation Kit. The technique involves amplifying specific polymorphic regions of DNA by Polymerase Chain Amplification (PCR). The amplified DNA fragments are then resolved through capillary electrophoresis.

2. Requirements

- ☆ Taq DNA polymerase (1U/reaction)
- ☆ Taq DNA polymerase buffer (1X)
- ☆ Primers (Forward and Reverse) (10 pmol/reaction each)
- ☆ Deoxynucleotide triphosphate (dNTPs) (2.5 mM)
- ☆ MgCl2 (if not included in Taq DNA polymerase buffer) (2.5 mM)
- ☆ Template DNA (100 ng/reaction)
- ☆ Polymer POP7
- ☆ Cathode buffer
- ☆ Anode buffer
- ☆ Capillary array

3. Protocol

Table 36.1: Prepare PCR Reaction Mixture

Chemical	*Stock*	*Volume/Reaction*	*Final Concentration*
PCR Buffer* (without $MgCl_2$)	10x	2.5 µl	1x
$MgCl_2$	25 mM	1.5 µl	1.5 mM
dNTP	10 mM of each	0.5 µl	200 µM of each dNTP
Forward Primer	10 µM	0.5 µl	0.2 µM
Reverse Primer	10 µM	0.5 µl	0.2 µM
Taq polymerase	5 U	0.2 µl	1 U
Milli-Q water (H_2O)	-	14.3 µl	-
DNA template	10 ng	5 µl	50 ng
Total reaction volume		25 µl	

Note: *QIAGEN PCR Buffer contains 15 mM $MgCl_2$. So, avoid additional $MgCl_2$ while using QIAGEN Kit and adjust volume of water accordingly.

1. Load the reaction mixture in PCR tube/plate.
2. Set up the reactions in a Thermocycler machine according to the following cycling conditions:

Cycle	*Temp.*	*Time*
1	95 °C	12 min
	94 °C	15 sec
10	50 °C	15 sec
	72 °C	30 sec
	89 °C	15 sec
20	50 °C	15 sec
	72 °C	30 sec
1	72 °C	30 min

- ✰ Prepare mixture of Hi-Di (Highly deionized) Formamide and internal marker ROX. Take 12 µl of HiDi Formamide and 0.5 µl of ROX for one sample. Dispense 12.5 µl of mix into 5 separate PCR tubes. Add 4 µl of PCR amplified samples to each tube and denature the samples at 95 °C for 5 min. Snap chill the samples on ice and load the samples into the '*ABI* 3500 - Genetic analyzer' system. Run the samples using '*ABI* 3500 Data Collection' software and analyze the data through '*ABI*' genotyping software package 'GeneMapper'.
- ✰ Load the 96-well plate in the Genetic Analyzer machine and run the '3500 Genetic Analyzer' (ABI) following manufacturer's instructions. Analyze the AFLP/SSR fragments using inbuilt software (ABI).

- Alternatively, resolve the amplified DNA products on 4 percent high resolution agarose (Sigma–Aldrich) or 3 percent metaphor agarose (Lonza) or 8 percent polyacrylamide gel electrophoresis (PAGE).
- Visualize and score the fragments in Gel Doc.

4. Real Time PCR Technology

The introduction of real-time PCR technology has significantly improved and simplified the quantification of nucleic acids, and this technology has become an invaluable tool for many scientists working in different disciplines. Especially in the field of molecular diagnostics, gene expression and copy number detection, this real-time PCR-based assays have gained importance.

The technology is based on the measurement of fluorescence during the PCR. The amount of emitted fluorescence is proportional to the amount of PCR product and enables the monitoring of the PCR reaction. The resulting PCR curve is used to define the exponential phase of the reaction, which is a prerequisite for accurate calculation of the initial copy number as well as for expression of the gene at the beginning of the reaction. The simplest and cheapest principle is based on intercalation of double-stranded DNA-binding dyes. This technology can be easily applied to already established PCR assays and does not need any additional fluorescence-labelled oligonucleotide. However, specific and nonspecific PCR products are both detected. Therefore, these assays require careful optimization of the PCR conditions and a clear differentiation between specific and nonspecific PCR products using melting-curve analysis.

4.1 Material Required

- Total RNA (Minimum of 500 ng to 5 µg)
- cDNA synthesis Kit
- SYBR Green/Taq Man Probe
- Real Time-PCR system

  Note: RNA isolated area should be RNAse free zone. DEPC treated materials be preferred for RNA isolation.

4.2 Total RNA Isolation from Phenol: Chloroform Method

- Take 0.3 g leaf tissues and grind to fine powder in liquid nitrogen
- Add 10 ml of extraction buffer (0.1 M tris HCl pH 9.0) containing 0.25 M sucrose, 0.2 m Nacl and 10 mM $MgCl_2$
- Add 750 µl of phenol (water saturated):chloroform and grind (1:1) (250 µl for 0.1 gm)
- Add 150 µl of 0.5 M sodium and ground (50 µl for 1.1 g tissue)
- Add 150 µl of 20 percent SDS and grind

- ☆ Transfer the ground tissues to the freshly sterilized eppendorf tube
- ☆ Add 144 µl of β- mercaptoethanol and shake @ 4 °C for 30 min
- ☆ Spin the tube @ 16,000 rpm for 30 minutes at the temperature of 15 °C
- ☆ Transfer the supernatant to another tube.
- ☆ Add eual volume of phenol: chloroform, vortexed and spun @ 15,000 rpm for 15 min at the temperature of 18 °C. Transfer only aqueous supernatant to another tube and equal volume of chloroform: isoamylalcohol (49: 1) is added and spin @ 15,000 rpm for 15 min.
- ☆ Transfer the aqueous phase to another tube and add 8 M lithium chloride ($LiCl_2$) to get final concentration of 3 M $LiCl_2$ and keept @ -70 °C for 4 h.
- ☆ Centrifuge the tube at 16,000 rpm for 30 min, discard the supernatant and wash the pellets once with 5 ml of 2M $LiCl_2$ and then with 70 percent alcohol by centrifuging at 15,000 rpm for 20 min, air dry the pellet and then dissolve in 150 µl of DEPC (diethyl pyrocarbonate) water.
- ☆ Estiamte purity and quantity by spectrophotometrically (OD value 260 nm).

4.3 Total RNA Isolation from Kit Method and RT-PCR Gene Expression Analysis

- ☆ Total RNA is isolated from the collected leaf samples using RNeasy Plant mini kit or from the traditional Phenol-chloroform method (Qiagen, USA), cDNA is synthesized from 1.0 µg of total RNA using High Capacity cDNA Reverse Transcriptase kit (Applied Biosystems, USA) according to manufacturer's instructions. cDNA is treating with RNAse H (1µl) for 20 min at 37 °C to remove any contaminating RNA.
- ☆ qRT-PCR is performed by using Power SYBR® Green PCR Master Mix (Applied Biosystems, USA). Reaction consisted of 150 ng of cDNA and 120 nM of each gene-specific primer in a final volume of 15 µL. Amplification is carried out for three technical replicates for each sample, including negative controls.
- ☆ ABI PRISM HT 7900 (Applied Biosystems, USA) can be used for the following thermal cycles: 50 °C for 2 min, 95 °C for 10 min; 40 cycles of 95 °C for 15 s, and 60 °C for 1 min. Expression levels are assessed based on the number of amplification cycles needed to reach a common fixed threshold (cycle threshold ΔCT) in the exponential phase of PCR. For relative quantification/expression, the $2^{-}\Delta\Delta^{CT}$ method between conditions in qRT-PCR is applied.

Table 36.2: qRT-PCR Master Mix

Reagent	*Amount*
cDNA	1 ul (1-100 ng)
Forward Primer (50-900 nM)	2 µl
Reverse primer (50-900 nM)	2 µl
2X SYBR green - (2x)	12.5 µl
H_2O	7.5 µl
Total volume	25 µl

4.4 Calculation of RT-PCR Gene Expression

Ct (sample) = ΔCt (gene of interest) – ΔCt (endogenous control)

ΔΔCt = ΔCt (sample) – ΔCt (Control sample)

Relative expression (R.E.) = (1 + PCR efficiency) – ΔΔCt

Or Relative quatification (RQ) = $2^{-\Delta\Delta Ct}$

Note: PCR efficiency = (10(–1/slope of standard curve) – 1

As standard curve the CT values were plotted against the logarithm of template concentration.

R.E+S.D. = (1 + PCR efficiency) (– ΔCt –S.D. dCt).

Chapter 37

Protoplast Isolation and Fusion in Potato

Jagesh K. Tiwari and Rasna Zinta

ICAR-Central Potato Research Institute, Shimla, Himachal Pradesh

1. Protoplast Isolation

- ☆ *In vitro plant material:* Three-week-old *in vitro*-grown microplants, raised from single nodal cuttings are used to isolate mesophyll protoplasts under sterile conditions
- ☆ *Microplants growth condition:* 16 h photoperiod/40 µmol $m^{-2}s^{-1}$/20 °C.
- ☆ *Pre-isolation (protoplast) incubation:* Cultivate *in vitro* plants at 20 °C for 48 h under a 16-h photoperiod in the dark (covered with black muslin cloth) prior to protoplast isolation to integrate cell cycles.
- ☆ *Protoplast isolation:* Mince young leaf tissues (1.0-2.0 g) in a Petri dish (BATTU2 = 90 × 15 mm) containing Protoplast Digestion Solution (PDS): 10.0 ml digestion solution for 1.0 g leaf tissue.
- ☆ *Incubation (for protoplast):* 16 h/dark/25 °C/optional: gyratory shaking at 40-50 rpm; not exceeding 50 rpm.

2. Post-isolation Handling

- ☆ Add 0.3 M KCl (filter-sterilize) (FS) to the digestion medium containing released protoplasts (protoplast suspension) in a 1:1 ratio (PDS:KCl).
- ☆ For example, add 15.0 ml of 0.3 M KCL to 15.0 ml digestion medium/solution followed by gentle shaking of PDS containing released protoplast

☆ Filter the suspension through a 40-µm nylon mesh, and collect in 10 ml centrifuge tubes; 60 µ can be used, but the debris and/or undigested tissues will be much.

3. Protoplast Purification

☆ Centrifuge the filtrate at 50 ×g (60 RCF) for 5 min, and then resuspend the pallet in 10 ml of 0.6 M sucrose (filter sterilized).

☆ Layer 1.0 ml of 0.3 M KCl onto this protoplast suspension, and centrifuge at 50 g for 5 min.

☆ Recover the live protoplasts (green) from the sucrose: KCl interface, and dilute with 10.0 ml of 0.3 M KCl.

☆ Centrifuge at 50 ×g for 5 min to pellet the protoplast. Resuspend the pellet in 0.5 M mannitol (sterile) to a final density of 1 x 10^6 protoplasts ml^{-1} for electrofusion.

Table 37.1. Protoplast Digestion Solution (PDS) Preparation

Sl.No.	*Chemicals*	*Strength*	*Medium Volume (500 ml)*
1.	MS Macronutrients		
	KNO_3	0.95 g/L	475.0 mg
	KH_2PO_4	0.085 g/L	42.5=43.0 mg
	$MgSO_4.7H_2O$	0.185 g/L	92.5=93.0 mg
	$CaCl_2.2H_2O$	0.660 g/L	330.0 mg
2.	MS Micronutrients (To be used as half strength)	× 1000	250 µl
3.	MS Fe-EDTA(MS stock 5) (To be used as full strength)	× 100	2.5 ml
4.	MS Vitamins (MS stock 6) (To be used as half strength)	× 1000	0.5 ml
5.	Myo-Inositol	100 mg/L	50.0 mg
6.	Polyvinylpyrrolidone (PVP)[AVM 10,000]	5.0 g/L	2.5 g
7.	MES (FW 195.24 for MES hydrate)	5.0 mM	488.0 mg
8.	D-Glucose (FW 180.16)	0.1 M	9.0 g
9.	D –Mannitol (FW-182.17)	0.4 M	36.434 g
10.	Cellulase 'ONOZUKA' RS or Cellulase ONOZUKA' R-10	1.0 percent	Not to be added here
11.	Macerozyme R-10	0.5 percent	Not to be added here

Notes:

☆ pH 5.7 to be adjusted using 0.1/1.0 N NaOH.

☆ Prepare the above digestion medium without digesting enzymes, adjust the pH and filter-sterilize.

☆ Store the medium in aliquots (20-25 ml) at -20 °C.

☆ Before using, thaw the medium, add enzymes at required amount to medium aliquots, centrifuge to dissolve and check/adjust the pH (5.7), if required.

☆ Filter-sterilized (0.2 μm filter) and use.

Table 37.2: Solutions Used in Protoplast Culture

Sl.No.	Chemicals	Strength	Medium Volume (500 ml)
1.	KCl (FW 74.55)	0.3 M	11.183 g
2.	Sucrose (FW 342.30)	0.6 M	102.69 g
3.	D-Mannitol (FW 182.17) (pH 7.0)	0.5 M	45.543 g
4.	Solution 2 (pH 7)		
	$CaCl_2.2H_2O$ (MW 147.02 g)	10 mM	220.53 mg
	D-Mannitol (FW 182.17 g) (pH 7)	0.5 M	150 ml
5.	Solution 3 (pH 7)		
	$CaCl_2.2H_2O$ (MW 147.02 g)	50 mM	1.1026 g
	D-Mannitol (FW 182.17 g) (pH 7)	0.5 M	150 ml
6.	Sodium Alginate solution		
	Na-Alginate (alginic acid sodium salt)	2.8 percent	2.8 g
	D-Mannitol (FW 182.17 g) (pH 7)	0.5 M	100 ml
7.	NaOH (FW 40)	1N	4.0 g/100 ml

Notes:

☆ Autoclave all solution at 121 °C for required duration depending on the volume, and then filter-sterilize using a 0.2 μm filter.

☆ Use 1N NAOH to prepare 0.1N NAOH and filter-sterilize the solution in small aliquots.

☆ pH adjustment of Mannitol (pH 7) may take longer time, adjust with 0.1N NaOH.

☆ Na-Alginate solution may take longer time while dissolving solution.

4. Protoplast Electrofusion

☆ Electrofusion medium: 0.5 M mannitol (FS)/pH 7.0-7.3/adjust pH with sterile 0.1 N NaOH.

☆ Symmetric fusion: 1:1 of each species under laminar work station

☆ Electrofusion settings (under laminar):

Method 1

CHAMBER: BTX Microslide Model 453/3.2mm gap

Alignment amplitude: 16 V

Alignment time:	25-30 s
Alignment field strength:	5V cm^{-1}
Electroporation amplitude:	260 V
DC pulse width:	60 µs
Number of pulses:	2
Electroporation field strength:	812V/cm

Method 2

CHAMBER:	BTX Microslide Model 453/3.2mm gap
Alignment amplitude:	32 V
Alignment time:	25-30 s
Alignment field strength:	100V/cm
Electroporation amplitude:	260 V
DC pulse width:	60 µs
Number of pulses:	1
Electroporation field strength:	812 V/cm

Method 3

CHAMBER:	Disposable cuvette P/N620 (2mm gap)
Alignment amplitude:	16 V
Alignment time:	25-30 s
Alignment field strength:	50 V/cm
Electroporation amplitude:	260 V
DC pulse with:	60 µs
Number of pulse:	2
Electroporation field strength:	812 V/cm

Notes:

- ☆ In a disposable cuvette, fuse protoplasts in an aliquot of 400 µl and allow to stand the fusion products for at least 30 min for post-fusion recovery.
- ☆ Post-fusion AC pulse can be employed maximum up to 9 s compress the fusion product.

5. Regeneration of Post-Fusion (Protoplast) Products

Protocol

- ☆ Dispense 50 µl Na-alginate 2.8 percent, prepared in 0.5 M mannitol in each box of 'castor rack' and mix well the 50 µl post-fusion products with sodium alginate to form film layer. (1:1 ratio *i.e.* 50 µl each protoplast and Na-alginate)

- Add ~5 ml Solution # 3 in each box and incubate at RT inside laminar for 30 min followed by add ~5 ml Solution # 2 and incubate same for 1.0 h
- Remove Solution # 2 and # 3 by Pasteur pipette and add 5 ml VKMG (VKM Glucose) liquid medium and tightly wrapped the 'castor rack' with parafilm.
- Incubate the castor racks (containing post-fusion products mixed with sodium alginate into VKMG) under dark at 25 °C for regeneration of microcalli (upto 4-6 months, depends upon the species).
- The first cell division of the protoplast usually occur at 3-4 days after fusion, examine the growth and development of post-fusion products cultures under microscope at regular intervals. Cell aggregates (microcalli) are formed at 4-6 months after the fusion.
- Before transfer the microcalli onto solid medium, dissolve the sodium alginate film surrounding post-fusion products/microcalli in Dissolving Solution.
- After dissolving sodium alginate, wash microcalli in Washing Solution.
- Dissolving solution: MS_{13K} Medium (Without hormones + Coconut water + Gelrite) + 20 mM or 50mM Na-Citrate dehydrate (pH=7.0)
- Washing solution: MS_{13K} Medium (Without hormones + Coconut water + Gelrite) (pH=5.8)
- Finally transfer microcalli onto MS_{13K} Medium (solid) in a Petri dish for development of micro-shoots/plants.
- Transfer of microshoots/plants onto the MS Medium for *in vitro* regeneration and multiplication of putative somatic hybrids' plantlets.

6. VKM Medium Preparation

6.1 Macro-elements

Chemical	Medium Volume (1 lit)
KNO_3	1480 mg/L
KH_2PO_4 (FW 136.09)	68 mg/L
$MgSO_4.7H_2O$ (FW 246.48)	984 mg/L
$CaCl_2.2H_2O$ (FW 147.02)	735 mg/L

No need to prepare stock solution, add directly during medium preparation.

6.2 VKM Stocks

Chemical	*Medium Volume (1 lit)*
VKM stock-I	1 ml/L
MS stock-5	10.0 ml/L
VKM stock II	10.0 ml/L
VKM stock III	10.0 ml/L
VKM stock IV	10.0 ml/L
VKM stock-V	10.0 ml/L
VKM stock-VI	400 µl/L

6.3 Complex Additives

Chemical	*Medium Volume (1 lit)*
Casein Hydrolysate	250 mg/L
Coconut water (**store at -20 °C)**	20 ml/L

Add directly during medium preparation. Coconut water to be added inside laminar.

6.4 Growth Hormones

Chemical	*Medium Volume (1 lit)*
2,4-D (FW 221) (0.2 mg/L) (store at 0 °C)*	200 µl/L
α-NAA (1.0 mg/L) (store at 0 °C)*	1000 µl/L = 1 ml/L
Zeatin (trans) (FW 219.2) (0.5 mg/L) (store at -20 °C)**	500 µl/L

Note:

*Separate stock of 1 mg/ml (dissolved in NaOH and water)

**Dissolved Zeatin-vial, supplied by manufacture, in 100 µl KOH and make up water volume to prepare stock of 1 mg/ml. (*e.g.* 10 mg/10 ml)

6.5 Sugar

Glucose (0.5 M) - 90.1g for 1 litre medium

Note:

- Adjust medium pH: 5.6-5.7 with 0.1 M KOH.
- VKM medium is used for the division and growth of protoplast.
- Please note, that-VKM medium must be filter-sterilized using 0.2 µ membrane filter (cellulose acetate/cellulose nitrate + 0.4 µ pre filter (glass fiber).
- NO AUTOCLAVING

7. VKMG (VKM Glucose) Medium

- ☆ VKM medium must be filter-sterilized using 0.2 μ membrane filter (cellulose acetate) + 0.4 μ prefilter (glass fibre)
- ☆ Weigh 9.01 g of 0.5 M Glucose in 100 ml VKM and adjust pH 5.6-5.7 with 0.1 M KOH

7.1 VKM Stock-I

Micro-elements	mg/L	Stock (×1000)
H_3BO_3 (FW 61.83)	3.0	300 mg/100 ml
$MnSO_4.H_2O$ (FW 169.02), OR	8.0	800 mg/100 ml
$MnSO_4.4H_2O$, OR	10.0	1000 mg/100 ml
$MnSO_4$ (anhyd)	7.0	700 mg/100 ml
$ZnSO_4.7H_2O$ (FW 287.54)	2.0	200 mg/100 ml
KI (FW 166.01) (Store in amber colour bottle)	0.75	75 mg/100 ml
$Na_2MoO_4.2H_2O$ (FW 241.95)	0.25	25 mg/100 ml
$CuSO_4.5H_2O$ (FW 249.68)	0.025	2.5 mg/100 ml*
$CoCl_2.6H_2O$ (FW 237.93) (store in amber colour bottle)	0.025	2.5 mg/100 ml*
1.0 ml VKM Stock-I for 1.0 L medium, Store at 2-8 °C		

Note: *Prepare 1 mg/ml separate stock of each and add 2.5 ml to prepare VKM Stock-I

7.2 MS Stock-5

Chemical	*mg/L*	*Stock (×100)*
$FeSO_4.7H_2O$ (FW 278.01)	27.8	278 mg/100 ml
$Na_2EDTA.2H_2O$ (FW 372.24)	37.2	373 mg/100 ml
10.0 ml MS Stock-5 for 1.0L medium		

7.3 VKM Stock-II

Sugar and Sugar Alcohol	*mg/L*	*Stock (×100)*
d-Mannitol (FW 182.17)	250.0	2.5g/100 ml
d-Sorbitol (FW 182.17)	250.0	2.5g/100 ml
Sucrose (FW 342.3)	250.0	2.5g/100 ml
Fructose (FW 180.2)	250.0	2.5g/100 ml
Ribose (FW 150.1)	250.0	2.5g/100 ml
Xylose (FW 150.1)	250.0	2.5g/100 ml
Mannose (FW 180.2)	250.0	2.5g/100 ml
Rhamnose (FW 182.2)	250.0	2.5g/100 ml

Sugar and Sugar Alcohol	*mg/L*	*Stock (×100)*
Cellobiose (FW 342.3)	250.0	2.5g/100 ml
m-Inositol (180.16)	100.0	1.0g/100 ml
Glucose*	90,000 (90 g)	9.0 g/100 ml
10 ml VKM Stock-II for 1.0 L medium, Store at 2-8 °C		

Note: *Glucose not to be added in the stock; to be added while preparing the medium at the end.

☆ Filter-Sterilize (FS) this stock and to be added inside the laminar work station (while preparing the medium)

7.4 VKM Stock-III

Organic Acid	*mg/L*	*Stock (×100)*
Sodium-pyruvate (FW 110.04) (Pyruvic acid-sodium salt)	20.0	200 mg/100ml
Citric acid (FW 192.12)	40.0	400 mg/100ml
Fumaric acid (FW 116.07)	40.0	400 mg/100ml
Malic acid (FW 134.09)	40.0	400 mg/100ml
10ml VKM Stock-III for 1.0 lit medium, Store at 0 °C		

Note: Filter-Sterilize this stock and to be added inside the LFCA Work station (while preparing the medium

7.5 VKM Stock-IV

Vitamin	*mg/L*	*Stock (×100)*
Calcium-D-pontothenic acid (FW 283.3)	1.0	10 mg/100ml
Choline chloride (FW 139.63)	1.0	10 mg/100ml
Ascorbic acid (FW 176.12)	2.0	20 mg/100ml
p-Aminobenzoic acid (FW 137.14)	0.02	0.2 mg*
Nicotinamide	1.0	10 mg/100ml
Pyridoxine-HCl (FW 205.6)	1.0	10 mg/100ml
Thiamine-HCl (FW 337.27)	10.0	100 mg/100ml
Biotin (FW 244.3) (Vit H)	0.01	0.10 mg**
10 ml VKM Stock-IV for 1.0 lit medium, Store at 0 °C		

Note: Filter-Sterilize and to be added inside the LFCA Work sterilize (while preparing the medium)

* 200 µl from 1mg/ml separate stock of p-Aminobenzoic acid.

** 100 µl from 1mg/ml separate stock of biotin in KOH/NaOH.

7.6 VKM Stock-V

Vitamin	*mg/L*	*Stock (×100)*
Vitamin A	0.01	0.1 mg
10 µl from stock of 10 mg/ml OR 100 µl from stock of 1 mg/ml for 100 ml VKM Stock-V. Dissolve in 100 percent ethanol only		
Vitamin B_{12}	0.02	0.2 mg
200.0 µl from 1mg/ml for 100 ml VKM Stock-V		
Vitamin D_3	0.01	0.1 mg
10 µl from stock of 10 mg/ml OR 100 µl from stock of 1 mg/ml for 100 ml VKM Stock-V. Dissolve in 100 percent ethanol only		
10ml VKM Stock-V for 1.0 lit medium, Store at -20 °C		

7.7 VKM Stock-VI

Vitamin	*mg/L*	*Stock*
Folic acid (FW 441.4)	0.4 mg	1 mg/ml
400 µl VKM Stock-VI for 1.0 lit medium, Store at 0 °C in amber colour bottle		

☆ This stock must be kept in an amber colour vial/or wrap Al-foil.

8. MS_{13K} Medium Preparation

8.1 Macro-elements

Chemical	*mg/L*	*Medium Volume (1000 ml)*
NH_4NO_3	1650	1650 mg/1000 ml
KNO_3	1900	1900 mg/1000 ml
$CaCl_2.2H_2o$	440	440 mg/1000 ml
$MgSO_4.7H_2O$	370	370 mg/1000 ml
KH_2PO_4	170	170 mg/1000 ml

Add directly during medium preparation.

8.2 Microelements

Chemical	*mg/L*	*Stock (×1000)*
H_3BO_3	6.2	620 mg/100 ml
$MnSO_4.H_2O$	16.9	1690 mg/100 ml
$ZnSO_{4.}7H2O$	10.59	1059 mg/100 ml
KI	0.83	83 mg/100 ml

Chemical	mg/L	Stock (×1000)
$Na_2MoO_4.2H_2O$	0.25	25 mg/100 ml
$CuSO_4.5H_2O$	0.025	2.5 mg/100 ml*
$CoCl_2.6H_2O$	0.025	2.5 mg/100 ml*
1 ml for 1.0 lit medium		

* Prepare 1 mg/ml stock and add 2.5 ml to prepare Microelements stock.

8.3 Iron–EDTA (or MS stock-V)

Chemical	mg/L	Stock (×100)
$Na_2EDTA.2H_2O$	37.2	373 mg/100 ml
$FeSO_4.7H_2O$	27.8	278 mg/100 ml
10.0 ml for 1 lit medium		

8.4 Vitamins (or MS stock-VI)

Chemical	mg/L	Stock (×100)
Nicotinamide	0.5	50 mg/100 ml
Pyridoxine-HCl	0.5	50 mg/100 ml
Thiamine-HCl	0.1	10 mg/100 ml
Glycine	2.0	200 mg/100 ml
1.0 ml for 1 lit medium		

8.5 Choline Chloride

Chemical	mg/L	Stock
Choline chloride	8.0	1 mg/ml

☆ 8 ml from stock of 1 mg/ml for 1 lit medium

8.6 Amino Acids

Chemical	mg/L	Stock (×100)
Arginine-HCl	6.24	62.4 mg/100ml
Asparaginic acid*	4.8	4.8 mg/100ml
Cystein**	1.2	12.0 mg/100ml
Glutamic acid	11.2	112.0 mg/100ml
Histidine	2.08	20.8 mg/100ml
Isoleucin	8.2	82.0 mg/100ml
Leucin	12.48	124.8 mg/100ml

Chemical	mg/L	Stock (×100)
Lysine	12.48	124.8 mg/100ml
Methionine	10.4	104.0 mg/100ml
Phenylalanine	4.0	40.0 mg/100ml
Proline	4.0	40.0 mg/100ml
Threonine	10.4	104.0 mg/100ml
Tryptophan	3.2	32.0 mg/100ml
Valine	10.4	104.0 mg/100ml
10.0 ml for 1 lit medium		

- ☆ Asparagine, isoleucine, tryptophan and valine take time to dissolve so use warm water

8.7 Other Chemicals

- ☆ Myo-Inositol: 100 mg for 1 lit medium
- ☆ Sucrose: 30.0 g for 1 lit medium
- ☆ Coconut milk/Water: 50 ml for 1 lit medium
- ☆ Growth Hormones (to be added by sterile - filtration)
- ☆ Gelrite: 2.5 g for 1 lit medium
- ☆ Prepare MS13K medium by mixing amount/volume of above.
- ☆ Adjust pH: 5.8; Autoclave: 121 °C for 20 min
- ☆ After autoclave, add the required amount of growth hormones by sterile filtration
- ☆ This medium is used for Callus regeneration and growth of post-fusion products.

8.7.1 Stock Solution of Growth Hormones

Chemical	mg/L	Stock (1mg/ml stock)
IAA	0.1 mg/l	100 µl/1 lit
Zeatin-riboside	2.0 mg/l	2 ml/1 lit

Solubility Parameters of Growth Hormones

Sl.No.	Hormones	Solubility	Stock Solution	Storage
1.	2,4-D (2,4 dichlorophenoxy acetic acid)	At 60 °C in water bath (ddH_2O)/100 percent EtOH/1N KOH or NaOH	20 mg/20 ml	0 °C
2.	α-NAA (Naphthalene acetic acid)	At 60 °C in water bath (ddH_2O)/1N KOH or NaOH	20 mg/20 ml	0 °C

Sl.No.	Hormones	Solubility	Stock Solution	Storage
3.	Zeatin mix isomer*	1N KOH or NaOH	10 mg/10 ml	-20 °C
4.	Zeatin riboside (ZR)*	1N KOH or NaOH	10 mg/10 ml	-20 °C
5.	BA (N^6Benzyl adenine)	1N KOH or NaOH	20 mg/10 ml	0 °C
6.	GA_3	100 percent EtOH/1N KOH or NaOH	20 mg/10 ml	0 °C

*Directly dissolve in the vial supplied by manufactures.

Solubility Parameters of Vitamins

Sl.No.	Vitamins	Solubility	Stock Solution	Storage
1.	4 (para) amino benzoic acid (PABA) (Vit Bx/Vit H1)	At 60 °C in water bath (ddH_2O)	20 mg/20 ml	2-8 °C (0 °C)
2.	Foliac acid (Vit M)	1N KOH	20 mg/20 ml	2-8 °C (0 °C)
3.	Biotin (Vit H/VitB_7)	At 60 °C in water bath (ddH_2O)/1N KOH	20 mg/20 ml	2-8 °C (0 °C)
4.	Vit A (Retinol acetate)*	100 percent EtOH (@ 25 mg/ml of EtOH)	20 mg/2 ml	-20 °C
5.	Vit B_{12} (Cyanocobalamine)	At 60 °C in water bath (ddH_2O)/1N KOH	20 mg/20 ml	2-8 °C (0 °C)
6.	VitD_3 (Cholcalciferol)*	100 percent EtOH (@ 10 mg/2ml of 100 percent EtOH)	20 mg/2 ml	2-8 °C (0 °C)
7.	Pyrodoxin-HCl (Vit B_6)**	ddH_2O	20 mg/20 ml	2-8 °C (0 °C)
8.	Thiamine-HCl (Vit B_1)**	ddH_2O	20 mg/20 ml	2-8 °C (0 °C)
9.	Nicotinamide (Vit B_3)**	ddH_2O	20 mg/20 ml	2-8 °C (0 °C)
10.	Nicotinic acid (Niacin)**(Vit B_3)	ddH_2O	20 mg/20 ml	2-8 °C (0 °C)

*May encounter difficulties in dissolving; ** Weigh directly while preparing the vitamins stock.

9. MS Medium Preparation for Micropropagation of Plants

9.1 MS Stock- 1

Chemical	Strength (x 50)	100 ml	250 ml	500 ml	1000 ml
NH_4NO_3 1650 mg/L	20 ml MS Stock 1 for 1L medium	8.250 g	20.625 g	41.250 g	82.500 g
KNO_3 1900 mg/L		9.500 g	23.750 g	47.500 g	95.000 g

9.2 MS Stock- 2

Chemical	*Strength (× 100)*	*100 ml*	*250 ml*	*500 ml*	*1000 ml*
$MgSO_4.7H_2O$ 370 mg/L	10 ml MS Stock 2 for 1L medium	3.700 g	9.250 g	18.500 g	37.000 g
$MnSO_4.H_2O$ 16.9 mg/L		169 mg	423 mg	845 mg	1690 mg
$ZnSO_4.7H_2O$ 8.6 mg/L		86 mg	215 mg	430 mg	860 mg
$CuSO_4.5H_2O$ 0.025 mg/L		0.25 mg (1.0 ml)	0.625 mg (2.5 ml)	1.25 mg (5.0 ml)	2.5 mg (10.0 ml)
Dissolve 25mg $CuSO_4.5H_2O$ in 100 ml dH_2O and then add required volume (within parenthesis) to MS 2 Stock					

9.3 MS Stock- 3

Chemical	*Strength (× 100)*	*100 ml*	*250 ml*	*500 ml*	*1000 ml*
$CaCl_2.2H_2O$ 440 mg/l	10 ml MS Stock 3 for 1L medium	4.400 g	11.000 g	22.000 g	44.000 g
KI 0.83 mg/L		8.3 mg	21.0 mg	41.5 mg	83.0 mg
$CoCl_2.6H_2O$ 0.025 mg/L		0.25 mg (1.0 ml)	0.625 mg (2.5 ml)	1.25 mg (5.0 ml)	2.5 mg (10.0 ml)
Dissolve 25mg $CoCl_2.6H_2O$ in 100 ml dH_2O and then add required volume (within parenthesis) to MS 3 Stock					

9.4 MS Stock- 4

Chemical	*Strength (× 100)*	*100 ml*	*250 ml*	*500 ml*	*1000 ml*
KH_2PO_4 170 mg/L	10 ml MS Stock 4 for 1L medium	1.700 g	4.250 g	8.500 g	17.000 g
H_3BO_3 6.2 mg/L		62.0 mg	155 mg	310 mg	620 mg
$NaMoO_4.2H_2O$ 0.25 mg/L		2.5 mg (1.0 ml)	6.25 mg (2.5 ml)	12.5 mg (5.0 ml)	25.0 mg (10.0 ml)
Dissolve 250mg $NaMoO_4.2H_2O$ in 100 ml dH_2O and then add required volume (within parenthesis) to MS 4 Stock					

9.5 MS Stock- 5

Chemical	*Strength (x 100)*	*100 ml*	*250 ml*	*500 ml*	*1000 ml*
$FeSO_4.7H_2O$ 27.8 mg/L	10 ml MS Stock 5 for 1L medium	278 mg	695 mg	1390 mg	2780 mg
$Na_2EDTA.2H_2O$ 37.3 mg/L		373 mg	933 mg	1865 mg	3730 mg
Store in amber colour bottle					

9.6 MS Stock- 6 (Vitamins)

Chemical	*Strength (x 1000)*	*100 ml*
Thiamine-HCl (0.1 mg/L)	1 ml MS Stock 6 for 1L medium	10.0 mg
Pyridoxine-HCl (0.5 mg/L)		50.0 mg
Nicotinic acid (0.5 mg/L)		50.0 mg
Glycine (2.0 mg/L)		200.0 mg
Store at 0 °C		

9.7 Weight and Add Directly

- ✰ Myo-Inositol: 100 mg/L
- ✰ Sucrose: 20.0 g/L
- ✰ pH: 5.8
- ✰ Gelrite: 2.0 g/L
- ✰ Autoclave-sterilize: 121 °C for 20 min

Chapter 38

Diagnostics of Potato Viruses

Ravinder Kumar, RK Tiwari, Chandan Maharana, T. Kochhar and Rajinder Kumar

ICAR-Central Potato Research Institute, Shimla, Himachal Pradesh

1. Serological Methods

Serological methods are most often applicable tests for detection and identification of bacterial/viral plant pathogens. These types of methods based on interaction between antigen-antibody. Usually, the detection of plant viruses by serological techniques is based on the capacity of the antigens (virus particles) to react *in vitro* with their specific antibodies. Antibodies, specific for a particular virus are adsorbed to a carrier. In case of plant viruses 'trapping' IgG is usually bound to a solid phase (polystyrene homologous antigen is detected by an antibody linked with an enzyme). ELISA (Enzyme-linked Immunosorbent assay) is simple to perform, highly sensitive, reliable assay (delectability varies from 0.01 ng/ml to 500 ng/ml of virus depending on the virus), and rapid when automated. It is suitable for testing a large number of samples at a time without requiring large quantities of expensive and difficult to produce antisera. Of the several variants of ELISA, the most commonly used ones is double antibody sandwich-(DAS) (Clark and Adams 1977; Goodwin and Banttari 1984). ELISA has the added advantage of its ability to detect ultra low concentration viruses, like PLRV and PALCV in crude extracts. The common direct double antibody antibody sandwich format of ELISA in polystyrene microtitre plates (Clark and Adams 1977) is the classical one and is widely used in routine testing of viruses. The reaction of colour intensities can be ready using an ELISA reader to quantify the relative antigen titres. It is recommended that at least two wells/sample be used and compared with A405 nm values obtained either through alkaline phosphatase-p-nitro-phenol) phosphate or A620 nm values for penicillinase-bromothymol blue systems for the healthy samples. A sample may be considered infected only if its mean absorbance values are greater than that of mean plus 2SD of the mean of healthy controls.

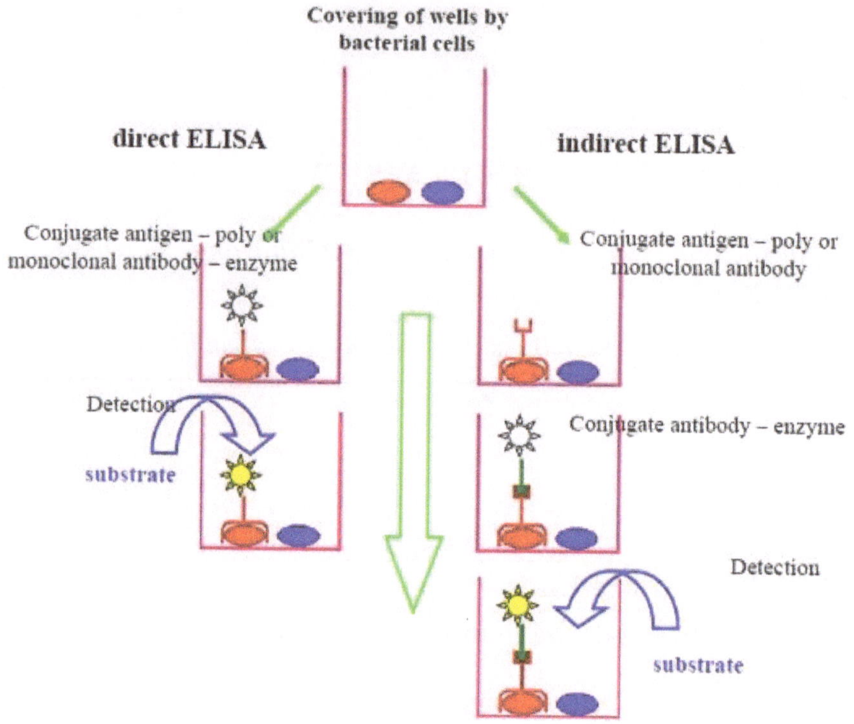

1.1 Direct and Indirect ELISA

Another effective variant is indirect ELISA. In this case, antibodies from two different animal species are required for trapping and detection, respectively. It helps to cut down the background. It is probably the best ELISA format for establishing the seroaffinity among strains of related viruses or even different viruses (Koening and Paul 1982). Besides, a single (universal) conjugate can be used in this format for detection of different viruses (Koenig and Paul, 1982). Adams and Barbara (1982) used in indirect ELISA based on $F(ab)^2$ fragments for trapping of antigens and the enzyme labelled protein-A for the detection to help in assessing serological relationship among virus strains but C 1q-indirect ELISA was found less sensitive for serologically different virus strains.

- ☆ **Buffers and other solutions:** PBS (Phosphate-buffered saline), pH 7.4, NaCl-8.0 g, KH_2PO_4-0.2 g, Na_2HPO_4-2.9 g, KCl-0.2 g, Adjust pH with either NaOH or HCL and make upto 1L. Coating or bicarbonate buffer, pH 9.6, Na_2CO_3-1.59 g, $NaHCO_3$-2.93 g make up to 1L with distilled water

- ☆ **Sample Buffer** pH 7.4, Na_2HPO_4-0.02 M, NaH_2PO_4-0.02 M, Adjust the pH and add 0.15 M NaCl
- ☆ **Extraction buffer/Conjugate buffer** pH 7.4, Sample buffer+0.05 percent Tween 20 + 2 per cent polyvinl pyrrolidon (PVP) + 0.2 percent, Egg albumin + 1 M Urea.
- ☆ **Substrate buffer** pH 9.8, Diethanolamine, 97 ml, $MgCl_2$-100 mg, Adjust pH 9.8 (with 6 N Hcl) and make up to 1 L with distilled water.
- ☆ **Alkaline phosphate (AP) and p-nitrophenly phosphatase system:** It was standardized by Clark and Adams (1977) and is the most common. It may be adopted with slight modifications. Therefore, the procedures involving das-ELISA with alkaline phosphatase, horse radish peroxidase and penicillinase are detailed hereunder.

1.2 DAS-ELISA

In DAS-ELISA, in first step, the specific antibodies are covered the wells of micoplate. Antibodies are attached to the bottom of each well and then the samples with bacterial cells are added. The plate is then gently rinsed to remove unbound bacteria cells. After washing, there are only cells which antigens reacted with antibodies. Next, the antibodies conjugated with enzyme are added (detection of pathogen can be possible when complex antibodies-antigens-antibodies-enzyme is obtained). In the last step, the substrate is added and then the plates can be read visually or by machine. The test is performed in micro-ELISA plates as follows:

- ☆ Add 100-200 μl purified diluted r-globulin in coating buffer to each well of the microtitre plate. Incubate at 37 °C.
- ☆ Wash by flooding wells with PBS- Tween-20. Leave at least 4-5 min. Repeat washing five times.
- ☆ Add 100-200 ul aliquots of the test sample to well. Leave 3 h at 37 °C or overnight at +4 °C.
- ☆ Wash plate 5 times as in 2 above.

1.3 DAS-ELISA (Direct)

- ☆ Add 100-200 μl aliquots of enzyme labelled r-globulin conjugate to each well. Incubate 3 at 37 °C.
- ☆ Wash plate 5 times as in 2 above.
- ☆ Add 100-200 ul freshly prepared substrate (p-nitro-phenol phosphate) @ 0.6 mg/ml in substrate buffer. Incubate at room temperature 10 min to 1 h.
- ☆ Stop reaction by adding 20 ul 3M NaOH to each well.

- ☆ Assess results by (a) Visual observation (b) measurement of absorbance at 405'nm.

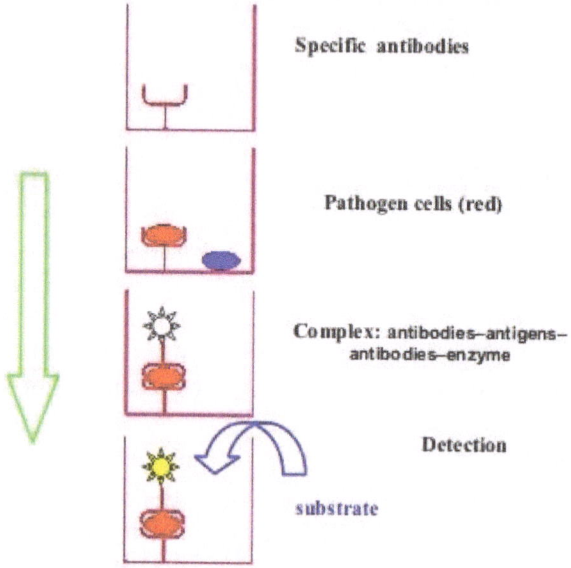

2. Electron Microscopy

Viral diseases can be detected or diagnosed by visual symptoms *viz.*, mosaic patterns on leaves, stunting of the plant, leaf malformations, and tuber malformations. The other method of detection is by variants of electron microscopy, protein based (ELISA) and nucleic acid based polymerase chain reaction. But all the above diagnostic techniques other than electron microscopy gives an indirect evidence of the causal agent (etiology) whereas, Transmission Electron Microscope (TEM) gives a direct access to see the causal agent. It is one of the most powerful scientific tools for carrying out detail structural studies of biological materials. In general, the purpose of any microscope is to form magnified image of a specimen so as to observe its maximum structural details (resolution). In TEM studies, negative staining makes it easy for detection of viruses from liquid samples. This led to the widespread application of TEM in the field of basics of virology and rapid diagnosis of viruses.

2.1 Material Required

Clean glass Petridis, Para film or dental wax, filter paper, copper grids of 400 mesh coated with carbon, Sorenen's phosphate buffer, micropipette with disposable tips, clean and dry tweezers with very tips, negative stain for example: 2 percent UA in DDW, eppendorf tubes of 0.5 ml capacity, wooden tooth picks and antiserum *etc.*

2.2 Leaf Dip Method

This method was originally described earlier and this method is mainly used for the detection of viruses particularly for those viruses that remain in very high concentration in leaf tissues. The protocols are mentioned below:

- ✰ Grind infected leaf of about 2-3 mm diameter in phosphate buffer 0.07 M, pH 6.5.
- ✰ Place 10 µl of homogenate on parafilm or waxed slide in a wet Petriplate.
- ✰ Place carbon coated copper grids (film side downward) on the surface of the droplet, ensuring that the grid surface is wet and allow it for 2-5 minutes.
- ✰ Pick up the grid by its edge with fine forceps and wash the grid with 10-15 drops of double distilled water (DDW) to remove the sap.
- ✰ The grids are stained with 2 percent aqueous uranyl acetate (UA). The excess stain is immediately drained off by using Whatman filter paper.
- ✰ Grids are then placed on sample holder and inserted in TEM.
- ✰ The grids are observed under TEM for its morphology.

2.3 Immunosorbent Electron Microscopy (ISEM)

It is a procedure in which EM support film is first coated with a layer of antibody which serves to trap the virus preparation. ISEM is used for two main purposes as follows:

- ✫ To trap increased number of particles on grid, by reducing the amount of host material.
- ✫ To estimate degree of serological relationship between viruses by making use of trapping response.

2.3.1 Trapping

Float the carbon coated grids (carbon side down) over drop of diluted antiserum (1:1000) in phosphate buffer (pH 6.5, 0.07 M). The drop can be put on a piece of parafilm and leave for 1 h at 37 °C.

- ✫ Wash the grid for 10-15 n phosphate buffer.
- ✫ Drain briefly and then place them over a drop of 10 µl leaf extract and leave for 30 min. at room temperature.
- ✫ Remove the grid and wash with approximately 10-15 drops of DDW.
- ✫ Stain with 2-4 drops of 2 percent freshly prepared uranyl acetate solution.
- ✫ Dry the grid by draining.
- ✫ Examine the grid under the TEM at 20,000 magnification.
- ✫ Count the number of particles from 10 different areas of treated and control grids.

2.3.2 Decoration

It is a process whereby virus particles are first individually observed to grid and grids are floated over specific antibody, layer of specific antibody halo is seen to have absorbed the stain and became electron dense. Milne and Luisoni in 1977 first introduced the decoration method to the Plant Virology.

- ✫ Virus particles are adsorbed to the grid.
- ✫ Grid is carefully washed with phosphate buffer.
- ✫ Put the grid over a drop of diluted antiserum at room temperature for 30 minutes.
- ✫ Wash the grid with 10-15 drops of DDW and stain with 2 percent UA and allow to dry.

2.3.3 Expected Results

We can see the size and shape (morphology) of the virus particles.

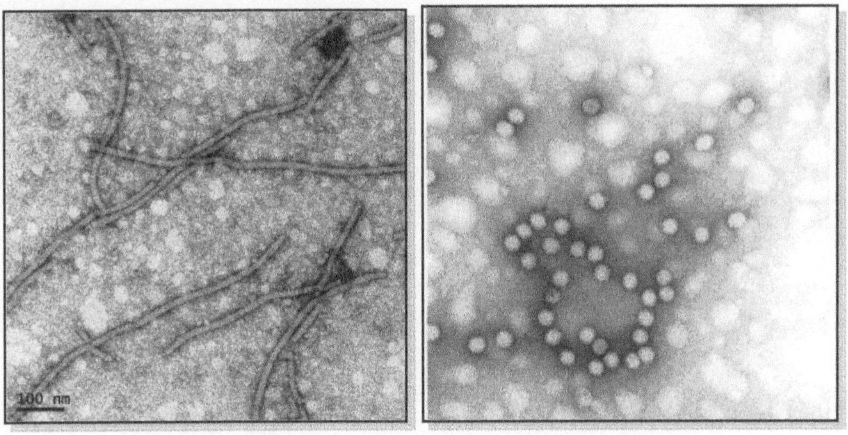

Elongated (PVA) and isometric (PLRV) *Potato Virus.*

COLOUR PLATES

Figure 2.1: Direct seeding tps crop. (pg. 14)

Figure 2.2: TPS Seedlings Transplant Crop. (pg.14)

Figure 2.2: TPS Seedlings Transplant Crop. (pg. 15)

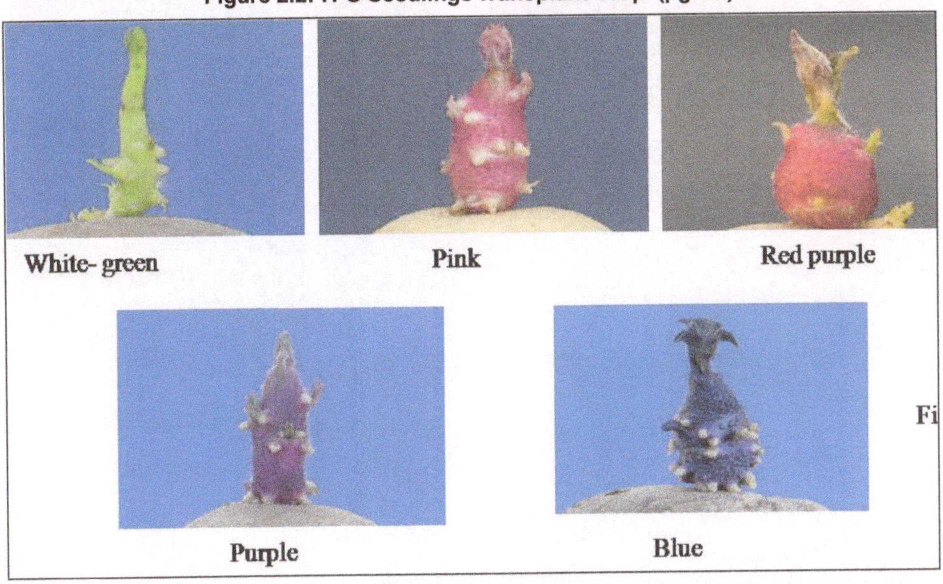

Figure 7.1 (pg. 63)

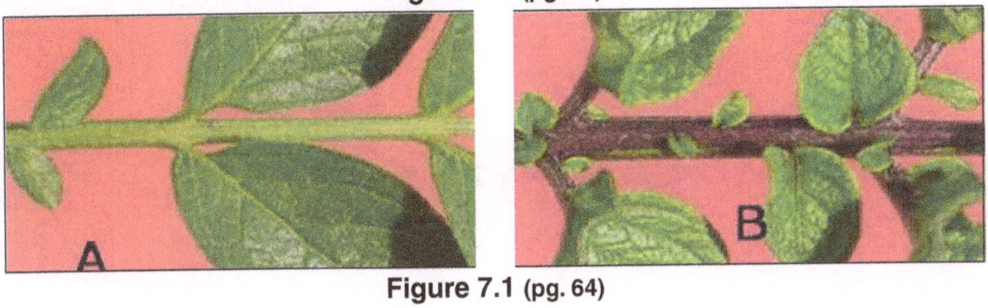

Figure 7.1 (pg. 64)

Figure 7.3 (pg. 65)

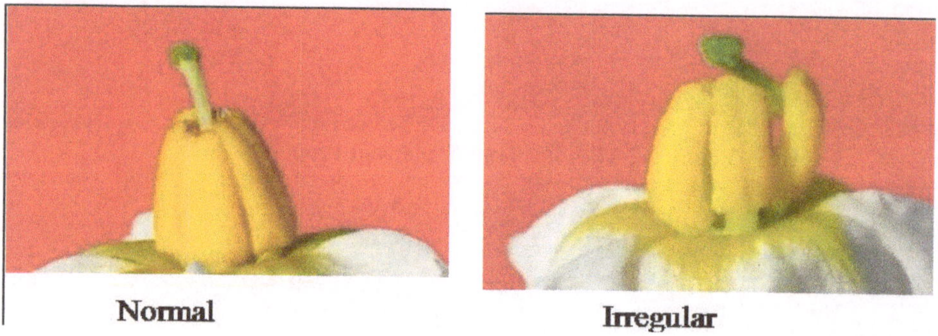

Figure 7.4 (pg. 65)

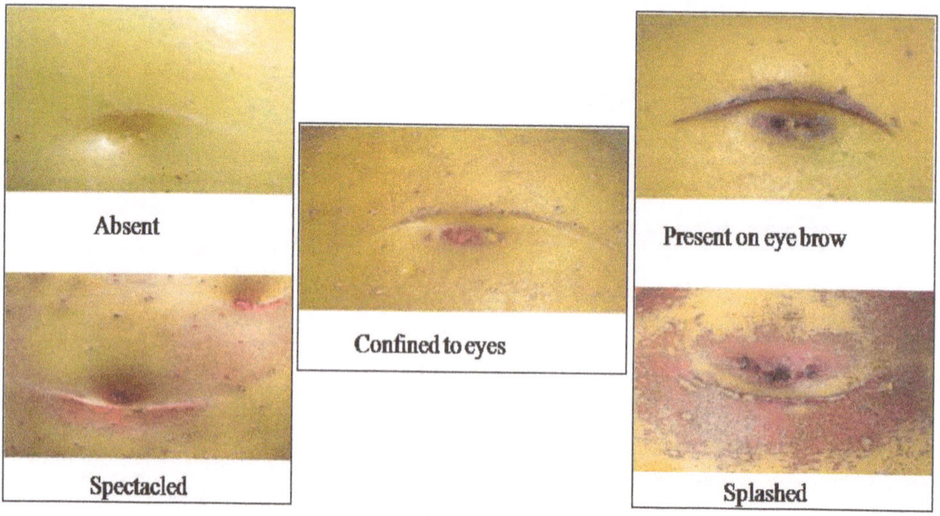

Figure 7.5 (pg. 66)

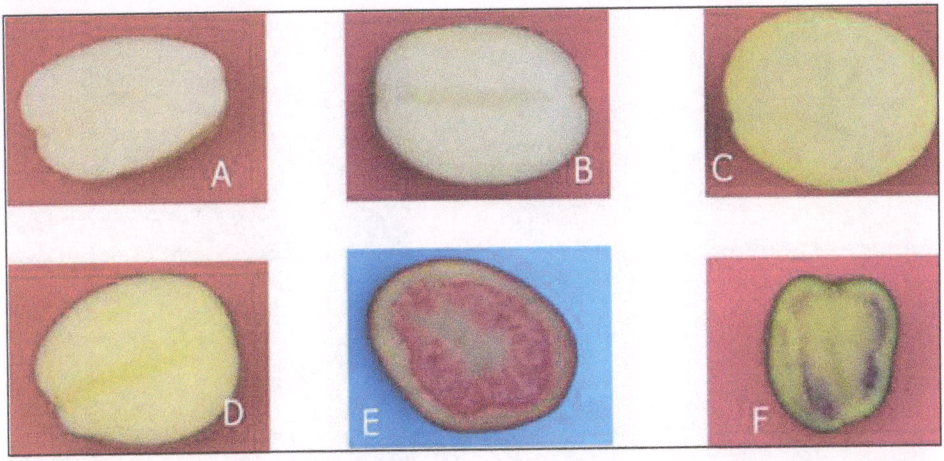

Figure 7.6 (pg. 66)

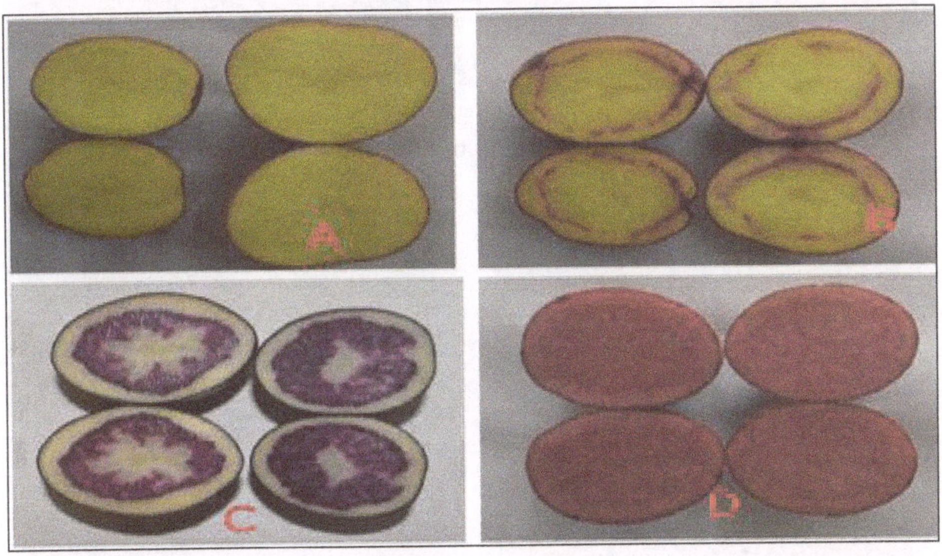

Figure 7.7 (pg. 67)

Compact Semi - compact Open

Figure 7.8 (pg. 67)

Figure 8.3. (i) Left panel: Somatic hybrids between C-13 (pg. 78)

Figure 10.1: PCN on Potato Root-Ball; PCN Infected Potato Field; (pg. 95)

Figure 10.2: Fenwick Can.(pg. 96)

Figure 10.3: Live and Dead Eggs and Juveniles Stained with Meldola's Blue Dye. (pg. 97)

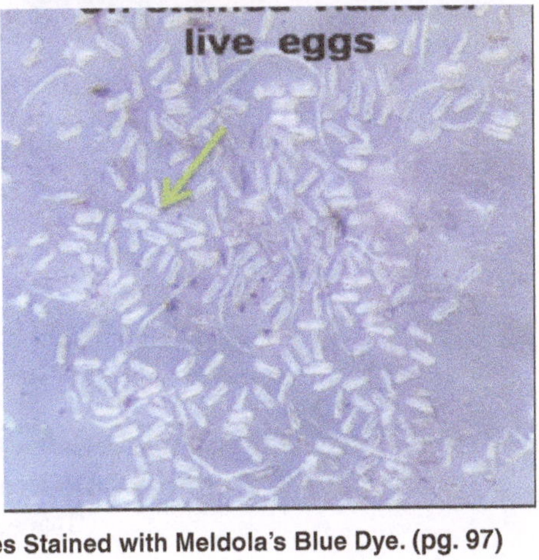

Figure 10.3: Live and Dead Eggs and Juveniles Stained with Meldola's Blue Dye. (pg. 97)

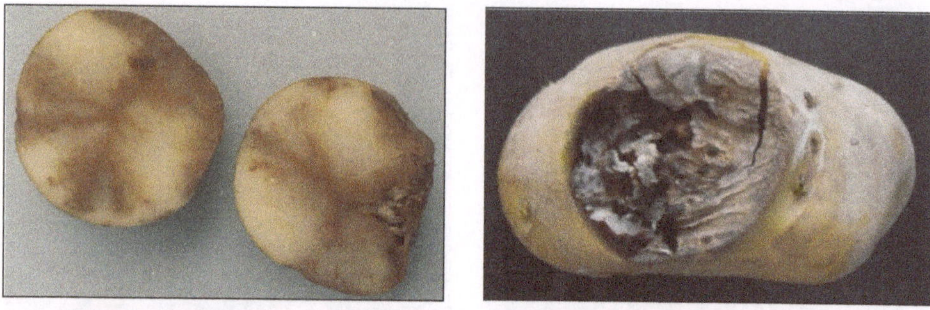

Figure 15.1: Tubers Infected with Late Blight and Dry Rot. (pg. 122)

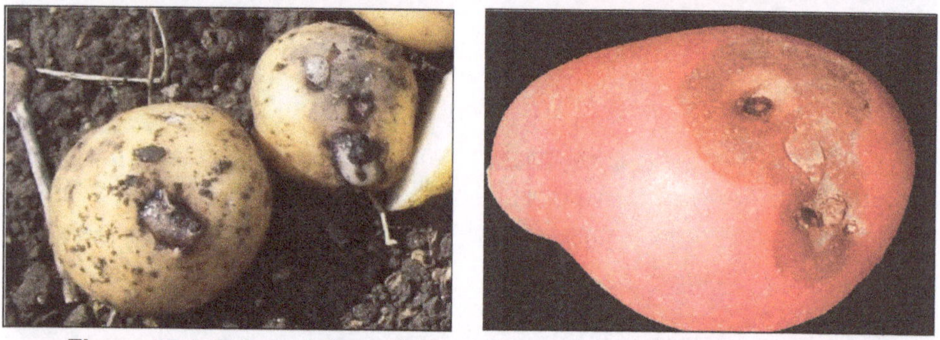

Figure 15.2: Tubers Infected with Brown Rot and Soft Rot. (pg. 123)

Figure 15.3: Uncured and a Cured Potato Tuber and Curing in Field (pg. 124)

Figure 15.4: A Traditional Cold Storage and tack of Potatoes Inside the Store. (pg. 126)

Figure 15.5: A Modified Potato Heap. (pg. 127)

Figure 15.6: A Commercial Cold Store using 10-12 °C Storage Technology, Potatoes and Chips. (pg. 128)

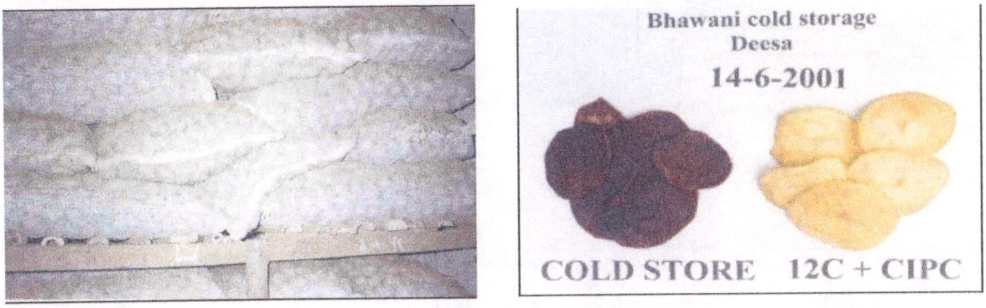

Figure 15.6: A Commercial Cold Store using 10-12 °C Storage Technology, Potatoes and Chips. (pg. 128)

Figure 18.1: Symptoms of Apical Leaf Curl Disease on Potato Plants. (pg. 152)

Figure 18.2: Primary Leaf Rolls Symptoms on Infected Potato Plants. (pg. 154)

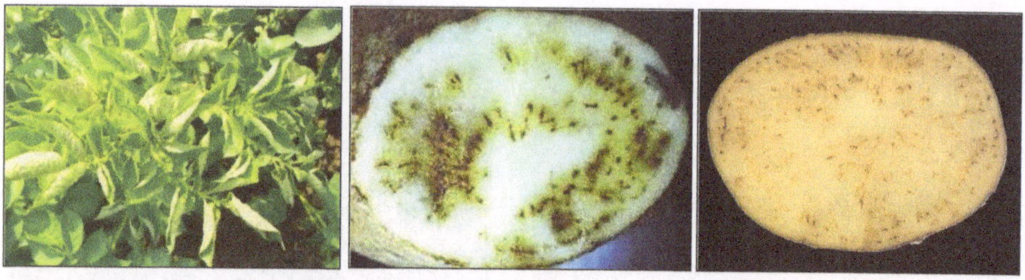

Figure 18.3: Leaf Roll Symptoms Due to Secondary Infection (Seed Borne) of PLRV. (pg. 155)

Figure 18.4: Potato Plants showing Symptoms of PVY. (pg. 158)

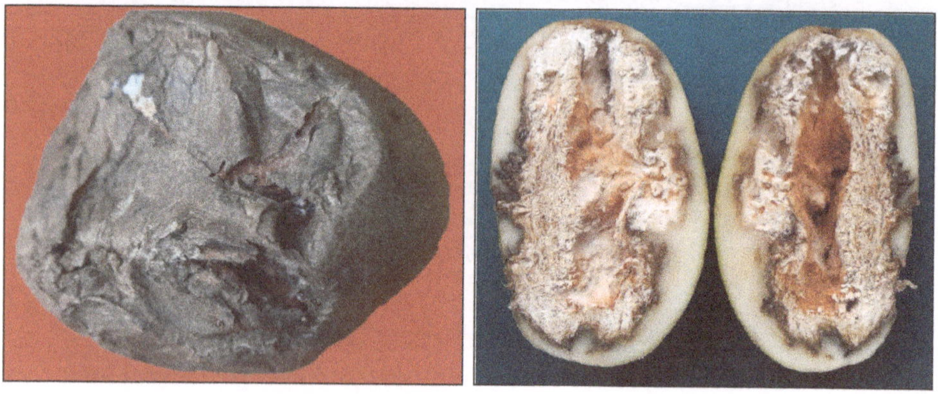

Figure 20.1: Symptoms of Fusarium Dry Rot on Tubers. (pg. 176)

Figure 20.2: Stem Canker (a), Rosetting of Plant Tops (b) and Black Scurf (c) Caused by *Rhizoctonia solani*. (pg. 178)

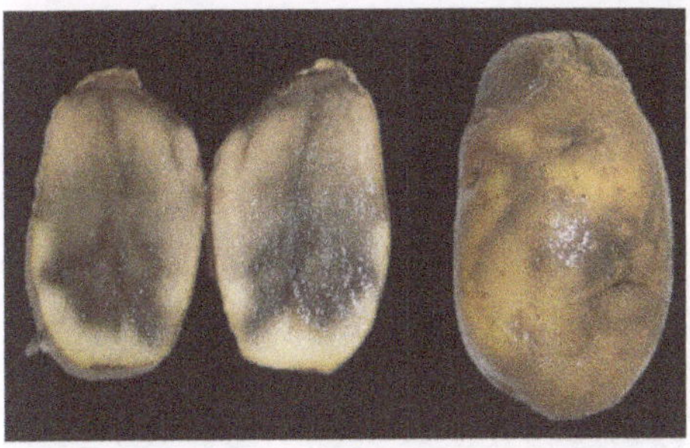

Figure 20.3: Charcoal Rot of Potato Tubers. (pg. 179)

Figure 20.4: Warty Outgrowths on Tubers of Potato.(pg. 180)

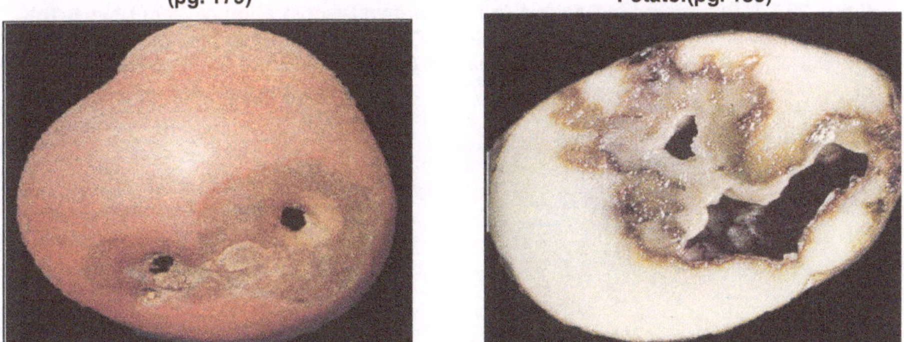

Figure 20.5: Soft Rot Symptoms (Left and Right) and Potato Tuber Cut to Show Tissues Affected by Soft Rot (Right). (pg. 181)

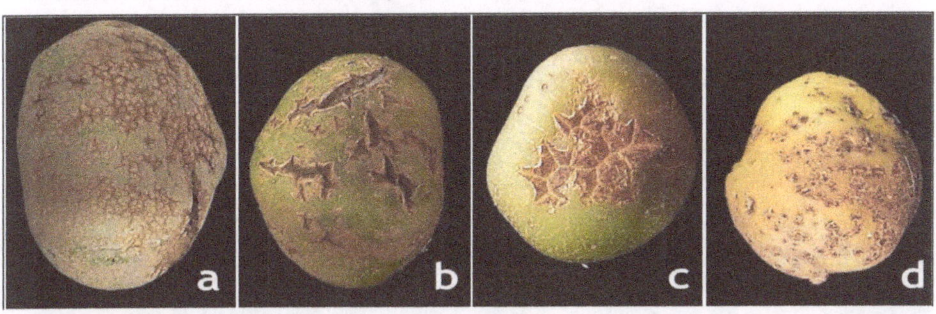

Figure 20.6: Various Types of Scab Symptoms Caused by *Streptomyces* Species on Potato Tubers. (pg. 183)

Figure 20.6: Various Types of Scab Symptoms Caused by *Streptomyces* Species on Potato Tubers. (pg. 183)

Figure 20.7.1: Symptoms of Bacterial Wilt (a); Brown Discoloration of Stem Tissues (b); Bacterial Streaming in Clear Water from Stem Cut Section of Potato (c) Infected with *R. solanacearum.* (pg. 185)

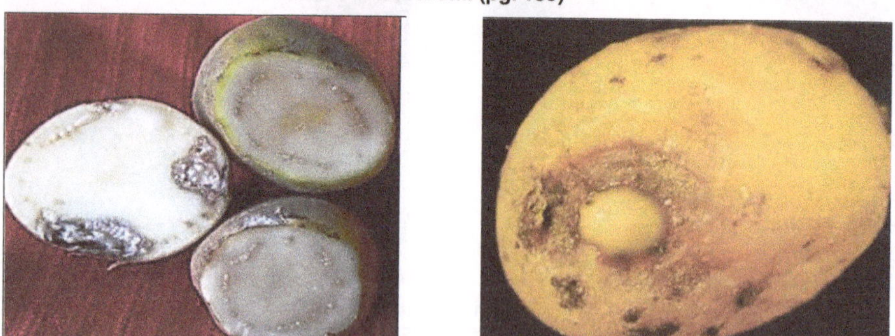

Figure 20.7.2: Vascular Browning of Tubers (Left) and Emergence of Bacterial Mass through Eye Due to *R. solanacearum.* (pg. 185)

Figure 20.8 (pg. 188)

Figure 20.8: Bacterial Wilt Symptoms (a); Brown Discoloration of Stem Tissues (b); and Bacterial Streaming in Clear Water from Stem of Potato Plant Infected with *R.solanacearum*; Symptoms of Common Scab on Potato Tubers (d,e). (pg. 188)

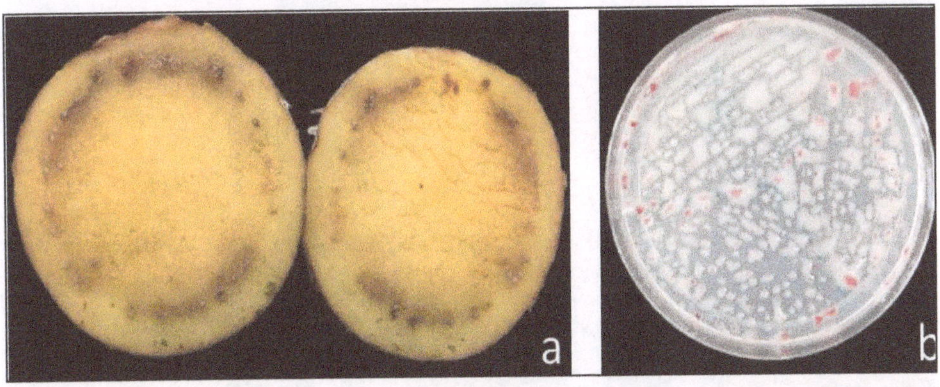

Figure 20.9: Grey-Brown Discoloration of Vascular Tissues and Bacterial Ooze in Potato Tuber Infected by *R. solanacearum* (a); Typical *R. solnacearum* Colonies on TZC Agar Medium (b). (pg. 188)

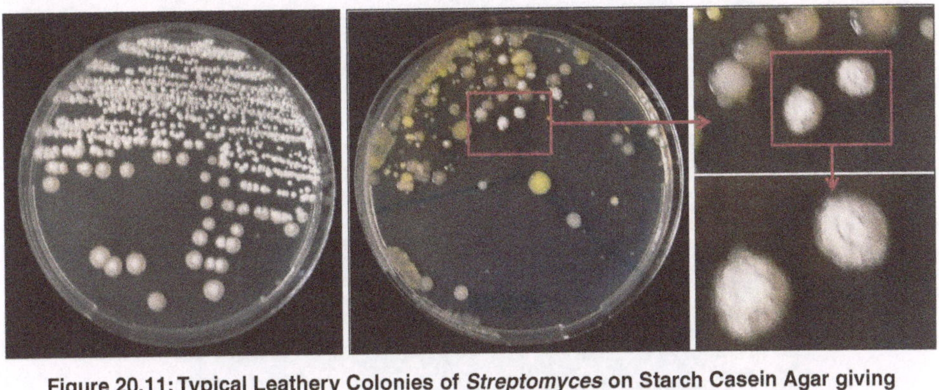

Figure 20.11: Typical Leathery Colonies of *Streptomyces* on Starch Casein Agar giving Powdery Appearance and Pure Culture of *Streptomyces* (Right). (pg. 193)

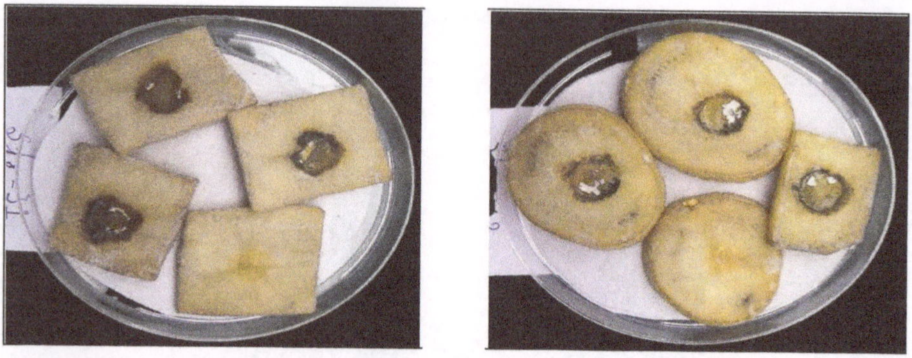

Figure 20.12: Pathogenicity Assay on Tuber Slices of cv. Kufri Chandermukhi: Necrotic Halos Produced by the Plugs from 5 Days Old Oat Meal Agar Cultures of *Streptomyces* strains, Slices without Halo are Control.(pg. 195)

Figure 22.2: Integration of Tissue Culture in Potato Seed Production.(pg. 209)

Figure 22.3: Aeroponics in Potato. Diagrammatic presentation of aeroponic system (upper) and mini-tubers as developed in aeroponic chamber (lower).**(pg. 211)**

Figure 22.4: Step-wise Seed Multiplication through Apical Rooted Cuttings. (pg. 216)

Figure 22.4: Step-wise Seed Multiplication through Apical Rooted Cuttings. (pg. 216)

Figure 22.4: Step-wise Seed Multiplication through Apical Rooted Cuttings. (pg. 216)

Figure 23.1: Green Manuring with Disc Harrow, Deep Tillage with Subsoiler Improves Infiltration and Final Field Preparation with Rotavator. (pg. 220)

Figure 23.2: Seed Bed Preparation with Power Tiller (*Source*: Kumar and Kumar 2018, CSISA, 2014). (pg. 221)

Figure 23.3: Tractor Operated Fertilizer Applicator-cum-Marker. (pg. 221)

Figure 23.4: Tractor Operated Fertilizer Applicator-cum-Marker. (pg. 222)

Figure 23.5: Potato Planting with Semi-automatic and Automatic Planters. (pg. 223)

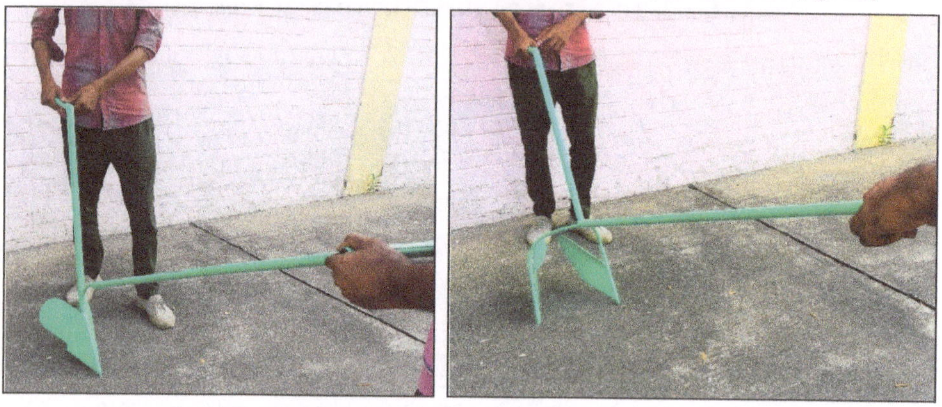

Figure 23.6: Ridger for Potato. (pg. 223)

Figure 23.7: CPRI-spade (*Patented Design-305966*). (pg. 224)

Figure 23.8: Interculturing and Fertilizer Top Dressing before Earthing Operations in Potato Crop. (pg. 224)

Figure 23.9: CPRI-Hand Tools for Weeding Operations (*Patented designs*). (pg. 225)

Figure 23.10: A View of Spraying Operation with Power Sprayers. (pg. 225)

Figure 23.11: CPRI Hand Tools for Potato Digging. (pg. 226)

Figure 23.11: CPRI Hand Tools for Potato Digging. (pg. 226)

Figure 23.13: Tractor Operated Potato Digger Elevator and Low Cost Blade Type Digger. (pg. 227)

Figure 23.14: Two Row Potato Combine Harvester. (pg. 228)

Figure 23.16: Spraying of Pesticides in Potato Crop using Drone. (pg. 229)

Figure 23.15: Rubber Belt Screen Type Potato Sorting and Grading Machine. (pg. 229)

Figure 25.4: Phenotype of Potato Plants under Low and High N in Aeroponics. (pg. 244)

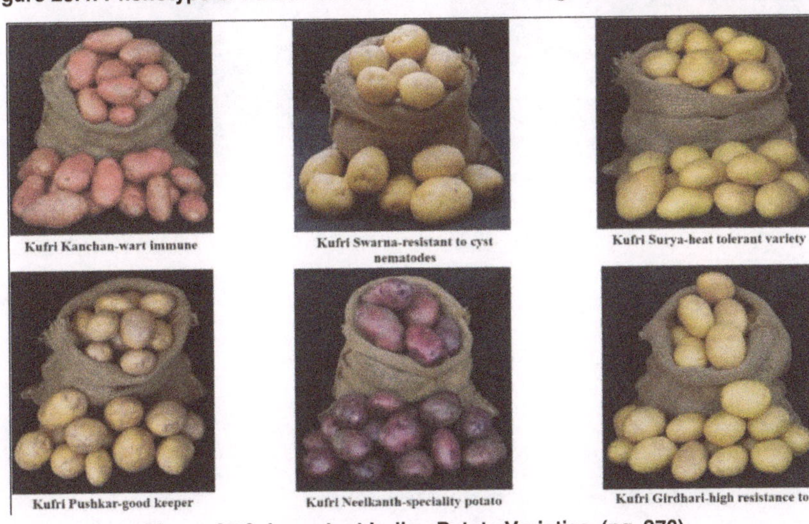

Figure 27.2: Important Indian Potato Varieties. (pg. 270)

(A) Kufri Surya (B) Kufri Lima (C) Kufri Kiran

Figure 28.1: Heat Tolerant Potato Varieties. (pg. 278)

(A) Kufri Thar-1 (B) Kufri Thar-2 (C) Kufri Thar-3

Figure 28.2: New Potato Varieties (with high water use efficiency) for Drought Prone Areas. (pg. 283)

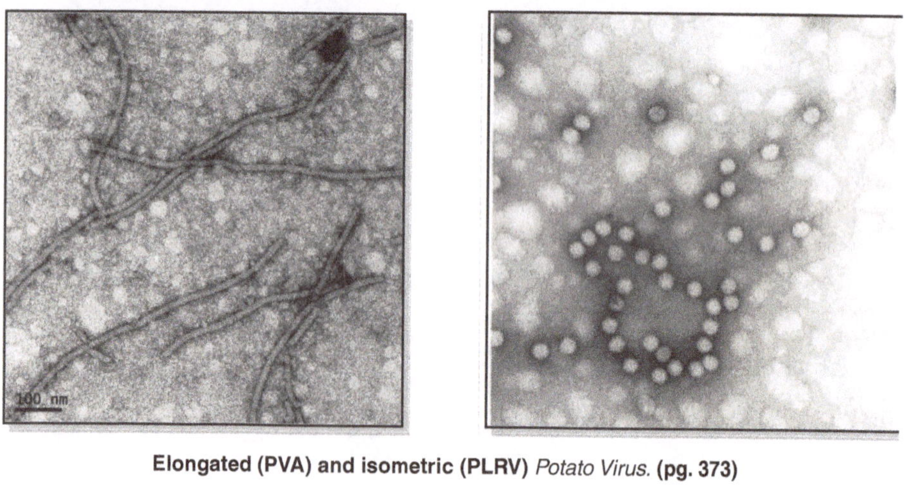

Elongated (PVA) and isometric (PLRV) *Potato Virus.* (pg. 373)

Suggested Readings

Ah-Hen K, Fuenzalida C, Hess S, Contreras A, Vega-Galvez A, Lemus-Mondaca L (2012) Antioxidant capacity and total phenolic compounds of twelve selected potato landrace clones grown in southern Chile. Chil. J. Agric. Res. 72 (1): 3-9.

Albishi T, John JA, Al-Khalifa AS, Shahidi F (2013) Phenolic content and antioxidant activities of selected potato varieties and their processing by-products. J Funct. Foods 5(2): 590-600.

Almekinders CJM (1992) The effect of photoperiod on flowering and TPS production in the warm tropics. Potato Res. 35: 433-442.

Arora RK, Sharma S (2014) Pre and post harvest diseases of potato and their management. In: Future Challenges in Crop Protection Against Fungal Pathogens (Eds.: Goyal A and Manoharachary C). Springer, New York. pp 149-183.

Arora RK, Sharma S, Singh BP (2014) Late blight disease of potato and its management. Potato J. 41: 16-40.

Bairwa A, Venkatasalam, EP, Priyank HM, Sharma S (2021) Introduction of potato cyst nematodes, life cycle and their management through bio-based amendments. In: Microbial Biotechnology in Crop Protection (Eds.: M Kaushal and R Prasad). Springer Nature, Singapore. pp 79-95.

Bandana, Sharma V, Raigond P, Singh B, Kaushik SK (2015) Ascorbic acid losses during storage of potato tubers. Potato J 42(1): 76-79.

Bansal KC, Nagarajan S (1987) Reduction of leaf growth by water stress and its recovery in relation to transpiration and stomatal conductance in potato (*S. tuberosum* L.) genotypes. Potato Res. 30: 497-506.

Basu PS, Minhas JS (1991) Heat tolerance and assimilate transport in different potato genotypes. J. Exp. Bot. 42: 861- 866.

Bhardwaj V, Kaushik SK, Chakrabarti SK, Pandey SK, Singh PH, Manivel P, Singh BP (2007) Combining resistance to late blight and PVY in potato. Potato J. 34: 41-42.

Bhardwaj V, Kaushik SK, Singh PH, Singh BP (2005) Tuber and foliage resistance to late blight in advanced potato hybrids. Potato J. 32: 131-32.

Bhatnagar A (2008) Insect associated with potato in Madhya Pradesh. Pest Manag. Hort. Ecosyst. 13 (2): 21-24.

Bhatnagar A, Sharma V, Singh BP (2016) Climate change and population buildup of whitefly on potato cultivars in North –Western India. Int. J. Agri. Stat. Sci., 12(1): 199-203.

Bhatnagar-Panwar M, Bhatnagar-Mathur P, Bhaaskarla VVA, Dumbala SR, Sharma KK (2013) Rapid, accurate and routine HPLC method for large-scale screening of pro-vitamin A carotenoids in oilseeds. J. Plant Biochem. Biotechnol. http: //dx.doi.rog/10.1007/s13562-013-0239-1

Bidani A, Nouri-Ellouz O, Lakhoua L, Sihachakr D, Cheniclet C, Mahjoub A, Drira N, Gargouri-Bouzid R (2007) Interspecific potato somatic hybrids between *Solanum berthaultii* and *Solanum tuberosum* L. showed recombinant plastome and improved tolerance to salinity. Plant Cell Tiss. Organ Cult. 91: 179-189.

Black CA (1965) Methods of soil analysis, Part 2, American Society of Agronomy, Madison, WI. pp 777-1572.

Botowicz B, Szczerbakowa A, Wielgat B (2005) RAPD analysis of the interspecific somatic hybrids *Solanum bulbocastanum* (+) *S. tuberosum*. Cell Mol. Biol. Lett. 10: 51-162.

Boogaard HL, CA Van Diepen, RP Rötter, JMCA Cabrera, HH Van Laar (1998) *WOFOST 7.1;* User's guide for the WOFOST 7.1 crop growth simulation model and WOFOST Control Center 1.5. Wageningen (Netherlands), DLO Win and Staring Centre. Technical Document.

Bradshaw JE, Bryan GJ, Ramsay G (2006) Genetic resources (including wild and cultivated *Solanum* species) and progress in their utilisation in potato breeding. Potato Res. 49: 49-65.

Buckseth T, Sharma AK, Pandey KK, Singh BP, Muthuraj R (2016) Methods of pre-basic seed potato production with special reference to aeroponics—A review. Sci. Hortic. 204: 79–87.

Burton WG (1989) The Potato. Longman Scientific and Technical, New York, USA. pp 742.

Caruso I, Castaldi L, Caruso G, Frusciante L, Carputo D (2008) Breeding potential of *Solanum tuberosum – S. commersonii* pentaploid hybrids: fertility studies and tuber evaluation. Euphytica 164: 357-363.

Chandla VK, Khurana SMP, Garg ID (2004) Aphids, their importance, Monitoring and Management in Seed Potato crop. Tech. Bulletin No. 61, CPRI, Shimla. pp 1-12

Cheng J, Saunders JA, Sinden SL (1995) Colorado potato beetle resistant somatic hybrid potato plants produced via protoplast electrofusion. In Vitro Cell Dev. Biol. 31: 90-95.

Chengappa PG (2004) Emerging trends in agro-processing in India. Ind. J. Agri. Econ. 59(1): 55-74.

CIP (2010) Laboratory Manual for *Phytophthora infestans* Work at CIP. International Potato center, Lima, Peru, 72p.

Clot CR, Polzer C, Prodhomme C, Schuit C, Engelen CJ, Hutten RC, van Eck HJ (2020) The origin and widespread occurrence of *Sli*-based self-compatibility in potato. Theor. Appl. Genet. 133(9): 2713-2728.

Collins WW (2000) The global initiative on late blight alliance for the future. In: (Eds.: SMP Khurana *et al.*). Potato, Global Research and Development, Vol. I. Indian Potato Association, CPRI, Shimla, India.

CPRI (2013) ICAR-CPRI Vision-2050. Central Potato Research Institute, Shimla, India.

CPRI Annual Report (2020) ICAR-Central Potato Research Institute, Shimla, India. 173p

Czosnek H, Hariton-Shalev A, Sobol I *et al.* (2017) The incredible journey of Begomoviruses in their whitefly vector. Viruses 9: . https: //doi. org/10.3390/v9100273.

Dalamu, Singh B, Barwal S, Raigond P, Sharma R, Joshi A (2015) Assessment of phytochemical diversity in Indian potato cultivars. Indian J Hort. 72(3): 447-450.

Dale MFB, Griffiths DW, Bain H (1998). Effect of bruising on the total glycoalkaloids and chlorogenic acid content of potato (*Solanum tuberosum*) tubers of five cultivars. J. Sci. Food Agric.77: 499-505.

De Bokx JA, van der Walt JPH (1987) Viruses of potatoes and seed potato production, Pudoc, Wageningen (Netherlands). pp 259.

Deuber H, Guignard C, Hoffmann L, Evers D (2012) Polyphenol and glycoalkaloid contents in potato cultivars grown in Luxembourg. Food Chem. 135: 2814-2824.

Diepen VCA, H Van Keulen, FWT Penning de Vries, IGAM Noij, J Goudriaan (1987) Simulated variability of wheat and rice in current weather conditions and in future weather when ambient CO_2 has doubled., Simulation Reports CABO-TT 14. CABO-DLO, WAU-TPE, Wageningen.

Dua VK, Govindakrishnan PM, Singh BP (2014) Calibration of WOFOST model for potato in India. Potato J. 41(2): 105-112.

Eggers EJ, van der Burgt A, van Heusden SA, de Vries ME, Visser RG, Bachem CW, Lindhout P (2021) Neofunctionalisation of the *Sli* gene leads to self-compatibility and facilitates precision breeding in potato. Nat. Commun. 12(1): 1-9.

Ellis D, Salas A, Chavez O, Gomez R, Anglin N (2020) *Ex situ* conservation of potato [*Solanum* Section Petota (Solanaceae)] genetic resources in genebanks. In: (Eds.: H Campos and O Ortiz). The Potato Crop. Springer Nature Switzerland. pp 109-138.

Evans K (2000) Management of potato nematodes. In: (Eds. SMP Khurana *et al.*). Potato, Global Research and Development, Vol. I, 335-344. IPA, Shimla.

Ewing EE (1981) Heat stress and tuberization stimulus. Am. Potato J. 58: 31-49.

Ezekiel R, Khurana SMP (2003) Nutritional and medicinal value of potatoes. Extension Bulletin no. 35. CPRI, Shimla. p 14.

FAO (2022) FAOSTAT Database. http: //www.fao.org/faostat/en/#data/QC Accessed 19 February 2022.

Farmers' rights under Protection of Plant Varieties and Farmers' Rights Act, 2001 Brochure. Guidelines for the conduct of test for Distinctiveness, Uniformity and Stability on Potato (*Solanum tuberosum* L.) November (2009). pp 33. *www.plantauthority.gov.in*

Fock I, Collonier C, Purwito A, Luisetti J, Sonvannavong V, Vedel F, Servaes A, Ambroise A, Kodja H, Ducreux G, Sihachakr (2001) Use of *S. stenotomum* for introduction of resistance to bacterial wilt in somatic hybrids of potato. Plant Physio. Biohem. 39: 899-908.

Garg ID, Khurana SMP (2003) Management of Viral Diseases. In: (Eds.: Khurana *et al.*). The Potato, Production and Utilization in Sub-Tropics. Mehta Publishers, New Delhi, pp 221-229.

Garg ID, Khurana SMP, Kumar S, Lakra BS (2001) Association of a geminivirus with potato apical leaf curl in India and its immuno-electron microscopic detection. J. Indian Potato Assoc. 28: 227-232.

Garg ID, Singh SP, Kushwah VS, Kumar D, Ashwani Kumar (2008) Seed Plot Technique. In: (Eds.: Pandey SK and Chakrabarti SK) Twenty Steps Towards Hidden Treasure. Malhotra Publishing House, New Delhi, India. pp 33-49.

Gaur PC, Pandey SK, Singh SV, Kumar D (1999) Indian Potato Varieties for Processing. Tech. Bull. 50, Central Potato Research Institute, Shimla. p 25.

Gaur PC, SK Pandey (1993) True Potato Seed (TPS) Technology. In: (Eds.: KL Chadha and JS Grewal). Advances in Horticulture, Vol. 7, 85-111. Malhotra Publishing House, New Delhi, India.

Ghosh SC, Asanuma K, Kusutani A, Toyota M (2000) Effect of moisture stress at different stages on the amount of total nonstructural carbohydrates, nitrate reductase activity and yield of potato. Japanese J. Tropical Agric. 44(3): 158-166.

Gopal J, Chahal GS, Minocha JL (2000) Progeny mean, heterosis and heterobeltiosis in *Solanum tuberosum* x *tuberosum* and *S. tuberosum* x *andigena* families under a short day sub-tropic environment. Potato Res. 43: 61-70.

Gopal J, Kumar V, Thakur SS (2004) Hybrid and open-pollinated true potato seed production: Some considerations particularly for developing short duration families. J. Indian Potato Assoc. 31: 13-20.

Gopal J, Rana MS (1988) Induction of flowering in potato in North-western plains of India. J. Indian Potato Asoc. 15: 91-93.

Grewal JS, KC Sud, RC Sharma (1990) Soil and plant tests for potato. Technical Bulletin No.29.Central Potato Research Institute, Shimla, India, pp 39.

Grewal, J.S., K.C. Sud and R.C. Sharma. 1990. Soil and Plant Tests for Potato, Tech. Bul. No. 29 CPRI, Shimla. pp 39.

Gupta VK, Luthra SK, Singh BP (2014) Breeding potatoes for value addition: A comprehensive approach for sustainable potato production. National Seminar on Emerging Problems of Potato. November 1-2, 2014, Central potato Research Institute, Shimla, HP, India. Abstracts: p 159.

Gupta VK, Luthra SK, Singh BP (2014) Potato processing varieties: Present status and future thrusts. National seminar on post-harvest management and processing of potato for increasing food security in India. 22 Sep, 2014, UAS Campus, Dharwad, Karnataka, India.

Harris PM (1992) The Potato Crop. In: (Eds. Khurana *et al.*) Chapman and Hall, London, UK, Mehta Publishers, New Delhi. pp 292-300.

Haverkort AJ, Struik PC, Visser RGF, Jacobsen E (2009) Applied biotechnology to control late blight in potato caused by *Phytophthora infestans*. Potato Res. 52: 249-264.

Hawkes JG (1958) Significance of wild species and primitive forms for potato breeding. Euphytica 7: 257-270.

Hawkes JG (1990) The Potato, Evolution, Biodiversity and Genetic Resources. Belhaven Press, London, pp 259.

Hawkes JG (1994) Origin of cultivated potatoes and species relationships. In: (Eds. Bradshaw JE and Mackay GR) Potato Genetics.CAB International, U.K., pp 3-42.

Haynes KG (2001) Variance components for yield and specific gravity in a diploid potato population after two cycles of recurrent selection. Amer. J. of Potato Res. 78: 69-75.

Horticulture Statistics at a Glance (2018) Government of India Ministry of Agriculture and Farmers' Welfare Department of Agriculture, Cooperation and Farmers' Welfare Horticulture Statistics Division.

Hosaka K, Hanneman R (1998) Genetics of self-compatibility in a self-incompatible wild diploid potato species Solanum chacoense. 2. Localization of an S locus inhibitor (*Sli*) gene on the potato genome using DNA markers. Euphytica 103: 265-271.

Howard HW (1970) Genetics of Potato (*Solanum tuberosum* L.). Logos Press Ltd., London, pp 20-24.

Jackson ML (1967) Soil chemical analysis, Prentice Hall of India Pvt. Ltd. New Delhi. pp 485.

Jadari R, Sihachakr D, Rossignol L, Ducreux G (1992) Transfer of resistance to *Verticillium dahliae* Kleb. from *Solanum torvum* S.W. into potato (*Solanum tuberosum* L.) by protoplast electrofusion. Euphytica 64: 39-47.

Jansky S (2006) Overcoming hybridization barriers in potato. Plant Breed. 125: 1-12.

Jansky SH, Charkowski AO, Douches DS, Gusmini G, Richael C, Bethke PC *et al.* (2016) Reinventing potato as a diploid inbred line-based crop. Crop Sci. 56: 1-11.

Kadian MS, Pande PC, Upadhaya MD (1990) TPS as a propagating material for north-western indo-gangetic plains. In: Commercial Adoption of TPS Technology: Prospects and Problems. Report of Group Meeting, Delhi, India. pp 40-48.

Kaspar K, Park JS, Brown CR, Mathison BD, Navarre DA, Chew BP (2013) Pigmented potato consumption improves immune response in men: a randomized controlled trial. Am. J. Advanced Food Sci. Technol. 1: 15-25.

Khan IA, SK Luthra, R Ezekiel (1994) Flowering induction in potato by extended photoperiod and application of growth regulators. In: (Eds.: Shekhawat *et al.*) Proceedings of "Potato: Present and future. Indian Potato Assoc., CPRI, Shimla. pp 46-48.

Khanna ML (1966) Breeding potato varieties tolerant to higher thermoperiods. Curr. Sci. 6: 143-144.

Khurana SMP, Garg ID, Singh BP, Gadewar AV (1999) Important diseases of potato in India and their management. In: (Eds. Upadhyay *et al.*). IPM Systems in Agriculture Aditya Books Pvt. Ltd., New Delhi, pp 263-298.

Khurana SMP, Minhas JS, Pandey SK (2003) The Potato: Production and Utilization in Sub-tropics. Mehta Publishers, New Delhi. 445p.

Khurana SMP, Singh MN (2000) Potato leafroll virus. In: (Ed. Khurana SMP) Diseases and Pests of Potato - a manual, CPRI, Shimla. pp 33-35.

Kim-Lee H, Moon JS, Hong YJ, Kim MS, Cho HM (2005) Bacterial wilt resistance in the progenies of the fusion hybrids between haploid of potato and *Solanum commersonii*. Amer. J. Potato Res. 82: 129-137.

Kozukue N, Misoo S, Yamada T, Kamijima O, Friedman M (1999) Inheritance of morphological characters and glycoalkaloids in potatoes of somatic hybrids between dihaploid *Solanum acaule* and tetraploid *Solanum tuberosum* J. Agric. Food Chem. 47: 4478-4483.

Krishna Prasad KS (2000) Root knot nematodes. *In*: (Eds. SMP Khurana). Diseases and Pests of potato-A manual, 41-44. Central Potato Research Institute, Shimla, HP, India.

Krishna Prasad KS (2006) Potato cyst nematodes and their management in the Nilgiris (India). CPRI Technical Bulletin No. 77. Central Potato Research Institute, Shimla, India, pp 20.

Ku SB, Edwards GE, Tanner CB (1977) Effect of light, carbon dioxide and temperature on photosynthesis, oxygen inhibition of photosynthesis and transpiration in potato. Plant Physiol. 59: 868-872.

Kumar A, Kumar D, Singh B (2007) Changes in processing quality of potato tubers occuring between haulm cutting and harvesting. Potato J. 34(3-4): 239-241.

Kumar D, Minhas JS (1994) Effect of water stress on leaf area, leaf water potential and tuber dry weight of potato cultivars Kufri Lalima and Kufri Bahar. In: (Eds.: Shekhawat *et al.*) Potato: Present and Future, IPA, CPRI Shimla, pp 270-272.

Kumar D, Minhas JS (2013) Evaluation of indigenous potato varieties, advanced clones and exotic genotypes against water deficit stress under sub-tropical environment. Indian J Plant Physiol. 18(3): 240-249.

Kumar D, Minhas JS (2015) Can water deficit be useful in potato? – Some issues. In: (Ed. Singh AL) Recent advances in crop physiology. Daya Publishing House. New Delhi Vol 2 pp 225-230.

Kumar D, Minhas JS, Gupta VK, Singh N, Rawal S *et al.* (2020) Kufri Thar-1: A new water-use efficient potato variety for low water availability areas in middle Gangetic plains, east coast plains and hills region in India. Potato J. 47(2): 116-125.

Kumar D, Minhas JS, Singh B (2003) Abiotic stress and potato production. In: (Eds.: Khurana *et al.*) The Potato-Production and Utilization in Sub-tropics. Mehta Publishers, New Delhi, pp 314-322.

Kumar M, Jatav MK, Dua VK, Trehan SP, Lal SS, Upadhyay NC (2014) Methodology to map spatial variability of available nutrients in area of intensively growing potato (*Solanum tuberosum*) using remote sensing and GIS. Indian J. Agric. Sci. 84 (3): 396-400.

Kumar NR, Singh BP, Khurana SMP, Pandey NK, Rana RK (2004) Destination of Indian Potatoes. *In:* (Eds. Khurana SMP, Singh BP, Luthra SK, Kumar NR, Kumar D and Kumar D) Processing and Export Potential of Indian Potatoes. IPA, Shimla. pp 38-43.

Kushwah VS, Govindakrishnan PM (2003) Cultural practices for seed and ware potato. *In:* (Eds. Khurana *et al.*) The Potato, Production and Utilization in Sub-tropics. Mehta Publishers, New Delhi, India.

Kushwah VS, Grewal JS (1992) Agro Techniques for Seed Potato Production. Technical Bulletin No. 37, Central Potato Research Institute, Shimla, India. pp 32.

Lakra BS (2002) Leaf curl: A threat to potato crop in Haryana. J. Myco.l Plant Pathol. 32: 367.

Lapwood DH (1967) Laboratory assessments of the susceptibility of potato tubers to infection by blight (*Phytophthora infestans*). Eur. Potato J. 10: 1127-1135.

Levy D (1983) Varietal differences in the response of potatoes to repeated short periods of water stress in hot climates. Potato Res. 26(4): 315-321.

Li PH (1985) Potato Physiology. Academic Press, London, UK: 586p

Lindhout P, Meijer D, Schotte T, Hutton RCB, Visser RGF, VanEck HJ (2011) Towards F_1 hybrid seed potato breeding. Potato Res. 54: 301-312.

Loebenstein G, Berger, PH, Brunt AA, Lawson RH (2001) Virus and Virus-like Diseases of Potatoes and Production of Seed-Potatoes. Kluwer Academic Publishers, Netherlands, pp 69-72.

Lowry OH, Rosenbrough RJ, Fern AL, Randall RJ (1951) Protein measurements with Folin-phenol reagent. J. Biol. Chem. 193: 263-75.

Luthra SK (2005) Suitability of induced flowering in potato under short day conditions. Seed Res. 33(2): 221-223.

Luthra SK, Khan IA (2000) Induction of flowering in potato under short day conditions. *In:* (Eds: Khurana *et al.*) Potato: Global Research and Development- Volume-I, Indian Potato Association, CPRI Shimla. pp 150-152.

Luthra SK, Pandey SK, Singh BP, Kang GS, Singh SV, Pande PC (2006) Potato Breeding in India. CPRI, Shimla Technical Bulletin No 74. pp 90.

Luthra SK, Rawal S, Gupta VK, Malik K (2018c) Performance of potato advanced hybrid MS/6-1947 under high temperature stress and water deficit conditions. Indian J. Agri Sci 88(7): 1070-1076.

Luthra SK, VK Gupta, Tiwari JK, Kumar V, Bhardwaj V *et al.* (2020) Potato Breeding in India. CPRI Technical Bulletin No 74 (revised), ICAR-Central Potato Research Institute, Shimla, Himachal Pradesh, India. pp 214.

Ma L, Zhang C, Zhang B, Tang F, Li F, Liao Q *et al.* (2021) A nonS-locus F-box gene breaks self-incompatibility in diploid potatoes. Nat. Commun. 12(1): 1-8.

Malcolmson JF (1976) Assessment of field resistance to blight (*P. infestans*) in potatoes. Trans. Br Mycol. Soc. 67: 321-325.

Mayo MA, D'Arcy CJ (1999) Family *Luteoviridae*: a re-classification of luteoviruses. *In:* (Eds. Smith HG and Barker H) The Luteoviridae, CAB International, Wallingford, UK. pp 15-22.

Midmore DJ (1992) Potato production in the tropics. In: (Eds. Harris, PM) The Potato Chapman and Hall, London, UK. pp 509-569.

Minhas JS, Kumar D, Joseph TA, Prasad KSK, Singh B (2001) Selection of heat tolerant potato genotypes and their performance under heat conditions. J. Indian Potato Assoc. 28: 132-134.

Minhas JS, Sukumaran NP (1988) Diurnal changes in net photosynthetic rate in potato in two environments. Potato Res. 31: 375-378.

Mojtahedi H, Brown CR, Santo GS (1995) Characterization of resistance in a somatic hybrid of *Solanum bulbocastanum* and *S. tuberosum* to *Meloidogyne chitwoodi*. J. Nematol. 27(1): 86-93.

Morris SC, Lee TH (1984) The toxicity and teratogenicity of Solanaceae glycoalkaloids, particularly those of potato (*Solanum tuberosum*). Food Technol. Australia 36: 118-124.

Mukherjee K, VermaY, Chakrabarti SK, Singh MN, Khurana SMP (2003) Cloning and sequencing of coat protein gene of an Indian Potato leafroll virus (PLRV) isolate and its similarity with other members of Luteoviridae. Virus Genes. 26: 247-253.

Nagarajan S, Minhas JS (1995) Inter-nodal elongation: A potential screening techniques for heat tolerance in potato. Potato Res. 38: 179-86.

Naik PS, Buckseth T (2018) Recent advances in virus elimination and tissue culture for quality potato seed production. In: (Eds.: Gosal SS, Wani SH) Biotechnologies of Crop Improvement, Volume 1. Springer Nature Switzerland AG. pp 131-158.

Naik PS, Karihaloo JL (2007) Micropropagation for Production of Quality Potato Seed in Asia-Pacific. Asia-Pacific Consortium on Agricultural Biotechnology, New Delhi, India pp 54.

Naik PS, Khurana SMP (2003) Micropropagation in potato seed production: need to revise seed certification standards. J. Indian Potato Assoc.30: 123-132.

Novy RG, Gillen AM, Whitworth JL (2007) Characterization of the expression and inheritance of potato leafroll virus (PLRV) and potato virus Y (PVY) resistance in three generations of germplasm derived from *Solanum etuberosum*. Theor. Appl. Genet. 114: 1161-1172.

Novy RG, Helgeson JE (1994) Somatic hybrids between *Solanum etuberosum* and diploid, tuber bearing *Solanum* clones. Theor. Appl. Genet. 89: 775-782.

Nyman M, Waara S (1997) Characterisation of somatic hybrids between *Solanum tuberosum* and its frost-tolerant relative *Solanum commersonii*. Theor. Appl. Genet. 95: 1127-1132.

Oberwalder B, Schilde-Rentschler L, Loffelhardt-Ruob B, Ninnemann H (2000) Differences between hybrids of *Solanum tuberosum* L. and *Solanum circaeifolium* Bitt. obtained from symmetric and asymmetric fusion experiments. Potato Res. 43: 71-82.

Orczyk W, Przetakiewicz J, Nadoloska-Orczyk A (2003) Somatic hybrids of *Solanum tuberosum*-application to genetics and breeding. Plant Cell Tissue Organ. Cult. 74: 1-13.

Pandey SK, Sarkar D Singh SV (2006) Potato processing in India: today and tomorrow. Potato J. 33(1-2): 11-19.

Pandit A, Rana Rajesh K, Pandey NK and Kumar NR (2003) Potato marketing in India. Indian Horticulture. 48 (2) July-September, 2003: 35-39.

Polgar Z, Wielgus SM, Horvath S, Helgeson JP (1999) DNA analysis of potato + *Solanum brevidens* somatic hybrid lines. Euphytica 105: 103–107.

Polzerová H, Patzak J, Greplová M (2011) Early characterization of somatic hybrids from symmetric protoplast electrofusion of *Solanum pinnatisectum* Dun. and *Solanum tuberosum* L. Plant Cell Tiss. Organ Cult. 104: 163-170.

Provisions of Plant Varieties and Farmers' Rights Act. 2001 Brochure Oct 2016. *www.plantauthority.gov.in*

Racz E, Parlagh-Huszar K, Kecskes T (1991) HPLC method for determination of ascorbic acid in fruits and vegetables. Periodica Polytechnica Chem. Engin. 351(2): 23-30.

Raigond P, Ezekiel R, Kaundal B (2014) Starch fractions of cooked potatoes at low temperature. Potato J 41(1): 58-67.

Raigond P, Kaundal B, Sood A, Devi S, Dutt S, Singh B (2018) Quantification of biguanide and related compounds (anti-diabetic) in vegetables and fruits. J. Food Compos. Anal. 74: 82-88.

Ramanujam S (1954) The potato in Indian agriculture. J. Soil Water Conserv. 3: 4-14.

Rana RK (2011) The Indian Potato Processing Industry-Global comparison and business prospects. Outlook on Agric. 40(3): 237-243.

Ranganna S (2007) Handbook of analysis and quality control for fruit and vegetable products. IInd ed. Tata Mc Graw Hill, New Delhi (India), pp 107.

Rasmussen JO, Nepper JP, Kirk HG, Tolstrup K, Rasmussen OS (1998) Combination of resistance to potato late blight in foliage and tubers by interspecific dihaploid protoplast fusion. Euphytica 102: 363-370.

Renia H (1995) True seed is a commercial reality in USA. Potato Rev. 5: 48-51.

Rokka VM, Tauriainen A, Pietilä L, Pehu E (1998) Interspecific somatic hybrids between wild potato *Solanum acaule* Bitt. and anther-derived dihaploid potato (*Solanum tuberosum* L.) Plant Cell Rep. 18: 82-88.

Sarkar D, Tiwari JK, Sharma S, Poonam, Sharma S, Gopal J, Singh BP, Luthra SK, Pandey SK, Pattanayak D (2011) Production and characterization of somatic hybrids between *Solanum tuberosum* L. and *S. pinnatisectum* Dun. Plant Cell Tiss. Organ Cult. 107: 427-440.

Serraf I, Sihachakr D, Ducreux G, Brown SC, Allot M, Barghi N, Rossignol L (1991) Interspecific somatic hybridization in potato by protoplast electrofusion. Plant Sci. 76: 115-126.

Shaner G, Finney RF (1977) The effect of nitrogen fertilization on the expression of slow mildewing resistance in knox wheat. Phytopathology 67: 1051-1056.

Sharma AK, Chandel RS, Kumar V, Singh BP (2012) Potato Cultivation in North Western Himalaya. Technical Bulletin No. 95, Central Potato Research Institute, Shimla, India. 72p.

Sharma S (2016) Fungal diseases of potato and their management. pp 154-169. In: (Eds.: Pandey *et al.*) Advances in Quality Potato Production and Post Harvest Management. Agrotech Publishing Academy, Udaipur, India.

Sharma S, Sagar V, Singh BP (2017) Important diseases of potato and their management. pp 433-456. In: (Eds. Gautam HR and Gupta SK) Diseases of Commercial Crops in India. Neoti Book Agency PVT. Ltd., New Delhi.

Singh A, Sunderasha S, Bhardwaj V, Singh B, Kumar D, Singh BP (2015) Expression profiling of potato cultivars with contrasting tuberization at elevated temperature using microarray analysis. Plant Physiol. Biochem. 97: 108-116.

Singh B, Sharma S, Singh BP (2010) Acrylamide in French fries of Indian potato varieties grown in north-western hills. In: Book of Abstracts of the "National conference of Plant Physiology" held at BHU, Varanasi, India during Nov 25-27 2010. Organised by ISPP, New Delhi. pp 339.

Singh BP, Nagesh M, Sharma Sanjeev, Sagar Vinay, Jeevalatha A, Sridhar J (2015) A Manual on Diseases and Pests of Potato. Technical Bulletin No. 101. ICAR-Central Potato Research Institute, Shimla, pp 90.

Singh BP, Rana RK, Govindakrishnan PM (2014) Vision 2050. Central Potato Research Institute, Shimla, India, pp 26.

Singh N, Sood MC (2013) Architecturing of planting geometry and drip layout systems for potato production under drip irrigation. Progress. Agric. 13 (2): 207-211.

Singh N, Sood MC (2013) Raised bed planting techniques for potato production under furrow and sprinkler irrigation methods. Ann. Hortic. 6 (1): 45-48.

Singh N, Sood MC, Lal SS (2014) Potato production under Drip and Sprinkler Irrigation. Technical Bulletin No. 97. Central Potato Research Institute, Shimla, India.

Singh PH, Singh BP, Singh L, Kaushik SK, Bhardwaj V (2004) Host resistance and aggressiveness of A1 and A2 mating types of *Phytophthora infestans* in potatoes. J. Indian Potato Assoc. 31: 39-42.

Singh RK, Buckseth T, Tiwari JK, Sharma AK, Singh V *et al.* (2019) Seed potato (*Solanum tuberosum*) production systems in India: A chronological outlook. Indian J. Agric. Sci. 89 (4): 578–87.

Singh S, Garg VK, Kumar S, Shekhawat GS (2000) Seed Production Manual. Central Potato Research Institute, Shimla, India. 91p.

Singh SP, Kumar M, Dua VK, Sharma SK, Sadawarti MJ, Roy S (2019) Leaf chlorophyll meter-A non-destructive method for scheduling nitrogen in potato crop. Potato J. 46(1): 73-80.

Sood MC, Singh N (2008) Micro-irrigation and fertigation. *In*: (Eds. SK Pandey and SK Chakrabarty). Twenty steps towards hidden treasure-Technologies that triggered potato revolution in India, 215- 226. Central Potato Research Institute, Shimla.

Sood S, Bhardwaj V, Sundaresha S (2020) Major paradigm shifts in potato breeding. In: (Eds. SS Gosal and SH Wani). Accelerated Plant Breeding, Vol. II, 1-16.

Sood S, Sundaresha S, Bhardwaj V, Kardile H, Chourasia KN (2021) Reinventing potato breeding through true potato seed based technology. Indian Hort.: 41-42.

Spooner DM, Ghislain M, Simon R, Jansky SH, Gavrilenko T (2014) Systematics, diversity, genetics, and evolution of wild and cultivated potatoes. Bot. Rev. 80: 283-383.

Spooner DM, Salas A (2006) Structure, Biosystematics, and Genetic resources. In: (Eds. J Gopal and SMP Khurana). Handbook of potato production, improvement and postharvest management, 1-40. Food Product Press, New York.

Stan M, Soran ML, Marutoiu C (2014) Extraction and HPLC determination of the ascorbic acid content of three indigenous spice plants. J. Analytical Chem. 69 (10): 998–1002.

Sud KC, Upadhayay NC, Trehan SP, Jatav MK, Kumar M, Lal SS (2008) Soil and Plant Tests for Judicious Use of Fertilizer in Potato. Technical Bulletin No. 91. Central Potato Research Institute, Shimla, India, pp 80.

Szczerbakowa A, Maciejewska U, Zimnoch-Guzowska E, Wielgat B (2003) Somatic hybrids Solanum nigrum (+) S. tuberosum: Morphological assessment and verification of hybridity Plant Cell Rep. 21: 577-584.

Szczerbakowa A, Tarwacka J, Oskiera M, Jakuczun H, Wielgat B (2010) Somatic hybridization between the diploids of *S.* x *michoacanum* and *S. tuberosum.* Acta Physiol. Plant 32: 867-873.

Talburt W F, Smith O (1975) Potato Processing. Third edition, pp 305-402. West port: The AVI Publ. Co., Inc.

Tareke E, Rydberg P, Karisson P, Eriksson S, Tornqvist M (2002) Analysis of acrylamide, a carcinogen formed in heated foodstuffs. J. Agric. Food Chem. 50: 4998-5006.

Tek AL, Stevenson WR, Helgeson JP, Jiang J (2004) Transfer of tuber soft rot and early blight resistances from *Solanum brevidens* into cultivated potato Theor. Appl. Genet. 109: 249-254.

Thieme R, Rakosy-Tican E, Gavrilenko T, Antonova O, Schubert J, Nachtigall M, Heimbach U, Thieme T (2008) Novel somatic hybrids (*Solanum tuberosum* L. + *Solanum tarnii*) and their fertile BC_1 progenies express extreme resistance to potato virus Y and late blight. Theor. Appl. Genet. 116: 691-700.

Thieme R, Rakosy-Tican E, Nachtigall M, Schubert J, Hammann T, Antonova O, Gavrilenko T, Heimbach U, Thieme T (2010) Characterization of the multiple resistance traits of somatic hybrids between *Solanum cardiophyllum* Lindl. and two commercial potato cultivars. Plant Cell Rep. 29: 1187-1201.

Tiwari JK, Plett D, Garnett T, Chakrabarti SK, Singh RK (2018) Integrated genomics, physiology and breeding approaches for improving nitrogen use efficiency in potato: translating knowledge from other crops. Funct. Plant Biol. 45: 587-605.

Tiwari JK, Poonam, Sarkar D, Pandey SK, Gopal J, Kumar SR (2010) Molecular and morphological characterization of somatic hybrids between *Solanum tuberosum* L. and *S. etuberosum* Lindl. Plant Cell Tiss. Organ Cult. 103: 175-187.

Tiwari JK, S Devi, T Buckesth, N Ali, RK Singh, R Zinta, VK Dua, SK Chakrabarti (2019) Precision phenotyping of contrasting potato (*Solanum tuberosum* L.) varieties in a novel aeroponics system for improving nitrogen use efficiency: in search of key traits and genes. J. Integ. Agric. 18: 2–12.

Tiwari JK, Sarkar D, Poonam, Sharma S, Gopal J (2011) *Solanum tuberosum* (+) *S. pinnatisectum* somatic hybrids: a new source of horizontal resistance to potato late blight in India. SOL Newsletter: 4.

Trehan SP (2009) Improving nutrient use efficiency by exploiting genetic diversity of potato Potato J 36: 121-135.

Trehan SP, Upadhayay NC, Sud KC, Kumar M, Jatav MK, Lal SS (2008) Nutrient Management in Potato. Technical Bulletin No. 90. Central Potato Research Institute, Shimla, India, pp 64.

Upadhya MD (1999) Present and future research for true potato seed technology. In: (Eds.: Khurana *et al.*). Potato, Global Research and Development, Vol.1, 641-655. Malhotra Publishing House, New Delhi, India.

Venkatasalam EP, Pandey KK, Singh V, Singh BP (2011) Seed potato production technology, CPRI, Shimla, Publication, pp 1-57.

Vos J (2009) Nitrogen responses and nitrogen management in potato. Potato Research 52: 305–317.

Yamada T, Misoo S, Ishii T, Ito Y, Takaoka K, Kamijima O (1997) Characterization of somatic hybrids between tetraploid *Solanum tuberosum* L. and dihaploid *S. acaule*. Breed Sci. 47: 229-236.

Zhang C, Yang Z, Tang D, Zhu Y, Wang P, Li D, Zhu G, Xiong X, Shang Y, Li C, Huang S (2021) Genome design of hybrid potato. Cell 184(15): 3873-3883.

Index

T

V

W

Y

Z

9 789359 191669

Printed at Repro India Ltd.